Lecture Notes in Mathematics 2253

More information about this subseries at http://www.springer.com/series/3114

Fondazione C.I.M.E., Firenze

C.I.M.E. stands for *Centro Internazionale Matematico Estivo*, that is, International Mathematical Summer Centre. Conceived in the early fifties, it was born in 1954 in Florence, Italy, and welcomed by the world mathematical community: it continues successfully, year for year, to this day.

Many mathematicians from all over the world have been involved in a way or another in C.I.M.E.'s activities over the years. The main purpose and mode of functioning of the Centre may be summarised as follows: every year, during the summer, sessions on different themes from pure and applied mathematics are offered by application to mathematicians from all countries. A Session is generally based on three or four main courses given by specialists of international renown, plus a certain number of seminars, and is held in an attractive rural location in Italy.

The aim of a C.I.M.E. session is to bring to the attention of younger researchers the origins, development, and perspectives of some very active branch of mathematical research. The topics of the courses are generally of international resonance. The full immersion atmosphere of the courses and the daily exchange among participants are thus an initiation to international collaboration in mathematical research.

C.I.M.E. Director (2002 – 2014)
Pietro Zecca
Dipartimento di Energetica "S. Stecco"
Università di Firenze
Via S. Marta, 3
50139 Florence
Italy
e-mail: zecca@unifi.it

C.I.M.E. Director (2015 –)
Elvira Mascolo
Dipartimento di Matematica "U. Dini"
Università di Firenze
viale G.B. Morgagni 67/A
50134 Florence
Italy
e-mail: mascolo@math.unifi.it

C.I.M.E. Secretary
Paolo Salani
Dipartimento di Matematica "U. Dini"
Università di Firenze
viale G.B. Morgagni 67/A
50134 Florence
Italy
e-mail: salani@math.unifi.it

CIME activity is carried out with the collaboration and financial support
of INdAM (Istituto Nazionale di Alta Matematica)

For more information see CIME's homepage: **http://www.cime.unifi.it**

Massimiliano Gubinelli • Panagiotis E. Souganidis •
Nikolay Tzvetkov

Singular Random Dynamics

Cetraro, Italy 2016

Franco Flandoli • Massimiliano Gubinelli •
Martin Hairer

Editors

Authors
Massimiliano Gubinelli
Hausdorff Center for Mathematics
University of Bonn
Bonn, Germany

Panagiotis E. Souganidis
Department of Mathematics
University of Chicago
Chicago, IL, USA

Nikolay Tzvetkov
Department of Mathematics
Universite de Cergy-Pontoise
Cergy-Pontoise Cedex, France

Editors
Franco Flandoli
Faculty of Sciences
Scuola Normale Superiore
Pisa, Italy

Massimiliano Gubinelli
Hausdorff Center for Mathematics
University of Bonn
Bonn, Germany

Martin Hairer
Department of Mathematics
Imperial College London
London, UK

ISSN 0075-8434   ISSN 1617-9692   (electronic)
Lecture Notes in Mathematics
C.I.M.E. Foundation Subseries
ISBN 978-3-030-29544-8   ISBN 978-3-030-29545-5   (eBook)
https://doi.org/10.1007/978-3-030-29545-5

Mathematics Subject Classification (2010): Primary: 35R60, 60H15, 37L50; Secondary: 35KXX, 35LXX, 35Qxx, 60Hxx

This Springer imprint is published by the registered company Springer Nature Switzerland AG.
The registered company address is: Gewerbestrasse 11, 6330 Cham, Switzerland

Preface

General Remarks

In recent years, a new paradigm emerged in the area of partial differential equations (PDEs): that probability may help to treat more singular problems than those that can be tackled classically by deterministic tools. The lectures given at the CIME summer school present two directions where this paradigm was developed into mature theories.

The first one are stochastic partial differential equations (SPDEs). Probability here is needed a priori, for the stochastic nature of the inputs. However, it does not play a merely technical role to define and treat stochastic terms. It is the basic tool which allows one to define, in suitable ways, nonlinear operations on distributions. More precisely, it selects special inputs that have all the necessary properties for the solution of very singular nonlinear equations.

The second one are deterministic PDEs with random initial conditions. The typical picture that emerges in certain classes of PDEs is that classical deterministic tools (like energy methods) allow to study the equation at sufficiently high level of regularity of initial conditions (and solutions); maybe more refined deterministic methods allow one to go beyond and treat certain classes of less regular initial conditions. But the most remarkable extension is made by considering solvability with random initial conditions, where probability selects suitable data.

A core question in these investigations is the interplay between rough inputs and nonlinearity. Fully nonlinear PDEs represent a sort of extreme class, where classical ideas of stochastic analysis cannot be applied directly. This is another direction we wanted to emphasise in the summer school.

Despite the different subjects, the main theme of interplay of randomness and nonlinear effects resulted in unexpected connections and analogies between the various approaches and results presented. For example, the space-time regularity of certain classes of Gaussian processes and some multilinear statistics thereof play clearly a main role in the theory of regularity structures as in the theory of energy solutions or in the analysis of the effect of random initial conditions on low regularity dispersive equations. Rough path theory is another common theme which showed up in regularity structures and in the analysis of viscosity solution of fully nonlinear SPDEs.

The Courses

Four courses of 6 h each were delivered to develop these ideas.

Massimiliano Gubinelli presented the approach to stochastic Burgers equation, related to the KPZ equation, based on the concept of energy solutions. Solving the stochastic Burgers equation is a priori very difficult since the square of white noise appears. Nevertheless, there is a way to define this operation under the regularisation of heat semigroup, as it appears in the variation of constant formulation. The series of lectures of Gubinelli presented the foundations of this approach with the basic existence result, a recent uniqueness result and ideas around the problem of weak universality of the KPZ equation.

The lectures of Martin Hairer were devoted to his recent theory of regularity structures, developed to deal with KPZ and other singular equations. This far-reaching theory leverages ideas from T. Lyons' rough path theory and generalises them to a multidimensional context to provide an analytic well-posedness theory for equations where the nonlinearities are apriori not well-defined. It does so by introducing tools to give a precise local description of a distribution far beyond what is possible with standard functional spaces and good enough to resolve the singularities in the nonlinear operations and decompose them into well-defined contributions. The stochastic quantisation equation for the Φ^4_3 Euclidean field theory has been the running example used in the lectures to illustrate the application of the general framework.

Martin Hairer wrote several introductions to his theory. In order to make his contribution to the volume original, we decided instead to write a review summarising the achievements in the field by him, his school and related groups and which could help the reader to find its way in the steadily increasing literature (see the Introduction).

Panagiotis E. Souganidis presented the theory, developed mainly in collaboration with P. L. Lions, about the pathwise weak solvability of fully nonlinear PDEs with rough time-dependent inputs, including Brownian motion. The lectures covered two classes of scalar fully nonlinear first- and second-order degenerate parabolic stochastic partial differential equations, including Hamilton-Jacobi equations and multidimensional scalar conservation laws. The lectures contained also the discussion of several examples and motivations, like the motion of interfaces and stochastic selection principle, or stochastic control and mean field games.

Nikolay Tzvetkov considered, as a main example of dispersive PDE, the nonlinear wave equation in space dimension 3. After basic definitions and properties, he described in detail some key result of the deterministic theory, for comparison with the probabilistic progresses. In particular, he showed well-posedness for relatively regular initial conditions by energy methods and for less regular ones by Strichartz estimates. Then, he presented the approach based on random initial data in classes of more singular functions. Here, new, probabilistic, Strichartz estimates can be proved and used perturbatively to solve the wave equation. Several additional results have been discussed, like large deviation bounds, continuous dependence, invariant measures and the more recent research direction of quasi-invariant measures, namely, the property of absolute continuity of the law at later time with respect to the law at time zero.

Final Remarks

The summer school was attended by 83 PhD students, young and more senior researchers, roughly one fourth of them from Italy and the remaining ones from a wide range of countries including France, Germany, Norway, Turkey, Greece, the United States, the United Kingdom and the Netherlands. All the junior participants (PhD or Master students, which formed the majority of the attendance) were fully supported by CIME and the European Mathematical Society with additional funds contributed by various other institutions: the Hausdorff Center for Mathematics in Bonn, the Laboratorio Ypatia delle Scienze Matematiche (joint project AMU-ECM-CNRS-INdAM), the ERC grant Dispeq of Nikolay Tzvetkov, the ERC grant of Martin Hairer and the IUF fellowship of Lorenzo Zambotti.

We thank the lecturers and all participants for their contributions to the success of the school. Moreover, we thank CIME staff for their efficient and continuous help.

Pisa, Italy   Franco Flandoli
Bonn, Germany   Massimiliano Gubinelli
London, UK   Martin Hairer

Contents

Chapter 1
Introduction

Franco Flandoli, Massimiliano Gubinelli, and Martin Hairer

One of the most remarkable recent progresses in stochastic analysis has been the invention of regularity structures (RS) as a tool to rigorously understand SPDEs which were previously only written down at a formal level without clear mathematical meaning.

A regularity structure allows to describe the local (in space-time) behaviour of a distribution by generalising the notion of polynomials to include basis elements which are themselves genuine distributions (possibly random), like for example certain functionals of white noise. Those basis elements are called a *model*. The family of local descriptions form naturally the section of a certain bundle over which a notion of parallel transport can be introduced in order to compare local descriptions in different points. This gives rise to natural subspaces of descriptions which changes "smoothly", therefore disconnecting the concept of regularity in the sense of classical functional spaces to that of regularity in the sense of allowing good analytic control. A Banach topology can be introduced under which the spaces of sufficiently regular sections of this bundle are called *modelled distributions*. In order to go from a section of this bundle of local regular descriptions to an actual distribution which delivers a numerical value when tested with a smooth test function we need to *reconstruct* such *global* description starting from the section of *local* descriptions provided by the modelled distribution. That this reconstruction

F. Flandoli
Faculty of Sciences, Scuola Normale Superiore, Pisa, Italy
e-mail: franco.flandoli@sns.it

M. Gubinelli (✉)
Hausdorff Center for Mathematics, University of Bonn, Bonn, Germany
e-mail: gubinelli@iam.uni-bonn.de

M. Hairer
Department of Mathematics, Imperial College London, London, UK
e-mail: m.hairer@imperial.ac.uk

© Springer Nature Switzerland AG 2019
F. Flandoli et al. (eds.), *Singular Random Dynamics*, Lecture Notes
in Mathematics 2253, https://doi.org/10.1007/978-3-030-29545-5_1

is possible (and unique under certain regularity conditions on the model and the modelled distribution) is the content of the *reconstruction* theorem which is the linchpin of the theory. In words, it states that there exists a unique distribution which is coherent with the family of local descriptions, somewhat like the fundamental theorem of calculus states that there exists only one primitive (modulo constants) which possesses a given derivative (which is a local description of the function itself).

The local description provided by the modelled distribution can then be used to analyse the behaviour of non-linear operations among distributions which are not controlled by standard functional spaces. In some sense, modelled distributions provide refined notions of regularity well beyond classical concepts like Sobolev or Besov regularity or even Hörmander's wave front sets, which are tailored to work for specific equations and capture the local structure of their solutions in a way effective to analyse nonlinear operations.

RS is composed of various facets which are designed to handle in quite large generality every natural class of singular SPDEs whose local behaviour can be understood "perturbatively". In the theory, this idea is encoded in the notion of *criticality*. Equations amenable to the RS analysis are those whose scaling behaviour is *subcritical*. Intuitively this means that, by blowing up locally the equation (or a smooth approximation), it should behave more and more like a linear equation with the non-linear terms gaining small factors due to the rescaling. The power of the method lies in the fact that this class still includes a large number of equations which are completely out of control with more classical theories. Let us give some well-known and relevant examples:

1. The (generalised) KPZ equation

$$\partial_t h(t, x) - \Delta h(t, x) + \Gamma(h(t, x))(\partial_x h(t, x))^2$$
$$= \sigma(h(t, x))\xi(t, x), \qquad (t, x) \in \mathbb{R}_+ \times \mathbb{T}, \qquad (1.1)$$

where here and in the following we will denote by ξ a space-time (or space) white noise.

2. The dynamic Φ_d^4 model for $d = 2, 3$

$$\partial_t \varphi(t, x) - \Delta \varphi(t, x) + \lambda(\varphi(t, x))^3 = \xi(t, x), \qquad (t, x) \in \mathbb{R}_+ \times \mathbb{T}^d. \qquad (1.2)$$

3. The Anderson model in two (and three) dimensions

$$\partial_t u(t, x) - \Delta u(t, x) - \xi(x)u(t, x) = 0, \qquad (t, x) \in \mathbb{R}_+ \times \mathbb{T}^{2,3}. \qquad (1.3)$$

All of these equations are purely formal, in order to be well defined they need suitable (infinite) counterterms which we haven't indicated explicitly. They have the heuristic intent of suggesting which kind of standard nonlinear PDE constitutes a suitable approximation once the noise terms are duly regularised.

Singular SPDEs (SSPDEs) are not just a mathematical challenge, devoid of physical applications. Quite the contrary, they are often the mathematical counterpart of very important and fundamental phenomenological physical theories which try to describe the large scale fluctuations of random extended systems. Here we understand 'large scale' in the sense that the mechanisms generating these random fluctuations live at scales which are much smaller than the typical scale of observation. From this point of view, one may compare SSPDEs with hydrodynamic limits and fluctuations around them in the theory of interacting particle systems, or with stochastic homogenisation theory. In these two examples however, there are only two scales which play a fundamental role: the microscopic scale and the macroscopic one, thus giving rise to Gaussian fluctuations. In SSPDEs, all the intermediate scales stay relevant, which gives rise to non-linear fluctuations and to non-Gaussian limiting random fields whose space-time dynamical features the SSPDEs are describing. Very much like in the martingale problem formulation of Markovian stochastic dynamics, the complicate nature of these fluctuations is reduced (via the non-linear dynamics) to its purest form, that of local uncorrelated Gaussian random fluctuations of white noise type.

In the following, we aim to give a reasoned guide to the current and fast evolving literature around singular SPDEs, RS and other related methods.

The very first success in handling SSPDEs has been Hairer's local solution theory for the KPZ equation [31] which was obtained using Lyons' rough path theory [49–51] and in particular, controlled paths [18, 22]. Very soon after, Hairer invented RS [33] and used them to give a local solution theory for the dynamical Φ_3^4 model (1.2). At the same time Gubinelli et al. introduced the notion of paracontrolled distributions [27] and later Kupiainen [47] applied renormalisation group ideas to give an alternative approach to the dynamical Φ_3^4 model. Later on, Otto and Weber also developed an alternative approach [58] to the study of equations described by regularity structures, see also the follow up paper by Otto et al. [59].

Nice introductions to some aspects of these theories can be found in the book of Friz and Hairer [18], in the papers of Mourrat et al. [57], in the ICM 2014 proceedings contribution of Hairer [32], in his course for the Brasilian School of Probability in 2015 [34] and in the more recent survey paper [36] by Hairer.

Starting with these initial developments many models were considered. Catellier and Chouk [11] gave a local solution theory for Φ_3^4 using paracontrolled distributions. Later, Mourrat and Weber constructed global space-time solutions for Φ_2^4 [56] and global solution in time on the torus for Φ_3^4 [55]. In this last work, they proved that the dynamic equation "comes down from infinity" in finite time, namely after a fixed time interval the solution can be bounded independently of the initial condition, a very strong property related to the coercive term $-\varphi^3$. More recently, Gubinelli and Hofmanová [25] proved global space-time existence and coming down from infinity for the Φ_3^4 model and for its elliptic analogue which lives naturally in $d = 5$. In doing so they developed a general technique to handle the full space situation in paracontrolled equations.

For Φ_d^4 many weak universality results are also known. Mourrat and Weber [54] proved that the Glauber dynamics for the Ising–Kac model (an Ising model with mesoscopic interactions of mean-field type) converges to the dynamic Φ_2^4 model, and this was later extended to more complex microscopic dynamics by Shen and Weber [62], showing that they can also converge to the dynamic Φ_2^6 model in some situations.

In $d = 3$, Hairer and Xu [45] gave the first weak universality result for Φ_3^4 by proving that a class of reaction diffusion equations with polynomial nonlinearities converge to it in a suitable crossover regime. Shen and Xu [63] examined the case of a regularised Φ_3^4 model with non-Gaussian noise and proved convergence to the Gaussian driven Φ_3^4 model under a space-time rescaling. Furlan and Gubinelli [19] proved the convergence of non-polynomial reaction-diffusion equations using Malliavin calculus. At the same time, an alternative approach to non-polynomial non-linearities was found by Hairer and Xu [46] who introduced a method based on proving probabilistic limit theorems for trigonometric polynomials of Gaussian random variables.

Dynamics can be used also to prove properties of the invariant measure, very much in the spirit of Parisi and Wu's original idea of *stochastic quantisation*. Hairer and Iberti [37] used the stationary Glauber dynamics to prove tightness of the Ising–Kac model in $d = 2$. Albeverio and Kusuoka [1] used a similar strategy in the Φ_3^4 case on the torus giving a proof of tightness of the Φ_3^4 measure which does not use the techniques of constructive quantum field theory. More recently, Gubinelli and Hofmanová [24] extended and simplified the analysis of Albeverio and Kusuoka to handle the Φ_3^4 model in the full space: they proved tightness and an integration by parts formula valid for any accumulation point.

As far as the KPZ equation is concerned, Gubinelli and Perkowski studied the KPZ equation on the torus proving convergence of discrete approximations, justifying rigorously the relation with the stochastic heat equation (SHE) and proving a variational description of the solution [26]. More recently, Perkowski and Rosati [60] proved global space-time existence for KPZ by relating it in a pathwise manner to the random directed polymer measure and also providing a renormalised stochastic control interpretation for it. A vectorial version of the KPZ equation motivated by phenomenological fluctuating hydrodynamic theory has been analysed by Kupiainen and Marcozzi [48] via the RG approach. It is interesting to note that the 3D Navier–Stokes equations driven by space-time white noise exhibit powercounting properties analogous to those of the KPZ equation, major differences being the presence of the pressure term and the vectorial nature of the equation. These hurdles were overcome by Zhu and Zhu [64].

Weak universality results for the KPZ equation started with the paper of Bertini and Giacomin [6] which showed that the logarithm of the solution of the stochastic heat equation describes the large scale fluctuations of the density of the weakly asymmetric simple exclusion process. This result predates by many years the invention of the modern approach to singular SPDEs and the KPZ equation in particular, but was one of the main motivations to develop a theory for such equations because it showed that there was a need of a theory capable of describing directly the dynamics of the fluctuations without resorting to taking the logarithm of the SHE (the so called Hopf–Cole transform), which is very problem-specific. Using regularity structures, Hairer and Quastel [42] were able to show that a large class of growth models in 1d (on a periodic domain) described by SPDEs with additive Gaussian noise converge to the KPZ equation in the weak asymmetry, large scale limit. Later on, Hairer and Shen [44] developed general techniques to handle non-Gaussian noise in the setup of Hairer and Quastel providing generalised central limits for these growth models. Indeed they showed that a large class of random fields with short range correlations driving a 1d weakly asymmetric growth model will generate the additive white noise perturbations appearing in KPZ, irrespective of their initial distribution. Cannizzaro and Matetski [10], using the adaptation to the discrete setting of RS theory developed by Hairer and Matetski [41], analyse a class of discrete KPZ equations and prove the convergence to their continuous counterparts. More recently, Matetski [53] developed a general strategy to handle noises defined via martingale properties, as those appearing in interacting particle systems and shows convergence to RS models driven by space-time Brownian motions, paving the way to very general weak universality statements.

Another class of models which played a fundamental role in the initial development of the theories of SSPDEs are the 2d and 3d Anderson models (1.3). A generalised version of the 2d model (on the torus) was used by Gubinelli et al. [27] to exemplify the application of paracontrolled distributions to singular SPDEs. The case of the full space was been treated by Hairer and Labbé in [39] for the 2d case and in [40] for the 3d case. A support theorem for the generalised Anderson model in 2d was obtained by Chouk and Friz [14].

A different point of view on the problem was taken by Allez and Chouk [2] who proved that Anderson's Hamiltonian operator $H = \Delta - \xi$ in $d = 2$ can be renormalised in such a way that it is self-adjoint and bounded below, although it has some rather unusual features. (For example, its domain does not contain any smooth function, except 0.) They also analyse some of its spectral properties. Perkowski and Martin [52] prove convergence of discrete versions of Anderson's Hamiltonian to its continuous counterpart via a small noise limit. In doing so they developed basic tools to apply paracontrolled calculus in a discrete setting. Non-linear Schrödinger evolutions (on a periodic 2d domain) associated with this operator were first considered by Debussche and Weber [17] and later extended to hyperbolic equations and to $d = 3$ by Gubinelli et al. [30] and to the full 2d space dispersive setting by Debussche and Martin [16].

These are some of the first results for equations which are not of parabolic type. Other results were obtained by Gubinelli et al. [29] for the 2d wave equation with polynomial nonlinearities, including also some weak universality results for 2d hyperbolic equations with additive space-time white noise. More recently the same authors solved in [28] also a 3d wave equation with quadratic nonlinearity by introducing a new paracontrolled decomposition involving certain random operators. A general solution theory for these nonlinear hyperbolic equation has not yet been developed, especially due to the fact that the notions of homogeneity and regularity developed for parabolic equations seems less amenable to capture the local behaviour of the solution and less efficient in controlling the non-linear interaction term.

In [43] Hairer and Shen gave a solution theory for the dynamical sine-Gordon model for certain values of the parameter α appearing in the equation. This model is interesting since it requires a multiplicative renormalisation and the scaling properties of the random objects needed to construct the RS models depend on a parameter. The Hairer–Shen solution reaches the level of "first order" expansion, well below the full range of values for α for which the equation is believed to be subcritical. While the analytic part of RS theory developed in [33] is capable to handle in complete generality the full subcritical regime of any non-linear SPDE with smooth coefficients, the renormalisation step and the analysis of the random RS model are usually done by hand and this become very rapidly a daunting task as the regularity of the stochastic objects reaches the critical scaling. For this reason Hairer and Shen limited their analysis to a relatively small range of values of α.

This limitation also plagues other very interesting applications of RS: for example the generalised KPZ equation (1.1) described in [35] and recently analysed in detail in [8]. The regularity structure needed to describe solutions to this equation contains a very large number of basis objects and it is unfeasible to prove by hand the renormalisation and the analytic properties needed in RS for all of them. In order to provide a complete "black box" theory which encompasses all these interesting but more complex situations one needs two new ingredients: a good understanding of the algebraic properties of the subtraction of diverging quantities, which is termed *negative renormalisation* and an equally good understanding of the centering procedure of the monomials in the RS model needed to leverage the notion of *homogeneity* (similar to physical scaling degree) which is crucial in the analytic estimates. This second subtraction procedure, which produces terms very much like a generalised Taylor remainder, is also called *positive renormalisation*. Positive and negative renormalisation do not commute and their interplay can be quite complex from a general point of view. In order to deal with this complexity, Bruned et al. [9] put in place a theory of co-interacting Hopf algebras which describe the abstract action of a large class of negative and positive renormalisation schemes on the RS proving that those actions are compatible with the requirements of the analytic side of RS theory. The use of Hopf algebras to describe negative renormalisation is reminiscent of the approach of Connes and Kreimer [15] to renormalisation of perturbative QFT, while their use for the positive renormalisation has been pioneered by Hairer and Kelly [38] in the description of the branched rough paths introduced in [23]. On the other hand, Chandra and Hairer [12] provided a multiscale probabilistic analysis of a general class of random fields with the aim to implement a specific negative renormalisation scheme applicable to a large class of RS models constructed from random fields (like white noise or more generally, stationary generalised random fields satisfying a hierarchy of cumulant bounds) and kernel convolutions. This so called *BHPZ renormalisation* (After Bogoliubov, Parasiuk, Hepp and Zimmermann who introduced the analogous renormalisation scheme to the analysis of perturbative QFT) provides a general scheme to turn the RS models arising in subcritical equations into well defined renormalised ones.

Finally Bruned et al. [7] provided the missing piece, namely a proof that the negative renormalisation provided by the BHPZ construction can be implemented via a transformation of the equation being solved by the introduction of appropriate *local* counterterms. This guarantees that for any subcritical equation there is a (finite) number of counterterms which one has to include in the equation so that, by suitably tuning these additional terms, regularised approximate solutions converge to a limiting object in a robust fashion. All together the four papers [7, 9, 12, 33] provide a complete, robust and systematic local well-posedness theory which is able to handle any smooth quasilinear parabolic equation driven by a general class of space-time random fields. Using this general theory Chandra et al. [13] were able to complete the analysis of the sine-Gordon model and provide a well-posedness theory in the full subcritical regime.

One interesting recent development is the analysis of the interplay between symmetries and renormalisation. To some extent, this grew out of the desire to gain a deeper understanding for the reason behind the exact cancellation between the two logarithmic divergencies appearing in [31]. The recent article [8] provides a very simple argument in support of the heuristic statement that 'if one can formally approximate a class of singular SPDEs in such a way that it exhibits a given symmetry, then it can be renormalised in a way that preserves this symmetry'. The generalised KPZ equation (1.1) exhibits two infinite-dimensional symmetries: on the one hand, this class of equations is formally invariant under changes of coordinates $h \mapsto \phi \circ h$ of the target space (which could be more than one-dimensional). On the other hand, Itô's isometry suggests that the law of the solution should only depend on $\sigma \sigma^\top$ rather than σ itself. Somewhat surprisingly, it turns out that *both* of these symmetries can be preserved simultaneously, which is quite unlike the case of SDEs where Stratonovich solutions satisfy the first symmetry but not the second one and Itô solutions satisfy the second one but not the first! Symmetry arguments are expected to play a crucial role in the stochastic quantisation of gauge theories which was recently performed in the simples case (2D, Abelian) in [61].

There have been also attempts to generalise paracontrolled distributions to handle more general equations by Bailleul and Bernicot who developed in [3] a paracontrolled calculus which can work in the manifold setting via heat kernel estimates and also higher order commutator estimates [4] to try to tackle less regular equations. However, paracontrolled calculus remains less general than RS theory as yet, the latter having reached a very complete state.

Let us mention also that through a series of paper published more or less at the same time, Bailleul et al. [5], Otto and Weber [58], Furlan and Gubinelli [20], Gerencsér and Hairer [21] developed extensions of the above theories to handle quasilinear equations. Quasilinear equations feature non-linearities also in the highest derivative term and, in the SSPDEs setting, these nonlinearities require appropriate renormalisation. The above papers provide more or less specific solutions to this problem. One key insight is that of Otto and Weber [58] who suggested to generalise the notion of the model (like in RS) to a parametric one in order to be able to perturb it in ways depending on the solution and obtain descriptions suitable to well approximate the local behaviour of the solutions to such quasilinear equations.

References

1. S. Albeverio, S. Kusuoka, The invariant measure and the flow associated to the Φ_3^4-quantum field model. Ann. Sc. Norm. Super. 56 (2019). https://doi.org/10.2422/2036-2145.201809_008
2. R. Allez, K. Chouk, The continuous Anderson Hamiltonian in dimension two (2015). arXiv e-prints, arXiv:1511.02718
3. I. Bailleul, F. Bernicot, Heat semigroup and singular PDEs. J. Funct. Anal. **270**(9), 3344–3452 (2016). With an appendix by F. Bernicot and D. Frey
4. I. Bailleul, F. Bernicot, Higher order paracontrolled calculus (2016). arXiv:1609.06966 [math]
5. I. Bailleul, A. Debussche, M. Hofmanová, Quasilinear generalized parabolic Anderson model equation. Stoch. Part. Differ. Equ. Anal. Comput. **7**(1), 40–63 (2019)
6. L. Bertini, G. Giacomin, Stochastic Burgers and KPZ equations from particle systems. Commun. Math. Phys. **183**(3), 571–607 (1997)
7. Y. Bruned, A. Chandra, I. Chevyrev, M. Hairer, Renormalising SPDEs in regularity structures (2017). arXiv:1711.10239 [math]
8. Y. Bruned, F. Gabriel, M. Hairer, L. Zambotti, Geometric stochastic heat equations (2019). arXiv e-prints, arXiv:1902.02884
9. Y. Bruned, M. Hairer, L. Zambotti, Algebraic renormalisation of regularity structures. Invent. Math. **215**(3), 1039–1156 (2019)
10. G. Cannizzaro, K. Matetski, Space-time discrete KPZ equation. Commun. Math. Phys. **358**(2), 521–588 (2018)
11. R. Catellier, K. Chouk, Paracontrolled distributions and the 3-dimensional stochastic quantization equation. Ann. Probab. **46**(5), 2621–2679 (2018)
12. A. Chandra, M. Hairer, An analytic BPHZ theorem for regularity structures (2016). arXiv: 1612.08138 [math-ph]
13. A. Chandra, M. Hairer, H. Shen, The dynamical sine-Gordon model in the full subcritical regime (2018). arXiv:1808.02594 [math-ph]
14. K. Chouk, P.K. Friz, Support theorem for a singular SPDE: the case of gPAM. Ann. Inst. Henri Poincaré Probab. Stat. **54**(1), 202–219 (2018)
15. A. Connes, D. Kreimer, Hopf algebras, renormalization and noncommutative geometry. Commun. Math. Phys. **199**(1), 203–242 (1998)
16. A. Debussche, J. Martin, Solution to the stochastic Schrödinger equation on the full space. Nonlinearity **32**(4), 1147–1174 (2019)
17. A. Debussche, H. Weber, The Schrödinger equation with spatial white noise potential. Electron. J. Probab. **23**, 16 (2018). Paper No. 28
18. P.K. Friz, M. Hairer, *A Course on Rough Paths: With an Introduction to Regularity Structures* (Springer, Berlin, 2014)
19. M. Furlan, M. Gubinelli, Weak universality for a class of 3d stochastic reaction–diffusion models. Probab. Theory Relat. Fields **173**, 1099–1164 (2019)
20. M. Furlan, M. Gubinelli, Paracontrolled quasilinear SPDEs. Ann. Probab. **47**(2), 1096–1135 (2019)
21. M. Gerencsér, M. Hairer, A solution theory for quasilinear singular SPDEs (2017). arXiv e-prints, arXiv:1712.01881
22. M. Gubinelli, Controlling rough paths. J. Funct. Anal. **216**(1), 86–140 (2004)
23. M. Gubinelli, Ramification of rough paths. J. Differ. Equ. **248**(4), 693–721 (2010)
24. M. Gubinelli, M. Hofmanová, A PDE construction of the Euclidean Φ_3^4 quantum field theory (2018). arXiv e-prints, arXiv:1810.017000

25. M. Gubinelli, M. Hofmanová, Global solutions to elliptic and parabolic ϕ^4 models in Euclidean space. Commun. Math. Phys. **368**(3), 1201–1266 (2019)
26. M. Gubinelli, N. Perkowski, KPZ reloaded. Commun. Math. Phys. **349**(1), 165–269 (2017)
27. M. Gubinelli, P. Imkeller, N. Perkowski, Paracontrolled distributions and singular PDEs. Forum Math. Pi **3**, e6, 75 (2015)
28. M. Gubinelli, H. Koch, T. Oh, Paracontrolled approach to the three-dimensional stochastic nonlinear wave equation with quadratic nonlinearity (2018). arXiv:1811.07808 [math]
29. M. Gubinelli, H. Koch, T. Oh, Renormalization of the two-dimensional stochastic nonlinear wave equations. Trans. Am. Math. Soc. **370**, 7335–7359 (2018)
30. M. Gubinelli, B. Ugurcan, I. Zachhuber, Semilinear evolution equations for the Anderson Hamiltonian in two and three dimensions. Stoch. Part. Differ. Equ. Anal. Comput., 1–68 (2019). https://doi.org/10.1007/s40072-019-00143-9
31. M. Hairer, Solving the KPZ equation. Ann. Math. **178**(2), 559–664 (2013)
32. M. Hairer, Singular stochastic PDEs, in *Proceedings of the International Congress of Mathematicians—Seoul 2014*, vol. 1 (Kyung Moon Sa, Seoul, 2014), pp. 685–670
33. M. Hairer, A theory of regularity structures. Invent. Math. **198**(2), 269–504 (2014)
34. M. Hairer, Introduction to regularity structures. Braz. J. Probab. Stat. **29**(2), 175–210 (2015)
35. M. Hairer, The motion of a random string (2016). arXiv:1605.02192 [math-ph]
36. M. Hairer, Renormalisation of parabolic stochastic PDEs. Jpn. J. Math. **13**(2), 187–233 (2018)
37. M. Hairer, M. Iberti, Tightness of the Ising-Kac model on the two-dimensional torus. J. Stat. Phys. **171**(4), 632–655 (2018)
38. M. Hairer, D. Kelly, Geometric versus non-geometric rough paths. Ann. Inst. Henri Poincaré Probab. Stat. **51**(1), 207–251 (2015)
39. M. Hairer, C. Labbé, A simple construction of the continuum parabolic Anderson model on $\mathbf{R}^2$. Electron. Commun. Probab. **20**(43), 11 (2015)
40. M. Hairer, C. Labbé, Multiplicative stochastic heat equations on the whole space. J. Eur. Math. Soc. **20**(4), 1005–1054 (2018)
41. M. Hairer, K. Matetski, Discretisations of rough stochastic PDEs. Ann. Probab. **46**(3), 1651–1709 (2018)
42. M. Hairer, J. Quastel, A class of growth models rescaling to KPZ. Forum Math. Pi **6**, e3, 112 (2018)
43. M. Hairer, H. Shen, The dynamical sine-Gordon model. Commun. Math. Phys. **341**(3), 933–989 (2016)
44. M. Hairer, H. Shen, A central limit theorem for the KPZ equation. Ann. Probab. **45**(6B), 4167–4221 (2017)
45. M. Hairer, W. Xu, Large-scale behavior of three-dimensional continuous phase coexistence models. Commun. Pure Appl. Math. **71**(4), 688–746 (2018)
46. M. Hairer, W. Xu, Large-scale limit of interface fluctuation models (2018). arXiv:1802.08192 [math-ph]
47. A. Kupiainen, Renormalization group and stochastic PDEs. Ann. Henri Poincaré **17**(3), 497–535 (2016)
48. A. Kupiainen, M. Marcozzi, Renormalization of generalized KPZ equation. J. Stat. Phys. **166**(3–4), 876–902 (2017)
49. T. Lyons, Differential equations driven by rough signals. Revista Matemática Iberoamericana **14**, 215–310 (1998)
50. T. Lyons, Z. Qian, *System Control and Rough Paths* (Oxford University Press, Oxford, 2002)
51. T.J. Lyons, M.J. Caruana, T. Lévy, *Differential Equations Driven by Rough Paths: Ecole d'Eté de Probabilités de Saint-Flour XXXIV-2004*, 1st edn. (Springer, Berlin, 2007)
52. J. Martin, N. Perkowski, Paracontrolled distributions on Bravais lattices and weak universality of the 2d parabolic Anderson model (2017). arXiv:1704.08653 [math]
53. K. Matetski, Martingale-driven approximations of singular stochastic PDEs (2018). arXiv e-prints, arXiv:1808.09429

54. J.-C. Mourrat, H. Weber, Convergence of the two-dimensional dynamic Ising-Kac model to Φ_2^4. Commun. Pure Appl. Math. **70**(4), 717–812 (2017)
55. J.-C. Mourrat, H. Weber, The dynamic Φ_3^4 model comes down from infinity. Commun. Math. Phys. **356**(3), 673–753 (2017)
56. J.-C. Mourrat, H. Weber, Global well-posedness of the dynamic ϕ^4 model in the plane. Ann. Probab. **45**(4), 2398–2476 (2017)
57. J.-C. Mourrat, H. Weber, W. Xu, Construction of Φ_3^4 diagrams for pedestrians, in *From Particle Systems to Partial Differential Equations*. Springer Proceedings in Mathematics and Statistics, vol. 209 (Springer, Cham, 2017), pp. 1–46
58. F. Otto, H. Weber, Quasilinear SPDEs via rough paths. Arch. Ration. Mech. Anal. **232**(2), 873–950 (2019)
59. F. Otto, J. Sauer, S. Smith, H. Weber, Parabolic equations with rough coefficients and singular forcing (2018). arXiv:1803.07884 [math]
60. N. Perkowski, T. C. Rosati, The KPZ equation on the real line (2018). arXiv:1808.00354 [math]
61. H. Shen, Stochastic quantization of an Abelian gauge theory (2018), arXiv:1801.04596 [math-ph]
62. H. Shen, H. Weber, Glauber dynamics of 2D Kac-Blume-Capel model and their stochastic PDE limits. J. Funct. Anal. **275**(6), 1321–1367 (2018)
63. H. Shen, W. Xu, Weak universality of dynamical Φ_3^4: non-Gaussian noise. Stoch. Part. Differ. Equ. Anal. Comput. **6**, 211–254 (2018)
64. R. Zhu, X. Zhu, Three-dimensional Navier-Stokes equations driven by space-time white noise. J. Differ. Equ. **259**(9), 4443–4508 (2015)

Chapter 2
Lectures on Energy Solutions
for the Stationary KPZ Equation

Massimiliano Gubinelli

Abstract These are a set of lectures delivered at the CIME-EMS Summer School in Applied Mathematics "Singular Random Dynamics" which have been held from 22 to 26 August 2016 in Cetraro, Italy. The goal of these lectures is to introduce the concept of *energy solution* for the Kadar–Parisi–Zhang equation and to discuss the application of this notion of solution to the analysis of the scaling limit of certain weakly-asymmetric growth processes.

2.1 Introduction

In these notes I will describe the martingale problem approach to the stationary KPZ equation introduced by Gonçalves and Jara [8, 9] and Gubinelli and Jara [11] under the name of *energy solutions*. Recent progress allowed to establish uniqueness for this formulation [14] and convergence results for various models [3, 5, 7, 10, 13]. These notes will be based on the material contained in the works [3, 11, 13, 14] and on the lecture notes [12] which covered the basic existence results. Here we will discuss also uniqueness and give an example of convergence results for a particular class of models. Note that recently a more general approach to uniqueness of the martingale problem has been introduced in [15] via a careful study of the formal generator of the process.

The equation we are interested in is the following SPDE, called the Kadar–Parisi–Zhang equation (KPZ)

$$\partial_t h(t, x) = \Delta h(t, x) + [(\partial_x h(t, x))^2 - \infty] + \xi(t, x), \qquad x \in \mathbb{T}, t \geqslant 0, \quad (2.1)$$

where $\mathbb{T}$ is the one dimensional torus, Δ the periodic Laplacian, ξ a space-time white noise and where the notation ∞ hints to the fact the quadratic terms needs a

M. Gubinelli (✉)
Hausdorff Center for Mathematics, University of Bonn, Bonn, Germany
e-mail: gubinelli@iam.uni-bonn.de

© Springer Nature Switzerland AG 2019
F. Flandoli et al. (eds.), *Singular Random Dynamics*, Lecture Notes
in Mathematics 2253, https://doi.org/10.1007/978-3-030-29545-5_2

renormalization procedure in order to be properly defined. This equation has been introduced (on the full line) by Kardar et al. in the celebrated paper [19] appeared in 1986 in the Physics literature in order to give a universal description of fluctuations of growing interfaces in one dimension. This topic generated a vast physics literature which is not our aim here to review. From the mathematical point of view the status of the equations is not a-priori very clear since it cannot be given a meaning via standard SPDE techniques due to difficulties to define the non-linear term (more on this later on). The first rigorous results on KPZ are contained in a paper by Bertini and Giacomin [2] where they (morally) proved the convergence of a the density field of an interacting particle system towards the solution $u(t, x)$ of the Stochastic Burgers Equation (SBE)

$$\partial_t u(t, x) = \Delta u(t, x) + \partial_x (u(t, x))^2 + \partial_x \xi(t, x), \qquad x \in \mathbb{T}, t \geqslant 0, \qquad (2.2)$$

which can be interpreted as the derivative of the KPZ equation. Again this SPDE is not well-posed (and not even well-defined) and their precise result is the convergence of the density field to a random field u which satisfies $u(t, x) = \partial_x \log Z(t, x)$ where Z is the solutions to the Stochastic Heat Equation (SHE) with multiplicative space-time white noise, namely the SPDE

$$\partial_t Z(t, x) = \Delta Z(t, x) + Z(t, x)\xi(t, x), \qquad x \in \mathbb{T}, t \geqslant 0, \qquad (2.3)$$

which is well-posed if understood via stochastic calculus for the cylindrical Wiener process W in $L^2(\mathbb{T})$ linked to the white noise ξ via the distributional relation $\xi = \partial_t \partial_x W(t, x)$. The result of Bertini and Giacomin shows that this indirect formulation of the KPZ equation is the physically correct one (since obtained via a scaling limit of a microscopic model) but it is still quite indirect and proofs of convergence to this kind of solutions are limited to the availability of the exponential transformation (called Hopf–Cole transformation) or some good approximation of it at the microscopic level. For the weakly asymmetric simple exclusion model (the one analysed by Bertini and Giacomin) this transformation can be implemented at the microscopic level, a fact discovered by Gärtner.

For a general review of the mathematical formulation and other results around the Hopf–Cole solutions and the related universality problem we refer to the nice and complete lecture notes of Quastel [21].

A first attempt at an intrinsic formulation of the KPZ/SBE equation is the one by Assign [1] where he manages to describe a *generalized* martingale problem which however is still too weak to allow interesting results. After this work, Jara and Gonçalves precised the notion of stationary martingale solution in their 2010 preprint [8] (whose published version is [9]). In a subsequent paper, Jara and myself gave a slightly different notion of solution based on the idea of identifying a general class of processes which would exhibit certain key path properties, similar to those observed for limiting points of the microscopic dynamics described e.g. [8]. Later on, in collaboration with Perkowski, we have established uniqueness of the refined formulation of the stationary martingale problem (here conventionally called *energy solution*). This last result opens the way to a large class of convergence theorems via the standard approach of stochastic compactness method:

(a) establish tightness of the sequence of macroscopic observables;
(b) prove that any limit point is an *energy solution*;
(c) use uniqueness to deduce the convergence of the whole sequence.

The topics covered in these notes are the following:

(a) Introduce the concept of *controlled process* which describe the possible limiting points of stationary stochastic dynamics which should converge to the SBE equation.
(b) Prove the regularising properties of these processes in the form of good estimates for additive functionals.
(c) Exploit these properties to define in this space a notion of solution to SBE and to prove uniqueness of solutions to SBE. Uniqueness will be the result of implementing the Hopf–Cole transformation at the level of controlled process (via a generalized Itô formula) and verifying that any energy solution is transformed into a solution of a (slightly modified SHE) which in turns enjoy pathwise uniqueness.
(d) Use the regularizing properties of controlled processes to prove that some rescaling of certain microscopic additive functionals converge to simpler macroscopic ones and use this to establish certain invariance principles, for example the convergence of a class of SPDEs to the SBE via suitable space-time rescalings.

2.1.1 Notations and Some Preliminaries

Let $\mathbb{T} = \mathbb{R}/(2\pi\mathbb{Z})$ be the one dimensional torus, $\mathscr{S}$ the space of C^∞ functions on $\mathbb{T}$ and $\langle f, g \rangle = \int_{\mathbb{T}} f(x)g(x)dx$ the $L^2(\mathbb{T})$ scalar product. For $k \in \mathbb{Z}$ we denote by $e_k(x) = \exp(-ikx)/(2\pi)^{1/2}$ the Fourier basis. Let $\Pi_0 : L^2(\mathbb{T}) \to L_0^2(\mathbb{T})$ the projection from $L^2(\mathbb{T})$ to $L_0^2(\mathbb{T}) = \{f \in L^2(\mathbb{T}) : \langle e_0, f \rangle = 0\}$, $\Pi_0^N : L^2(\mathbb{T}) \to L_0^2(\mathbb{T})$ the projection on the span of $(e_k)_{0<|k|\leqslant N}$ and by $\Pi^N : L^2(\mathbb{T}) \to L^2(\mathbb{T})$ the projection on the span of $(e_k)_{0\leqslant|k|\leqslant N}$. Let H^α the standard Sobolev spaces on $\mathbb{T}$ of index $\alpha \in \mathbb{R}$ defined as

$$H^\alpha := H^\alpha(\mathbb{T}) := \left\{\rho \in \mathscr{S}' : \|\rho\|_{H^\alpha}^2 := \sum_{k\in\mathbb{Z}}(1 + |k|^2)^\alpha|\rho(e_k)|^2 < \infty\right\}.$$

and $H_0^\alpha = \Pi_0 H^\alpha$ their mean-zero counterparts. The space of distributions $\mathscr{S}' = \mathscr{S}'(\mathbb{T}^d)$ is the set of linear maps f from $\mathscr{S} = C^\infty(\mathbb{T}^d, \mathbb{C})$ to $\mathbb{C}$, such that there exist $k \in \mathbb{N}$ and $C > 0$ with

$$|\langle f, \varphi \rangle| := |f(\varphi)| \leqslant C \sup_{|\mu|\leqslant k} \|\partial^\mu \varphi\|_{L^\infty(\mathbb{T}^d)}$$

for all $\varphi \in \mathscr{S}$. The Fourier transform $\mathscr{F} f : \mathbb{Z}^d \to \mathbb{C}$,

$$\mathscr{F} f(k) = \hat{f}(k) = \langle f, e_k \rangle,$$

with $e_k = e^{-i\langle k, \cdot \rangle} / (2\pi)^{d/2}$, is defined for all $f \in \mathscr{S}'$, and it satisfies $|\mathscr{F} f(k)| \leqslant |P(k)|$ for a suitable polynomial P. Conversely, if $(g(k))_{k \in \mathbb{Z}^d}$ is at most of polynomial growth, then its inverse Fourier transform

$$\mathscr{F}^{-1} g = \sum_{k \in \mathbb{Z}} g(k) e_k^*$$

defines a distribution (here $e_k^* = e^{i\langle k, \cdot \rangle} / (2\pi)^{d/2}$ is the complex conjugate of e_k). Parseval formula

$$\langle f, \varphi^* \rangle_{L^2(\mathbb{T})} = \int_{\mathbb{T}} f(x) \varphi(x)^* \mathrm{d}x = \sum_k \hat{f}(k) \hat{\varphi}(k)^*$$

extends from $f, \varphi \in L^2(\mathbb{T})$ to $f \in \mathscr{S}'$ and $\varphi \in \mathscr{S}$. Moreover for $f \in \mathscr{S}'$, $\varphi \in \mathscr{S}$ and for $u, v : \mathbb{Z} \to \mathbb{C}$ with u of polynomial growth and v of rapid decay

$$\mathscr{F}(f\varphi)(k) = (2\pi)^{-d/2} \sum_{\ell} \hat{f}(k - \ell) \hat{\varphi}(\ell)$$

and

$$\mathscr{F}^{-1}(uv)(x) = (2\pi)^{d/2} \langle \mathscr{F}^{-1} u, (\mathscr{F}^{-1} v)(x - \cdot) \rangle.$$

Linear maps on $\mathscr{S}'$ can be defined by duality: if $A : \mathscr{S} \to \mathscr{S}$ is such that for all $k \in \mathbb{N}$ there exists $n \in \mathbb{N}$ and $C > 0$ with $\sup_{m \leqslant k} \|\partial_x^m (A\varphi)\|_{L^\infty} \leqslant C \sup_{m \leqslant n} \|\partial_x^m \varphi\|_{L^\infty}$, then we set $\langle {}^t A f, \varphi \rangle = \langle f, A\varphi \rangle$. Differential operators are defined by $\langle \partial_x^n f, \varphi \rangle = (-1)^n \langle f, \partial_x^n \varphi \rangle$. Any $\varphi : \mathbb{Z} \to \mathbb{C}$ growing at most polynomially defines a *Fourier multiplier*

$$\varphi(\mathrm{D}) : \mathscr{S}' \to \mathscr{S}', \qquad \varphi(\mathrm{D}) f = \mathscr{F}^{-1}(\varphi \mathscr{F} f).$$

Denote $C_T V = C([0, T], V)$ the space of continuous functions from $[0, T]$ to the Banach space V endowed with the supremum norm and with $C_T^\gamma V = C^\gamma([0, T], V)$ the subspace of γ-Hölder continuous functions in $C_T V$ with the γ-Hölder norm.

We will need also a fixed family of smoothing operators indexed by $L \geqslant 1$. Let $q : \mathbb{T} \to \mathbb{R}_+$ be an even smooth function of compact support around 0 and such that $\int_{\mathbb{T}} q(x)\mathrm{d}x = 1$. Let $\delta^L(y) = Lq(Ly)$, $\delta_x^L(y) = \delta^L(x - y)$, $\rho_x^L = \delta_x^L - 1/(2\pi)$ and $\mathscr{I}_0^L f = \rho^L * f$. We let also $\rho_x^\infty(y) = \delta_x(y) - 1/(2\pi)$ and then $\mathscr{I}_0 f = \rho^\infty * f = \Pi_0 f$.

2.1.2   White Noise

Fix a complete probability space $(\Omega, \mathcal{F}, \mathbb{P})$ where is defined a spatial white noise η on $\mathbb{T}$, i.e. η is a centred Gaussian process indexed by $L^2(\mathbb{T})$, with covariance

$$\mathbb{E}[\eta(f)\eta(g)] = \int_{\mathbb{T}} f(x)g(x)\mathrm{d}x.$$

We can choose a version of η such that $\eta \in \mathscr{S}'$ almost surely. Indeed letting

$$Q_\lambda := \sum_{k \in \mathbb{Z}} \exp(\lambda |\eta(e_k)|^2/2)/(1 + |k|^2)$$

we have

$$\sup_{k \in \mathbb{Z}} |\eta(e_k)| \leqslant 2\lambda^{-1} \log(1 + |k|^2) + 2\lambda^{-1} \log Q_\lambda$$

and moreover $\mathbb{E}[Q_\lambda] < \infty$ for any $\lambda < 1$. This implies easily the existence of the random distribution η and gives precise informations about the regularity of such a distribution. In particular it holds that $\eta \in H^{-1/2-\varepsilon}(\mathbb{T})$ a.s. for all $\varepsilon > 0$ and we will let μ be the law of η as a random variable taking values in $H^{-1/2-\varepsilon}(\mathbb{T})$ for some fixed and small $\varepsilon > 0$.

The space-time white noise on $\mathbb{R}_+ \times \mathbb{T}$ is similarly defined as the centred Gaussian process ξ indexed by $L^2(\mathbb{R}_+ \times \mathbb{T})$ with covariance

$$\mathbb{E}[\xi(f)\xi(g)] = \int_{\mathbb{R}_+ \times \mathbb{T}} f(t, x)g(t, x)\mathrm{d}t\mathrm{d}x.$$

A nice realisation for this process, the one we will use below, is to define $M_t(\varphi) = \sqrt{2}\xi(\mathbb{I}_{[0,t]}\varphi)$ for $\varphi \in L^2(\mathbb{T})$ and $t \geqslant 0$ and observe that $(M_t(\varphi))_{t \geqslant 0, \varphi \in L^2(\mathbb{T})}$ is a Gaussian random field with covariance

$$\mathbb{E}[M_t(\varphi)M_s(\psi)] = 2(t \wedge s)\langle \varphi, \psi \rangle_{L^2(\mathbb{T})}.$$

In particular, for every $\varphi \in \mathscr{S}$ the stochastic process $(M_t(\varphi))_{t \geqslant 0}$ is a Brownian motion with covariance $2\|\varphi\|_{L^2(\mathbb{T})}^2$. We will use this fact to have a rigorous interpretation of the white noise ξ appearing in our equation. Moreover the notation M stresses the fact that $(M_t(\varphi))_t$ is a martingale in its natural filtration and more generally in the filtration $\mathcal{F}_\bullet = (\mathcal{F}_t)_t$ with $\mathcal{F}_t = \sigma(M_s(\varphi) : s \leqslant t, \varphi \in L^2(\mathbb{T}))$, $t \geqslant 0$.

2.2 The Ornstein–Uhlenbeck Process

Let X be a solution to

$$X_t(\varphi) = X_0(\varphi) + \int_0^t X_s(\Delta\varphi)ds + \partial_x M_t(\varphi) \tag{2.4}$$

for all $t \geqslant 0$ and $\varphi \in \mathscr{S}$. This equation has *at most* one solution (for fixed X_0). Indeed, the difference D between two solutions should satisfy $D_t(\varphi) = \int_0^t D_s(\Delta\varphi)ds$, which means that D is a distributional solution to the heat equation. Taking $\varphi = e_k$ we get $D_t(e_k) = -k^2 \int_0^t D_s(e_k)ds$ and then by Gronwall's inequality $D_t(e_k) = 0$ for all $t \geqslant 0$. To obtain the existence of a solution, observe that

$$X_t(e_k) = X_0(e_k) - k^2 \int_0^t X_s(e_k)ds + \partial_x M_t(e_k)$$

and that $\partial_x M_t(e_0) = 0$. For all $k \neq 0$ the process $\beta_t(k) = M_t(e_k)$ is a complex valued Brownian motion (i.e. real and imaginary part are independent Brownian motions with the same variance). The covariance of β is given by

$$\mathbb{E}[\beta_t(k)\beta_s(m)] = 2(t \wedge s)\delta_{k+m=0}$$

and moreover $\beta_t(k)^* = \beta_t(-k)$ for all $k \neq 0$ (where $\cdot^*$ denotes complex conjugation). In other words, $(X_t(e_k))_{t,k}$ is a complex-valued Ornstein–Uhlenbeck process [18, Example 5.6.8] which solves a linear one-dimensional SDE and has an explicit representation given by

$$X_t(e_k) = e^{-k^2 t} X_0(e_k) - ik \int_0^t e^{-k^2(t-s)}d_s\beta_s(k).$$

This is enough to determine $X_t(\varphi)$ for all $t \geqslant 0$ and $\varphi \in \mathscr{S}$. Moreover $(X_t(e_k) : t \in \mathbb{R}_+, k \in \mathbb{Z})$ is a complex Gaussian random field. If we take $X_0 \sim \eta$ where η is a space white noise, independent of $(\beta_s(k))_{s \geqslant 0, k \in \mathbb{Z}}$ we have that X_t has mean zero and covariance

$$\mathbb{E}[X_t(e_k)X_s(e_m)] = \delta_{k+m=0}e^{-|t-s|k^2}$$

as well as

$$\mathbb{E}[X_t(e_k)X_s(e_m)^*] = \delta_{k=m}e^{-|t-s|k^2}.$$

In particular, $\mathbb{E}[|X_t(e_k)|^2] = 1$. Note that $X_t(e_k) \sim \mathcal{N}_{\mathbb{C}}(0, 1)$ for all $k \in \mathbb{Z}_0$ and $t \in \mathbb{R}$, where we write

$$U \sim \mathcal{N}_{\mathbb{C}}(0, \sigma^2)$$

if $U = V + iW$, where V and W are independent random variables with distribution $\mathcal{N}(0, \sigma^2/2)$. The random distribution X_t then satisfies $X_t(\varphi) \sim \mathcal{N}(0, \|\varphi\|^2_{L^2(\mathbb{T})})$. That is, the white noise on $\mathbb{T}$. It is also possible to deduce that the white noise on $\mathbb{T}$ is indeed the invariant measure of the Ornstein–Uhlenbeck process, that it is the only one, and that it is approached quite fast [18]. Next we examine the Sobolev regularity of X.

Lemma 1 *Let* $\alpha < -1/2$ *almost surely* $X \in CH^\alpha$.

Proof Let $\alpha = -1/2 - \varepsilon$ and consider

$$\|X_t - X_s\|^2_{H^\alpha} = \sum_{k \in \mathbb{Z}} (1 + |k|^2)^\alpha |X_t(e_k) - X_s(e_k)|^2.$$

Let us estimate the $L^{2p}(\Omega)$ norm of this quantity for $p \in \mathbb{N}$ by writing

$$\mathbb{E}\|X_t - X_s\|^{2p}_{H^\alpha} = \sum_{k_1,\ldots,k_p \in \mathbb{Z}} \prod_{i=1}^p (1 + |k_i|^2)^\alpha \mathbb{E} \prod_{i=1}^p |X_t(e_{k_i}) - X_s(e_{k_i})|^2.$$

By Hölder inequality, we get

$$\mathbb{E}\|X_t - X_s\|^{2p}_{H^\alpha} \lesssim \sum_{k_1,\ldots,k_p \in \mathbb{Z}} \prod_{i=1}^p (1 + |k_i|^2)^\alpha \prod_{i=1}^p (\mathbb{E}|X_t(e_{k_i}) - X_s(e_{k_i})|^{2p})^{1/p}.$$

Note now that $X_t(e_{k_i}) - X_s(e_{k_i})$ is a Gaussian random variable, so that there exists a universal constant C_p for which

$$\mathbb{E}|X_t(e_{k_i}) - X_s(e_{k_i})|^{2p} \leqslant C_p (\mathbb{E}|X_t(e_{k_i}) - X_s(e_{k_i})|^2)^p.$$

Moreover,

$$X_t(e_k) - X_s(e_k) = (e^{-k^2(t-s)} - 1)X_s(e_k) + ik \int_s^t e^{-k^2(t-r)} \mathrm{d}_r \beta_r(k),$$

leading to

$$\mathbb{E}|X_t(e_k) - X_s(e_k)|^2 \lesssim (k^2(t-s))^\kappa,$$

for any $\kappa \in [0, 1]$ and $k \neq 0$, while for $k = 0$ we have $\mathbb{E}|X_t(e_0) - X_s(e_0)|^2 = 0$. Let us introduce the notation $\mathbb{Z}_0 = \mathbb{Z} \setminus \{0\}$. Therefore,

$$\mathbb{E}\|X_t - X_s\|_{H^\alpha}^{2p} \lesssim \sum_{k_1,\ldots,k_p \in \mathbb{Z}_0} \prod_{i=1}^{p}(1 + |k_i|^2)^\alpha \prod_{i=1}^{p} \mathbb{E}|X_t(e_{k_i}) - X_s(e_{k_i})|^2$$

$$\lesssim (t-s)^{\kappa p} \sum_{k_1,\ldots,k_p \in \mathbb{Z}_0} \prod_{i=1}^{p}(1 + |k_i|^2)^\alpha (k_i^2)^\kappa (|X_0(e_{k_i})|^2 + 1)$$

$$\lesssim (t-s)^{\kappa p} \left[\sum_{k \in \mathbb{Z}_0}(1 + |k|^2)^\alpha (k^2)^\kappa \right]^p$$

If $\alpha < -1/2 - \kappa$, the sum on the right hand side is finite and we obtain an estimation for the modulus of continuity of $t \mapsto X_t$ in $L^{2p}(\Omega; H^\alpha)$:

$$\mathbb{E}\|X_t - X_s\|_{H^\alpha}^{2p} \lesssim (t-s)^{\kappa p}.$$

Kolmogorov's continuity criterion allows us to conclude that almost surely $X \in CH^\alpha$.   □

Note that the regularity of the Ornstein–Uhlenbeck process does not allow us to form the quantity X_t^2 point-wise in time. One can show that $X_t^2(e_k)$ does not make sense as a random variable. So we should expect that, at fixed time, the regularity of the Ornstein–Uhlenbeck process is like that of the space white noise and this is a way of understanding our difficulties in defining X_t^2 since this will be, modulo smooth terms, the square of the space white noise.

A different matter is to make sense of the time-integral of $\partial_x X_t^2$. Let us give it a name and call it $J_t(\varphi) = \int_0^t \partial_x X_s^2(\varphi)ds$. For $J_t(e_k)$, the computation of its variance gives a quite different result.

Lemma 2 *Almost surely, $J \in C^{1/2-}H^{-1/2-}$.*

Proof We have now

$$\mathbb{E}[|J_t(e_k)|^2] = \frac{1}{\pi}k^2 \int_0^t \int_0^t \sum_{\ell+m=k} \mathbb{E}[X_s(e_\ell)X_{s'}(e_{-\ell})]\mathbb{E}[X_s(e_m)X_{s'}(e_{-m})]dsds'.$$

If $s > s'$, we have

$$\mathbb{E}[X_s(e_\ell)X_{s'}(e_{-\ell})] = e^{-\ell^2(s-s')},$$

and therefore

$$\mathbb{E}[|J_t(e_k)|^2] = \frac{k^2}{2\pi} \int_0^t \int_0^t \sum_{\ell+m=k} e^{-(\ell^2+m^2)|s-s'|} \mathrm{d}s\mathrm{d}s'$$

$$\leqslant \frac{1}{\pi} k^2 t \sum_{\ell+m=k} \int_0^\infty e^{-(\ell^2+m^2)r} \mathrm{d}r = \frac{1}{\pi} k^2 t \sum_{\ell+m=k} \frac{1}{\ell^2+m^2}$$

Now for $k \neq 0$

$$\sum_{\ell+m=k} \frac{1}{\ell^2+m^2} \sim \int_{\mathbb{R}} \frac{\mathrm{d}x}{x^2+(k-x)^2} \sim \frac{1}{|k|}.$$

So finally $\mathbb{E}[|J_t(e_k)|^2] \lesssim |k|t$. From which is easy to conclude that at fixed t the random field J_t belongs almost surely to $H^{-1/2-}$. Redoing a similar computation in the case $J_t(e_k) - J_s(e_k)$, we obtain $\mathbb{E}[|J_t(e_k) - J_s(e_k)|^2] \lesssim |k| \times |t-s|$. To go from this estimate to a path-wise regularity result of the distribution $(J_t)_t$, following the line of reasoning of Lemma 1, we need to estimate the p-th moment of $J_t(e_k) - J_s(e_k)$. We already used in the proof of Lemma 1 that all moments of a Gaussian random variable are comparable. By Gaussian hypercontractivity (see Theorem 3.50 of [17]) this also holds for polynomials of Gaussian random variables, so that

$$\mathbb{E}[|J_t(e_k) - J_s(e_k)|^{2p}] \lesssim_p (\mathbb{E}[|J_t(e_k) - J_s(e_k)|^2])^p.$$

From here we easily derive that almost surely $J \in C^{1/2-}H^{-1/2-}$ which is the space of $1/2-$-Holder continuous functions with values in $H^{-1/2-}$. □

This shows that $\partial_x X_t^2$ exists as a space-time distribution but not as a continuous function of time with values in distributions in space. The key point in the proof of Lemma 2 is the fact that the correlation $\mathbb{E}[X_s(e_\ell)X_{s'}(e_{-\ell})]$ of the Ornstein–Uhlenbeck process decays quite rapidly in time.

The construction of the process J does not solve our problem of constructing $\int_0^t \partial_x u_s^2 \mathrm{d}s$ since we need similar properties for the full solution u of the non-linear dynamics (or for some approximations thereof), and all we have done so far relies on explicit computations and the specific Gaussian features of the Ornstein–Uhlenbeck process. But at least this give us a hint that indeed there could exist a way of making sense of the term $\partial_x u(t,x)^2$, even if only as a space-time distribution, and that in doing so we should exploit some decorrelation properties of the dynamics.

To deal with the full solution u, we need a replacement for the Gaussian computations based on the explicit distribution of X that we used above. This will be provided, in the current setting, by stochastic calculus along the time direction. Indeed, note that for each $\varphi \in \mathscr{S}$ the process $(X_t(\varphi))_{t\geqslant 0}$ is a semi-martingale in the filtration $(\mathscr{F}_t)_{t\geqslant 0}$.

Before proceeding with these computations, we need to develop some tools to describe the Itô formula for functions of the Ornstein–Uhlenbeck process. This will also serve us as an opportunity to set up some analysis on Gaussian spaces.

2.3 Gaussian Computations

For cylindrical functions $F : \mathscr{S}' \to \mathbb{R}$ of the form $F(\rho) = f(\rho(\varphi_1), \ldots, \rho(\varphi_n))$ with $\varphi_1, \ldots, \varphi_n \in \mathscr{S}$ and $f : \mathbb{R}^n \to \mathbb{R}$ at least C_b^2, we have by Itô's formula

$$d_t F(X_t) = \sum_{i=1}^{n} F_i(X_t) dX_t(\varphi_i) + \frac{1}{2} \sum_{i,j=1}^{n} F_{i,j}(X_t) d\langle X(\varphi_i), X(\varphi_j) \rangle_t,$$

where $\langle \rangle_t$ denotes the quadratic covariation of two continuous semimartingales and where $F_i(\rho) = \partial_i f(\rho(\varphi_1), \ldots, \rho(\varphi_n))$ and $F_{i,j}(\rho) = \partial_{i,j}^2 f(\rho(\varphi_1), \ldots, \rho(\varphi_n))$, with ∂_i denoting the derivative with respect to the i-th argument. Now recall that

$$dX_t(\varphi_i) = X_t(\Delta\varphi_i) dt + d\partial_x M_t(\varphi_i)$$

is a continuous semimartingale, and therefore

$$d\langle X(\varphi_i), X(\varphi_j) \rangle_t = d\langle \partial_x M(\varphi_i), \partial_x M(\varphi_j) \rangle_t = \langle \partial_x \varphi_i, \partial_x \varphi_j \rangle_{L^2(\mathbb{T})} dt,$$

and then

$$d_t F(X_t) = \sum_{i=1}^{n} F_i(X_t) d\partial_x M_t(\varphi_i) + L_0 F(X_t) dt,$$

where L_0 is the second-order differential operator defined on cylindrical functions F as

$$L_0 F(\rho) = \sum_{i=1}^{n} F_i(\rho) \rho(\Delta\varphi_i) + \sum_{i,j=1}^{n} F_{i,j}(\rho) \langle \partial_x \varphi_i, \partial_x \varphi_j \rangle_{L^2(\mathbb{T})}. \qquad (2.5)$$

Another way to describe the generator L_0 is to give its value on the functions $\rho \mapsto \exp(\rho(\psi))$ for $\psi \in \mathscr{S}$, which is

$$L_0 e^{\rho(\psi)} = e^{\rho(\psi)}(\rho(\Delta\psi) - \langle \psi, \Delta\psi \rangle_{L^2(\mathbb{T})}).$$

If F, G are two cylindrical functions (which we can take of the form $F(\rho) = f(\rho(\varphi_1), \ldots, \rho(\varphi_n))$ and $G(\rho) = g(\rho(\varphi_1), \ldots, \rho(\varphi_n))$ for the same $\varphi_1, \ldots, \varphi_n \in \mathscr{S}$), we can check that

$$L_0(FG) = (L_0 F)G + F(L_0 G) + \mathcal{E}(F, G), \tag{2.6}$$

where the quadratic form $\mathcal{E}$ is given by

$$\mathcal{E}(F, G)(\rho) = 2 \sum_{i,j} F_i(\rho) G_j(\rho) \langle \partial_x \varphi_i, \partial_x \varphi_j \rangle_{L^2(\mathbb{T})}. \tag{2.7}$$

In particular, the quadratic variation of the martingale obtained in the Itô formula for F is given by

$$d \left\langle \int_0^\cdot \sum_{i=1}^n F_i(X_s) d \partial_x M_s(\varphi_i) \right\rangle_t = \mathcal{E}(F, F)(X_t) dt.$$

Lemma 3 (Gaussian Integration by Parts) $(Z_i)_{i=1,\ldots,M}$ *is an M-dimensional Gaussian vector with zero mean and covariance* $(C_{i,j})_{i,j=1,\ldots,M}$ *iff for all $g \in C_b^1(\mathbb{R}^M)$ we have*

$$\mathbb{E}[Z_k g(Z)] = \sum_\ell C_{k,\ell} \mathbb{E}\left[\frac{\partial g(Z)}{\partial Z_\ell}\right].$$

As a first application of this formula let us show that $\mathbb{E}[L_0 F(\eta)] = 0$ for every cylindrical function, where η is a space white noise with mean zero, i.e. $\eta(\varphi) \sim \mathcal{N}(0, \|\varphi\|^2_{L^2(\mathbb{T})})$ for all $\varphi \in L_0^2(\mathbb{T})$, and $\eta(1) = 0$. Here we write $L_0^2(\mathbb{T})$ for the subspace of all $\varphi \in L^2(\mathbb{T})$ with $\int_\mathbb{T} \varphi dx = 0$. Indeed, note that by polarization $\mathbb{E}[\eta(\varphi_i) \eta(\Delta \varphi_j)] = \langle \varphi_i, \Delta \varphi_j \rangle_{L^2(\mathbb{T})}$, leading to

$$\mathbb{E} \sum_{i,j=1}^n F_{i,j}(\eta) \langle \partial_x \varphi_i, \partial_x \varphi_j \rangle_{L^2(\mathbb{T})} = -\mathbb{E} \sum_{i,j=1}^n F_{i,j}(\eta) \langle \varphi_i, \Delta \varphi_j \rangle_{L^2(\mathbb{T})}$$

$$= - \sum_{i,j=1}^n \langle \varphi_i, \Delta \varphi_j \rangle_{L^2(\mathbb{T})} \mathbb{E} \frac{\partial}{\partial \eta(\varphi_i)} F_j(\eta)$$

$$= - \sum_{j=1}^n \mathbb{E}[\eta(\Delta \varphi_j) F_j(\eta)],$$

so that $\mathbb{E}[L_0 F(\eta)] = 0$ (here we interpreted $\partial_j f$ as a function of $n+1$ variables, with trivial dependence on the $(n+1)$-th one). In combination with Itô's formula, this indicates that the white noise law should indeed be a stationary distribution for

X (convince yourself of it!). From now on we fix the initial distribution $X_0 \sim \eta$, which means that $X_t \sim \eta$ for all $t \geqslant 0$.

As another application of the Gaussian integration by parts formula, we get

$$\frac{1}{2}\mathbb{E}[\mathcal{E}(F, G)(\eta)] = -\sum_{i,j} \mathbb{E}[F_i(\eta)G_j(\eta)]\langle \varphi_i, \Delta\varphi_j \rangle_{L^2(\mathbb{T})}.$$

$$= -\sum_{i,j} \mathbb{E}[(F(\eta)G_j(\eta))_i]\langle \varphi_i, \Delta\varphi_j \rangle_{L^2(\mathbb{T})}$$

$$+ \sum_{i,j} \mathbb{E}[F(\eta)G_{ij}(\eta)]\langle \varphi_i, \Delta\varphi_j \rangle_{L^2(\mathbb{T})}$$

$$= -\sum_{j} \mathbb{E}[F(\eta)G_j(\eta)\eta(\Delta\varphi_j)]$$

$$+ \sum_{i,j} \mathbb{E}[F(\eta)G_{ij}(\eta)]\langle \varphi_i, \Delta\varphi_j \rangle_{L^2(\mathbb{T})}$$

$$= -\mathbb{E}[(FL_0 G)(\eta)].$$

Combining this with (2.6) and with $\mathbb{E}[L_0(FG)(\eta)]=0$, we obtain $\mathbb{E}[(FL_0 G)(\eta)] = \mathbb{E}[(GL_0 F)(\eta)]$. That is, L_0 is a symmetric operator with respect to the law of η.

Consider now the operator D, defined on cylindrical functions F by

$$DF(\rho) = \sum_i F_i(\rho)\varphi_i \tag{2.8}$$

so that DF takes values in $\mathscr{S}'$, the continuous linear functionals on $\mathscr{S}$, D is independent of the specific representation of F, that is if

$$F(\rho) = f(\rho(\varphi_1), \ldots, \rho(\varphi_n)) = g(\rho(\psi_1), \ldots, \rho(\psi_m))$$

for all $\rho \in \mathscr{S}'$, then

$$\sum_i \partial_i f(\rho(\varphi_1), \ldots, \rho(\varphi_n))\varphi_i = \sum_j \partial_j g(\rho(\psi_1), \ldots, \rho(\psi_m))\psi_m.$$

A way to show this is to consider that for all $\theta \in \mathscr{S}$,

$$\langle DF(\rho), \theta \rangle = \frac{d}{d\varepsilon}F(\rho + \varepsilon\theta)|_{\varepsilon=0}.$$

By Gaussian integration by parts we get

$$\mathbb{E}[F(\eta)\langle\psi, DG(\eta)\rangle] + \mathbb{E}[G(\eta)\langle\psi, DF(\eta)\rangle] = \sum_i \mathbb{E}[(FG)_i(\eta)\langle\psi, \varphi_i\rangle]$$

$$= 2\mathbb{E}[\eta(\psi)(FG)(\eta)],$$

and therefore

$$\mathbb{E}[F(\eta)\langle\psi, DG(\eta)\rangle] = \mathbb{E}[G(\eta)\langle\psi, -DF(\eta) + \eta F(\eta)\rangle].$$

So if we consider the space $L^2(\mathrm{Law}(\eta))$ with inner product $\mathbb{E}[F(\eta)G(\eta)]$, then the adjoint of D is given by $D^* F(\rho) = -DF(\rho) + \rho F(\rho)$. Let $D_\psi F(\rho) = \langle\psi, DF(\rho)\rangle$ and similarly for $D_\psi^* F(\rho) = -D_\psi F(\rho) + \rho(\psi)F(\rho)$. If $(f_n)_{n\geqslant 1}$ is an orthonormal basis of $L^2(\mathbb{T})$ then

$$L_0 = \sum_n D_{f_n}^* D_{\Delta f_n}.$$

Moreover, we have

$$[D_\theta, D_\psi^*]F(\rho) = (D_\theta D_\psi^* - D_\psi^* D_\theta)F(\rho) = \langle\psi, \theta\rangle_{L^2(\mathbb{T})} F(\rho),$$

whereas $[D_\theta^*, D_\psi^*] = 0$. Therefore,

$$[L_0, D_\psi^*] = \sum_n [D_{e_n}^* D_{\Delta e_n}, D_\psi^*] = \sum_n D_{e_n}^* [D_{\Delta e_n}, D_\psi^*] + \sum_n [D_{e_n}^*, D_\psi^*] D_{\Delta e_n}$$

$$= \sum_n D_{e_n}^* \langle\psi, \Delta e_n\rangle_{L^2(\mathbb{T})} = D_{\Delta\psi}^*.$$

So if ψ is an eigenvector of Δ with eigenvalue λ, then $[L_0, D_\psi^*] = \lambda D_\psi^*$. Let now $(\psi_n)_{n\in\mathbb{N}}$ be an orthonormal eigenbasis for Δ with eigenvalues $\Delta\psi_n = \lambda_n\psi_n$ and consider the functions

$$H(\psi_{i_1}, \ldots, \psi_{i_n}) : \mathscr{S}' \to \mathbb{R}, \qquad H(\psi_{i_1}, \ldots, \psi_{i_n})(\rho) = (D_{\psi_{i_1}}^* \cdots D_{\psi_{i_n}}^* 1)(\rho).$$

Then

$$L_0 H(\psi_{i_1}, \ldots, \psi_{i_n}) = L_0 D_{\psi_{i_1}}^* \cdots D_{\psi_{i_n}}^* 1$$

$$= D_{\psi_{i_1}}^* L_0 D_{\psi_{i_2}}^* \cdots D_{\psi_{i_n}}^* 1 + \lambda_{i_1} D_{\psi_{i_1}}^* \cdots D_{\psi_{i_n}}^* 1 \qquad (2.9)$$

$$= \cdots = (\lambda_{i_1} + \cdots + \lambda_{i_n}) H(\psi_{i_1}, \ldots, \psi_{i_n}), \qquad (2.10)$$

where we used that $L_0 1 = 0$. These functions are eigenfunctions for L_0 and the eigenvalues are all the possible combinations of $\lambda_{i_1} + \cdots + \lambda_{i_n}$ for $i_1, \ldots, i_n \in \mathbb{N}$. We have immediately that for different n these functions are orthogonal in $L^2(\text{law}(\eta))$. They are actually orthogonal as soon as the indices i differ since in that case there is an index j which is in one but not in the other and using the fact that $D^*_{\psi_j}$ is adjoint to D_{ψ_j} and that $D_{\psi_j} G = 0$ if G does not depend on ψ_j we get the orthogonality. The functions $H(\psi_{i_1}, \ldots, \psi_{i_n})$ are polynomials and they are called *Wick polynomials*.

Lemma 4 *For all $\psi \in \mathscr{S}$, almost surely*

$$(e^{D^*_\psi} 1)(\eta) = e^{\eta(\psi) - \|\psi\|^2/2}.$$

Proof If F is a cylindrical function of the form $F(\rho) = f(\rho(\varphi_1), \ldots, \rho(\varphi_m))$ with $f \in \mathscr{S}(\mathbb{R}^m)$, then

$$\mathbb{E}[F(\eta)(e^{D^*_\psi} 1)(\eta)] = \mathbb{E}[e^{D_\psi} F(\eta)] = \mathbb{E}[F(\eta + \psi)] = \mathbb{E}[F(\eta) e^{\eta(\psi) - \|\psi\|^2/2}],$$

where the second step follows from the fact that if we note $\Psi_t(\eta) = F(\eta + t\psi)$ (note that every $\psi \in \mathscr{S}$ can be interpreted as an element of $\mathscr{S}'$) we have $\partial_t \Psi_t(\eta) = D_\psi \Psi_t(\eta)$ and $\Psi_0(\eta) = F(\eta)$ so that $\Psi_t(\eta) = (e^{t D_\psi} F)(\eta)$ for all $t \geqslant 0$ and in particular for $t = 1$. The last step is simply a Gaussian change of variables. Indeed if we take $\varphi_1 = \psi$ and $\varphi_k \perp \psi$ for $k \geqslant 2$ we have

$$\mathbb{E}[F(\eta + \psi)] = \mathbb{E}[f(\eta(\psi) + \langle \psi, \psi \rangle, \eta(\varphi_2), \ldots, \eta(\varphi_m))]$$

since $(\eta + \psi)(\varphi_k) = \eta(\varphi_k)$ for $k \geqslant 2$. Now observe that $\eta(\psi)$ is independent of $(\eta(\varphi_2), \ldots, \eta(\varphi_m))$ so that

$$\mathbb{E}[f(\eta(\psi) + \langle \psi, \psi \rangle, \eta(\varphi_2), \ldots, \eta(\varphi_m))]$$

$$= \int_{\mathbb{R}} \frac{e^{-z^2/2\|\psi\|^2}}{\sqrt{2\pi \|\psi\|^2}} \mathbb{E}[f(z + \langle \psi, \psi \rangle, \eta(\varphi_2), \ldots, \eta(\varphi_m))]$$

$$= \int_{\mathbb{R}} \frac{e^{-z^2/2\|\psi\|^2}}{\sqrt{2\pi \|\psi\|^2}} e^{z - \|\psi\|^2/2} \mathbb{E}[f(z, \eta(\varphi_2), \ldots, \eta(\varphi_m))] = \mathbb{E}[F(\eta) e^{2\eta(\psi) - \|\psi\|^2}].$$

To conclude the proof, it suffices to note that $\mathbb{E}[F(\eta)(e^{D^*_\psi} 1)(\eta)] = \mathbb{E}[F(\eta) e^{2\eta(\psi) - \|\psi\|^2}]$ implies that $(e^{D^*_\psi} 1)(\eta) = e^{\eta(\psi) - \|\psi\|^2/2}$. $\qquad\qquad\square$

Theorem 1 *The Wick polynomials $\{H(\psi_{i_1}, \ldots, \psi_{i_n})(\eta) : n \geqslant 0, i_1, \ldots, i_n \in \mathbb{N}\}$ form an orthogonal basis of $L^2(\text{law}(\eta))$.*

Proof Taking $\psi = \sum_i \sigma_i \psi_i$ in Lemma 4, we get

$$e^{\sum_i \sigma_i \eta(\psi_i) - \sum_i \sigma_i^2 \|\psi_i\|^2/2} = (e^{D_\psi^*} 1)(\eta) = \sum_{n \geqslant 0} \frac{((D_\psi^*)^n 1)(\eta)}{n!}$$

$$= \sum_{n \geqslant 0} \sum_{i_1, \ldots, i_n} \frac{\sigma_{i_1} \cdots \sigma_{i_n}}{n!} H(\underbrace{\psi_{i_1}, \ldots, \psi_{i_n}}_{n \text{ times}})(\eta),$$

which is enough to show that any random variable in L^2 can be expanded in a series of Wick polynomials showing that the Wick polynomials are an orthogonal basis of $L^2(\text{law}(\eta))$ (but they are still not normalized). Indeed assume that $Z \in L^2(\text{Law}(\eta))$ but $Z \perp H(\psi_{i_1}, \ldots, \psi_{i_n})(\eta)$ for all $n \geqslant 0$, $i_1, \ldots, i_n \in \mathbb{N}$, then

$$0 = e^{\sum_i \sigma_i^2 \|\psi_i\|^2/2} \mathbb{E}[Z(e^{D_\psi^*} 1)(\eta)] = e^{\sum_i \sigma_i^2 \|\psi_i\|^2/2} \mathbb{E}[Z e^{\sum_i \sigma_i \eta(\psi_i) - \sum_i \sigma_i^2 \|\psi_i\|^2/2}]$$

$$= \mathbb{E}[Z e^{\sum_i \sigma_i \eta(\psi_i)}].$$

Since the σ_i are arbitrary, this means that Z is orthogonal to any polynomial in η (consider the derivatives in $\sigma \equiv 0$) and then that it is orthogonal also to $\exp(i \sum_i \sigma_i \eta(\psi_i))$. So let $f \in \mathscr{S}(\mathbb{R}^M)$ and $\sigma_i = 0$ for $i > m$, and observe that

$$0 = (2\pi)^{-m/2} \int d\sigma_1 \cdots d\sigma_m \mathscr{F} f(\sigma_1, \ldots, \sigma_m) \mathbb{E}[Z e^{i \sum_i \sigma_i \eta(\psi_i)}]$$

$$= \mathbb{E}[Z f(\eta(\psi_1), \ldots, \eta(\psi_M))],$$

which means that Z is orthogonal to all the random variables in L^2 which are measurable with respect to the σ-field generated by $(\eta(\psi_n))_{n \geqslant 0}$. This implies $Z = 0$. That is, Wick polynomials form a basis for $L^2(\mu)$. $\square$

Example 1 The first few (un-normalized) Wick polynomials are

$$H(\psi_i)(\rho) = D_{\psi_i}^* 1(\rho) = \rho(\psi_i),$$

$$H(\psi_i, \psi_j)(\rho) = D_{\psi_i}^* D_{\psi_j}^* 1 = D_{\psi_i}^* \rho(\psi_j) = -\delta_{i=j} + \rho(\psi_i)\rho(\psi_j),$$

and

$$H(\psi_i, \psi_j, \psi_k)(\rho) = D_{\psi_i}^*(-\delta_{j=k} + \rho(\psi_j)\rho(\psi_k))$$

$$= -\delta_{j=k}\rho(\psi_i) - \delta_{i=j}\rho(\psi_k) - \delta_{i=k}\rho(\psi_j) + \rho(\psi_i)\rho(\psi_j)\rho(\psi_k).$$

Some other properties of Wick polynomials can be derived using the commutation relation between D and D*. By linearity $D^*_{\varphi+\psi} = D^*_\varphi + D^*_\psi$, so that using the symmetry of H we get

$$H_n(\varphi + \psi) := H\underbrace{(\varphi + \psi, \ldots, \varphi + \psi)}_{n} = \sum_{0 \leqslant k \leqslant n} \binom{n}{k} H(\underbrace{\varphi, \ldots, \varphi}_{k}, \underbrace{\psi, \ldots, \psi}_{n-k}).$$

Then note that by Lemma 4 we have

$$(e^{D^*_\varphi} 1)(\eta)(e^{D^*_\psi} 1)(\eta) = e^{\eta(\varphi) - \|\varphi\|^2/2} e^{\eta(\psi) - \|\psi\|^2/2} = e^{\eta(\varphi+\psi) - \|\varphi+\psi\|^2/2 + \langle\varphi,\psi\rangle}$$

$$= (e^{D^*_{\varphi+\psi}} 1)(\eta) e^{\langle\varphi,\psi\rangle}.$$

Expanding the exponentials,

$$\sum_{m,n} \frac{H_m(\varphi)}{m!} \frac{H_n(\psi)}{n!} = \sum_{r,\ell} \frac{H_r(\varphi + \psi)}{r!} \frac{(2\langle\varphi,\psi\rangle)^\ell}{\ell!}$$

$$= \sum_{p,q,\ell} \frac{H(\overbrace{\varphi, \ldots, \varphi}^{p}, \overbrace{\psi, \ldots, \psi}^{q})}{p! q!} \frac{(\langle\varphi,\psi\rangle)^\ell}{\ell!},$$

and identifying the terms of the same homogeneity in φ and ψ respectively we get

$$H_m(\varphi) H_n(\psi) = \sum_{p+\ell=m} \sum_{q+\ell=n} \frac{m! n!}{p! q! \ell!} H(\overbrace{\varphi, \ldots, \varphi}^{p}, \overbrace{\psi, \ldots, \psi}^{q})(\langle\varphi,\psi\rangle)^\ell. \quad (2.11)$$

This gives a general formula for such products. By polarization of this multilinear form, we can also get a general formula for the products of general Wick polynomials. Indeed taking $\varphi = \sum_{i=1}^m \kappa_i \varphi_i$ and $\psi = \sum_{j=1}^n \lambda_j \psi_j$ for arbitrary real coefficients $\kappa_1, \ldots, \kappa_m$ and $\lambda_1, \ldots, \lambda_n$, we have

$$H_m(\sum_{i=1}^m \kappa_i \varphi_i) H_n(\sum_{j=1}^n \lambda_j \psi_j)$$

$$= \sum_{i_1, \ldots, i_m} \sum_{j_1, \ldots, j_n} \kappa_{i_1} \cdots \kappa_{i_m} \lambda_{j_1} \cdots \lambda_{j_m} H(\varphi_{i_1}, \ldots, \varphi_{i_m}) H(\psi_{j_1}, \ldots, \psi_{j_n}).$$

Deriving this with respect to all the κ, λ parameters and setting them to zero, we single out the term

$$\sum_{\sigma \in S_m, \omega \in S_n} H(\varphi_{\sigma(1)}, \ldots, \varphi_{\sigma(m)}) H(\psi_{\omega(1)}, \ldots, \psi_{\omega(n)})$$

$$= m! n! H(\varphi_1, \ldots, \varphi_m) H(\psi_1, \ldots, \psi_n),$$

where S_k denotes the symmetric group on $\{1, \ldots, k\}$, and where we used the symmetry of the Wick polynomials. Doing the same also for the right hand side of (2.11) we get

$$H(\varphi_1, \ldots, \varphi_m) H(\psi_1, \ldots, \psi_n)$$

$$= \sum_{p+\ell=m} \sum_{q+\ell=n} \frac{1}{p! q! \ell!} \sum_{i,j} H(\overbrace{\varphi_{i_1}, \ldots, \varphi_{i_p}}^{p}, \overbrace{\psi_{j_1}, \ldots, \psi_{j_q}}^{q}) \prod_{r=1}^{\ell} (\langle \varphi_{i_{p+r}}, \psi_{j_{q+r}} \rangle),$$

where the sum over i, j runs over $i_1, \ldots, i_m$ permutation of $1, \ldots, m$ and similarly for $j_1, \ldots, j_n$. Since $H(\varphi_{i_1}, \ldots, \varphi_{i_p}, \psi_{j_1}, \ldots, \psi_{j_q})(\eta)$ is orthogonal to 1 whenever $p + q > 0$, we obtain in particular

$$\mathbb{E}[H(\psi_1, \ldots, \psi_n)(\eta) H(\psi_1, \ldots, \psi_n)(\eta)] = \frac{1}{n!} \sum_{i,j} \prod_{r=1}^{n} (\langle \psi_{i_r}, \psi_{j_r} \rangle)$$

$$= \sum_{\sigma \in S_n} \prod_{r=1}^{n} (\langle \psi_r, \psi_{\sigma(r)} \rangle).$$

In conclusion, we have shown that the family

$$\left\{ \left(\sum_{\sigma \in S_n} \prod_{r=1}^{n} (\langle \psi_r, \psi_{\sigma(r)} \rangle) \right)^{-1/2} H(\psi_{i_1}, \ldots, \psi_{i_n})(\eta) : n \geqslant 0, i_1, \ldots, i_n \in \mathbb{N} \right\}$$

is an orthonormal basis of $L^2(\text{law}(\eta))$.

Remark 1 In our problem it will be convenient to take the Fourier basis as basis in the above computations. Let $e_k(x) = \exp(ikx)/\sqrt{2\pi} = a_k(x) + i b_k(x)$ where $(a_k)_{k \in \mathbb{N}}$ and $(b_k)_{k \in \mathbb{N}}$ form together a real valued orthonormal basis for $L^2(\mathbb{T})$. Then $\rho(e_k)^* = \rho(e_{-k})$ whenever ρ is real valued, and we will denote $D_k = D_{e_k} = D_{a_k} + i D_{b_k}$ and similarly for $D_k^* = D_{a_k}^* - i D_{b_k}^* = -D_{-k} + \rho(e_{-k})$. In this way, D_k^* is the adjoint of D_k with respect to the Hermitian scalar product on $L^2(\Omega; \mathbb{C})$ and the Ornstein–Uhlenbeck generator takes the form

$$L_0 = \sum_{k \in \mathbb{N}} (D_{\partial_x a_k}^* D_{\partial_x a_k} + D_{\partial_x b_k}^* D_{\partial_x b_k}) = \frac{1}{2} \sum_{k \in \mathbb{Z}} k^2 D_k^* D_k \qquad (2.12)$$

(convince yourself of the last identity by observing that $D_k^* D_k + D_{-k}^* D_{-k} = 2(D_{a_k}^* D_{a_k} + D_{b_k}^* D_{b_k})$!). Similarly,

$$\mathcal{E}(F, G) = \sum_{k \in \mathbb{Z}} k^2 (D_k F)^* (D_k G). \qquad (2.13)$$

2.4   The Itô Trick

We are ready now to start our computations. Recall that we want to analyze $J_t(\varphi) = \int_0^t \partial_x X_s^2(\varphi) ds$ using Itô calculus with respect to the Ornstein–Uhlenbeck process. We want to understand J_t as a correction term in Itô's formula. If we can find a function G such that $L_0 G(X_t) = \partial_x X_t^2$, then we get from Itô's formula

$$\int_0^t \partial_x X_s^2 ds = G(X_t) - G(X_0) - M_{G,t},$$

where M_G is a martingale depending on G. Of course, G will not be a cylindrical function but we only defined L_0 on cylindrical functions. So to make the following calculations rigorous we would again have to replace $\partial_x X_t^2$ by $\partial_x \Pi_n X_t^2$ and then pass to the limit. As before we will perform the calculations already in the limit $N = +\infty$, in order to simplify the computations and to not obscure the ideas through technicalities. The next problem is that the point-wise evaluation $\int_0^t \partial_x X_s^2(x) ds$ does not make any sense because the integral will only be defined as a space distribution. So we will consider

$$G : \mathscr{S}' \to \mathscr{S}'.$$

Note however that we can reduce every such G to a function from $\mathscr{S}'$ to $\mathbb{C}$ by considering $\rho \mapsto G(\rho)(e_k)$ for all $k \in \mathbb{Z}_0$.

Now for a fixed k, we have

$$\partial_x X_t^2(e_k) = ik \sum_{\ell+m=k} X_t(e_\ell) X_t(e_m) = ik \sum_{\ell+m=k} H_{\ell,m}(X_t), \qquad (2.14)$$

where $H_{\ell,m}(\rho) = (D_{-\ell}^* D_{-m}^* 1)(\rho) = \rho(e_\ell)\rho(e_m) - \delta_{\ell+m=0}$ is a second order Wick polynomial so that $L_0 H_{\ell,m} = -(\ell^2 + m^2) H_{\ell,m}$ by (2.9). Therefore, it is enough to take

$$G(X_t)(e_k) = -ik \sum_{\ell+m=k} \frac{H_{\ell,m}(X_t)}{\ell^2 + m^2}. \qquad (2.15)$$

This corresponds to the distribution $G(X_t)(\varphi) = -\int_0^\infty \partial_x (e^{s\Delta} X_t)^2(\varphi) ds$ (check it!). Then

$$G(X_t)(\varphi) = G(X_0)(\varphi) + M_{G,t}(\varphi) + J_t(\varphi),$$

where $M_{G,t}(\varphi)$ is a martingale with quadratic variation

$$d\langle M_{G,*}(\varphi), M_{G,*}(\varphi)\rangle_t = \mathcal{E}(G(*)(\varphi), G(*)(\varphi))(X_t) dt.$$

We can estimate

$$\mathbb{E}[|J_t(\varphi) - J_s(\varphi)|^{2p}] \lesssim_p \mathbb{E}[|M_{G,t}(\varphi) - M_{G,s}(\varphi)|^{2p}] + \mathbb{E}[|G(X_t)(\varphi) - G(X_s)(\varphi)|^{2p}].$$

To bound the martingale expectation, we will use the following Burkholder inequality:

Lemma 5 *Let m be a continuous local martingale with $m_0 = 0$. Then for all $T \geqslant 0$ and $p > 1$,*

$$\mathbb{E}[\sup_{t \leqslant T} |m_t|^{2p}] \leqslant C_p \mathbb{E}[\langle m \rangle_T^p].$$

Applying Burkholder's inequality, we obtain

$$\mathbb{E}[|J_t(\varphi) - J_s(\varphi)|^{2p}] \lesssim_p \mathbb{E}\left[\left|\int_s^t \mathcal{E}(G(*)(\varphi), G(*)(\varphi))(X_r)\mathrm{d}r\right|^p\right]$$

$$+\mathbb{E}[|G(X_t)(\varphi) - G(X_s)(\varphi)|^{2p}]$$

$$\leqslant (t-s)^{p-1} \int_s^t \mathbb{E}[|\mathcal{E}(G(*)(\varphi), G(*)(\varphi))(X_r)|^p]\mathrm{d}r$$

$$+\mathbb{E}[|G(X_t)(\varphi) - G(X_s)(\varphi)|^{2p}]$$

$$= (t-s)^p \mathbb{E}[|\mathcal{E}(G(*)(\varphi), G(*)(\varphi))(\eta)|^p]$$

$$+\mathbb{E}[|G(X_t)(\varphi) - G(X_s)(\varphi)|^{2p}],$$

using that $X_r \sim \eta$. Now

$$D_m G(\rho)(e_k) = -ik \frac{\rho(e_{k-m})}{(k-m)^2 + m^2},$$

and therefore

$$\mathcal{E}(G(*)(e_k), G(*)(e_k))(\rho) = \sum_m m^2 D_{-m} G(\rho)(e_{-k}) D_m G(\rho)(e_k)$$

$$= k^2 \sum_{\ell+m=k} m^2 \frac{|\rho(e_\ell)|^2}{(\ell^2 + m^2)^2} \lesssim k^2 \sum_{\ell+m=k} \frac{|\rho(e_\ell)|^2}{\ell^2 + m^2},$$

which implies that

$$\mathbb{E}[|\mathcal{E}(G(*)(e_k), G(*)(e_k))(\eta)|] \lesssim k^2 \mathbb{E} \sum_{\ell+m=k} \frac{|\eta(e_\ell)|^2}{\ell^2 + m^2} \lesssim k^2 \sum_{\ell+m=k} \frac{1}{\ell^2 + m^2} \lesssim |k|.$$

A similar computation gives also that

$$\mathbb{E}[|\mathcal{E}(G(*)(e_k), G(*)(e_k))(\eta)|^p] \lesssim |k|^p.$$

Further, we have

$$\mathbb{E}[|G(X_t)(e_k) - G(X_s)(e_k)|^2] \lesssim k^2 \sum_{\ell+m=k} \mathbb{E}\left[\frac{|H_{\ell,m}(X_t) - H_{\ell,m}(X_s))^2}{(\ell^2 + m^2)^2}\right]$$

$$\lesssim k^2 |t - s| \sum_{\ell+m=k} \frac{m^2}{(\ell^2 + m^2)^2} \lesssim |k||t - s|.$$

Given that G is a second order polynomial of a Gaussian process we can apply once more Gaussian hypercontractivity to obtain

$$\mathbb{E}[|J_t(e_k) - J_s(e_k)|^{2p}] \lesssim_p (t - s)^p |k|^p.$$

The advantage of the Itô trick with respect to the explicit Gaussian computation is that it goes over to the non-Gaussian case.

2.5 An Approximation Scheme

Our aim now is to devise suitable approximation for the SBE and try to modify the Itô trick to get enough compactness to be able to extract limits. The properties of these limit points will suggest a suitable notion of solution.

For any $N \geqslant 1$ consider solutions u^N to the SPDE

$$\partial_t u^N = \Delta u^N + \partial_x \Pi^N (\Pi^N u^N)^2 + 2\partial_x \xi.$$

These are generalized functions such that

$$du_t^N(e_k) = -k^2 u_t^N(e_k)dt + [\partial_x \Pi^N (\Pi^N u^N)^2](e_k)dt + ik d\beta_t(k)$$

for $k \in \mathbb{Z}$ and $t \geqslant 0$. We take u_0 to be the white noise with covariance $u_0(\varphi) \sim \mathcal{N}(0, \|\varphi\|^2)$. The point of our choice of the non-linearity is that this (infinite-dimensional) system of equations decomposes into a finite dimensional system for $(v^N(k) = \Pi^N u^N(e_k))_{k:|k| \leqslant N}$ and an infinite number of one-dimensional equations for each $u^N(e_k)$ with $|k| > N$. Indeed if $|k| > N$ we have $[\partial_x \Pi^N (\Pi^N u^N)^2](e_k) = 0$ so $u_t(e_k) = X_t(e_k)$ the Ornstein–Uhlenbeck process with initial condition $X_0(e_k) = u_0(e_k)$ which renders it stationary in time (check it). The equation for $(v^N(k))_{|k| \leqslant N}$ reads

$$dv_t^N(k) = -k^2 v_t^N(k)dt + b_k(v_t^N)dt + ik d\beta_t(k), \qquad |k| \leqslant N, t \geqslant 0$$

where

$$b_k(v_t^N) = ik \sum_{\ell+m=k} \mathbb{I}_{|\ell|,|k|,|m| \leqslant N} v_t^N(\ell) v_t^N(m).$$

This is a standard finite-dimensional ODE having global solutions for all initial conditions which gives rise to a nice Markov process. The fact that solutions do not blow up even if the interaction is quadratic can be seen by computing the evolution of the norm

$$A_t = \sum_{|k| \leqslant N} |v_t^N(k)|^2$$

and by showing that

$$dA_t = 2 \sum_{|k| \leqslant N} v_t^N(-k) dv_t^N(k)$$

$$= -2 \sum_{|k| \leq N} k^2 |v_t^N(k)|^2 dt$$

$$+2 \sum_{|k| \leqslant N} v_t^N(-k) b_k(v_t^N) dt + 2ik \sum_{|k| \leqslant N} v_t^N(-k) d\beta_t(k).$$

Since A is nonnegative, we increase its absolute value by omitting the first contribution. But now

$$\sum_{|k| \leqslant N} v_t^N(-k) b_k(v_t^N) = 2i \sum_{k,\ell,m:\ell+m=k} \mathbb{I}_{|\ell|,|k|,|m| \leqslant N} k v_t^N(\ell) v_t^N(m) v_t^N(-k)$$

$$= -2i \sum_{k,\ell,m:\ell+m+k=0} \mathbb{I}_{|\ell|,|k|,|m| \leqslant N} (k) v_t^N(\ell) v_t^N(m) v_t^N(k)$$

and by symmetry of this expression it is equal to

$$= -\frac{2}{3} i \sum_{k,\ell,m:\ell+m+k=0} \mathbb{I}_{|\ell|,|k|,|m| \leqslant N} (k+\ell+m) v_t^N(\ell) v_t^N(m) v_t^N(k) = 0,$$

so $|A_t| \leq |A_0 + M_t|$ where $dM_t = 2 \sum_{|k| \leqslant N} \mathbb{I}_{|k| \leqslant N} (ik) v_t^N(-k) d\beta_t(k)$. Now

$$\mathbb{E}[M_T^2] \lesssim \int_0^T \sum_{|k| \leqslant N} k^2 |v_t^N(k)|^2 dt \lesssim N^2 \int_0^T A_t dt$$

and then by martingales inequalities

$$\mathbb{E}[\sup_{t\in[0,T]} (A_t)^2] \leqslant 2\mathbb{E}[A_0^2] + 2\mathbb{E}[\sup_{t\in[0,T]} (M_t)^2] \leqslant 2\mathbb{E}[A_0^2] + 8\mathbb{E}[M_T^2]$$

$$\leqslant 2\mathbb{E}[A_0^2] + CN^2 \int_0^T \mathbb{E}(A_t)dt.$$

As a consequence, Gronwall's inequality gives

$$\mathbb{E}[\sup_{t\in[0,T]} (A_t)^2] \lesssim e^{CN^2T}\mathbb{E}[A_0^2],$$

from where we can deduce (by a continuation argument) that almost surely there is no blowup at finite time for the dynamics. The generator L^N for the Galerkin dynamics is given by

$$L^N F(\rho) = L_0 F(\rho) + B^N F(\rho),$$

where

$$B^N F(\rho) = \sum_k \mathbb{I}_{|k|\leqslant N} (\partial_x (\Pi^N \rho)^2)(e_k) D_k F(\rho).$$

And again the non-linear drift B^N is antisymmetric with respect to the invariant measure of L_0 by a computation similar to that for the full drift B. Next, using Echeverría's criterion [4] the invariance of the white noise follows from its infinitesimal invariance which can be checked at the level of the generator L^N.

2.5.1   Time Reversal

In order to carry over the Ito trick's computation to the full process u^N solution of the non-linear dynamics we need to replace the generator of X with that of u^N and to have a way to handle the boundary terms. The idea is now to reverse the Markov process u^N in time, which will allow us to kill the antisymmetric part of the generator and at the same time kill the boundary terms. Indeed observe that we have the Itô formula

$$d_t F(u_t^N) = \sum_{i=1}^n F_i(u_t^N)\partial_x dM_t(\varphi_i) + L^N F(u_t^N)dt,$$

where L^N is now the full generator of the approximate non-linear dynamics. Formally, the non-linear term B^N is antisymmetric with respect to the invariant measure of L_0. Indeed since B^N is a first order operator

$$\mathbb{E}[(B^N F(\eta))G(\eta)] = \mathbb{E}[(B^N(FG)(\eta))] - \mathbb{E}[F(\eta)(B^N G(\eta))]$$

$$= -\mathbb{E}[F(\eta)(B^N G(\eta))] \qquad (2.16)$$

provided $\mathbb{E}[B^N F(\eta)] = 0$ for any cylinder function F. Let us show this.

$$\mathbb{E}[B^N F(\eta)] = \sum_{k:|k|\leqslant N} \mathbb{E}[(\partial_x(\Pi^N\eta)^2)(e_k)D_k F(\eta)]$$

$$= - \sum_{k:|k|\leqslant N} \mathbb{E}[(D_k(\partial_x(\Pi^N\eta)^2)(e_k))F(\eta)]$$

$$+ \sum_{k:|k|\leqslant N} \mathbb{E}[D_k[(\partial_x(\Pi^N\eta)^2)(e_k)F(\eta)]]$$

But now we get from (2.14)

$$D_k(\partial_x(\Pi^N\eta)^2)(e_k) = \sqrt{2}ik\eta(e_0)\mathbb{I}_{|k|\leqslant N} = \mathbb{I}_{|k|\leqslant N}\pi^{-1/2}ik\langle\eta, 1\rangle = 0,$$

where we used that $\langle\eta, 1\rangle = 0$. Gaussian integration by parts then gives

$$\mathbb{E}[B^N F(\eta)] = \sum_{k:|k|\leqslant N} \mathbb{E}[D_k[(\partial_x(\Pi^N\eta)^2)(e_k)F(\eta)]]$$

$$= \sum_{k:|k|\leqslant N} \mathbb{E}[\eta(e_k)(\partial_x(\Pi^N\eta)^2)(e_k)F(\eta)]$$

$$= \mathbb{E}[\langle\eta, \Pi^N\partial_x(\Pi^N\eta)^2\rangle F(\eta)] = \frac{1}{3}\mathbb{E}[\langle1, \partial_x(\Pi^N\eta)^3\rangle F(\eta)] = 0$$

since $\langle1, \partial_x(\Pi^N\eta)^3\rangle = -\langle\partial_x 1, (\Pi^N\eta)^3\rangle = 0$.

The dynamics of u^N backwards in time has a Markovian description. If $(y_t)_{t\geqslant 0}$ is a stationary Markov process on a Polish space, with semigroup $(P_t)_{t\geqslant 0}$ and stationary distribution μ and P_t^* the adjoint of P_t in $L^2(\mu)$, then (P_t^*) is a semigroup of operators on $L^2(\mu)$ (that is $P_0^* = \mathrm{id}$ and $P_{s+t}^* = P_s^* P_t^*$ as operators on $L^2(\mu)$). Moreover if $y_0 \sim \mu$, then for all $T > 0$ the process $\hat{y}_t = y_{T-t}, t \in [0, T]$, is also Markov, with semigroup $(P_t^*)_{t\in[0,T]}$, and that μ is also an invariant distribution for (P_t^*).

Now if we reverse the process u^N in time letting $\hat{u}_t^N = u_{T-t}^N$, we have by stationarity

$$\mathbb{E}[F(\hat{u}_t^N)G(\hat{u}_0^N)] = \mathbb{E}[F(u_{T-t}^N)G(u_T^N)] = \mathbb{E}[F(u_0^N)G(u_t^N)].$$

So if we denote by $\hat{L}^N$ the generator of $\hat{u}^N$:

$$\mathbb{E}[\hat{L}^N F(\hat{u}_0^N) G(\hat{u}_0^N)] = \frac{d}{dt}\bigg|_{t=0} \mathbb{E}[F(\hat{u}_t^N) G(\hat{u}_0^N)]$$

$$= \frac{d}{dt}\bigg|_{t=0} \mathbb{E}[F(u_0^N) G(u_t^N)] = \mathbb{E}[LG(u_0^N) F(u_0^N)],$$

which means that $\hat{L}^N$ is the adjoint of L^N in $L^2(\mu)$, that is

$$\hat{L}^N F(\rho) = L_0^N F(\rho) - B^N F(\rho) = L_0^N F(\rho) - \sum_{k:|k|\leqslant N} (\partial_x (\Pi^N \rho)^2)(e_k) D_k F(\rho).$$

In other words, the reversed process solves

$$\hat{u}_t^N(\varphi) = \hat{u}_0^N(\varphi) + \int_0^t \hat{u}_s^N(\Delta\varphi)ds + \int_0^t \langle (\Pi^N \hat{u}_s^N)^2, \partial_x\varphi\rangle ds - \int_0^t \hat{\xi}_s(\partial_x\varphi)ds$$

for a different space-time white noise $\hat{\xi}$. Then Itô's formula for $\hat{u}^N$ gives

$$d_t F(\hat{u}_t^N) = \sum_{i=1}^n F_i(\hat{u}_t^N) d\partial_x \hat{M}_t(\varphi_i) + \hat{L}F(\hat{u}_t^N)dt,$$

where for all test functions φ, the process $\partial_x \hat{M}(\varphi)$ is a continuous martingale in the filtration of $\hat{u}^N$ with covariance

$$d\langle \partial_x \hat{M}(\varphi), \partial_x \hat{M}(\psi)\rangle_t = 2\langle \partial_x\varphi, \partial_x\psi\rangle_{L^2(\mathbb{T})}dt.$$

Combining the Itô formulas for u^N and $\hat{u}^N$, we get

$$F(u_T^N)(\varphi) = F(u_0^N)(\varphi) + \partial_x M_{F,T}^N(\varphi) + \int_0^T L^N F(u_s^N)(\varphi)ds$$

and

$$F(u_0^N)(\varphi) = F(\hat{u}_T^N)(\varphi) = F(\hat{u}_0^N)(\varphi) + \partial_x \hat{M}_{F,T}^N(\varphi) + \int_0^T \hat{L}^N F(\hat{u}_s^N)(\varphi)ds$$

$$= F(u_T^N)(\varphi) + \partial_x \hat{M}_{F,T}^N(\varphi) + \int_0^T \hat{L}^N F(u_s^N)(\varphi)ds.$$

Summing up these two equalities gives

$$0 = \partial_x M_{F,T}^N(\varphi) + \partial_x \hat{M}_{F,T}^N(\varphi) + \int_0^T (\hat{L}^N + L^N) F(u_s^N)(\varphi)ds,$$

that is

$$2 \int_0^T L_0 F(u_s^N)(\varphi) \mathrm{d}s = -\partial_x M_{F,T}^N(\varphi) - \partial_x \hat{M}_{F,T}^N(\varphi).$$

An added benefit of this forward–backward representation is that the only term which required a lot of informations about u^N, that is the boundary term $F(u_t^N)(\varphi) - F(u_s^N)(\varphi)$ does not appear at all now. As above if $2L_0 F_N(\rho) = \partial_x(\Pi^N \rho)^2$, we end up with

$$\int_0^T \partial_x(\Pi^N u_s^N)(\varphi) \mathrm{d}s = -\partial_x M_{F_N,T}^N(\varphi) - \partial_x \hat{M}_{F_N,T}^N(\varphi). \tag{2.17}$$

Setting $\mathcal{B}_t^N(\varphi) = \int_0^t \partial_x(\Pi_N u_s)^2(\varphi) \mathrm{d}s$ we can now show that

$$\mathbb{E}[|\mathcal{B}_t^N(e_k) - \mathcal{B}_s^N(e_k)|^{2p}] \lesssim_p (t-s)^p |k|^p$$

and letting $\mathcal{B}_t^{N,M} = \mathcal{B}_t^N - \mathcal{B}_t^M$ we get

$$\mathbb{E}[|\mathcal{B}_t^{N,M}(e_k) - \mathcal{B}_s^{N,M}(e_k)|^{2p}] \lesssim_p (|k|/N)^{\varepsilon p}(t-s)^p |k|^p$$

for all $1 \leqslant N \leqslant M$. From this we derive that

$$(\mathbb{E}[\|\mathcal{B}_t^{N,M} - \mathcal{B}_s^{N,M}\|_{H^\alpha}^{2p}])^{1/2p} \lesssim_{p,\alpha} N^{-\varepsilon/2}(t-s)^{1/2}$$

for all $\alpha < -1-\varepsilon$. This estimate allows to prove compactness of the approximations $\mathcal{B}^N$ and then convergence to a limit $\mathcal{B}$ in $L^{2p}(\Omega; C^{1/2-}H^{-1-})$.

These uniform estimates are the key to prove tightness of the triplet $(u^N, \mathcal{B}^N, M)$ and obtain limit points $(u, \mathcal{B}, M)$ which all will share the following properties:

1. the law of u_t is the white noise μ for all $t \in [0, T]$;
2. For any test function $\varphi \in \mathscr{S}$ the process $t \mapsto \mathcal{B}_t(\varphi)$ is a.s. of zero quadratic variation, $\mathcal{B}_0(\varphi) = 0$ and the pair $(u(\varphi), \mathcal{B}(\varphi))_{0 \leqslant t \leqslant T}$ satisfies the equation

$$u_t(\varphi) = u_0(\varphi) + \int_0^t u_s(\Delta\varphi) \mathrm{d}s + \mathcal{B}_t(\varphi) + \partial_x M_t(\varphi)$$

where $(M_t(\varphi))_{0 \leqslant t \leqslant T}$ is a martingale with respect to the filtration generated by $(u, \mathcal{B})_{0 \leqslant t \leqslant T}$ with quadratic variation $[M(\varphi)]_t = 2t \|\varphi\|_{L^2(\mathbb{T})}^2$;
3. the reversed processes $\hat{u}_t = u_{T-t}$, $\widehat{\mathcal{B}}_t = -(\mathcal{B}_T - \mathcal{B}_{T-t})$ satisfies the same equation with respect to its own filtration (the backward filtration of u).

The requirement for zero quadratic variation for $\mathcal{B}(\varphi)$ is due to the fact that we want to be able to perform the Itô trick at the level of the limit points and that we cannot expect the limiting drift $(\mathcal{B}_t(\varphi))_t$ to have finite variation, that is limit points

u will not be semimartingales but only *Dirichlet processes*: sums of martingales and zero variation processes. Luckily in this setting it is still possible to derive an Itô formula and everything goes through as described above, as we will see below.

For the moment the only property which is still not clear is this zero quadratic variation for the drift. Indeed the only information the previous estimates gives on $t \mapsto \mathcal{B}_t(\varphi)$ is that it belongs to $C_T^{1/2-}(\mathbb{R})$ which is not a sufficient regularity. An interpolation procedure solves the problem. Indeed it is possible to prove that $t \mapsto \mathcal{B}_t(\varphi)$ belongs to $C_T^{3/4-}(\mathbb{R})$ which is now enough to prove the zero quadratic variation property. See [11] for additional details on the limiting procedure and the interpolation and [23] for details on how to implement the Itô trick on the level of diffusions.

2.6 Controlled Processes and Energy Solutions

Here we define stationary energy solutions of (2.2), taking inspiration by the description of the limit points for the approximation scheme of the previous section. We introduce first a class of processes u which at fixed time are distributed like the (zero-mean) white noise on $\mathbb{T}$ but for which the nonlinear term $\partial_x u^2$ is defined as a space-time distribution. In this class of processes it makes then sense to look for solutions of the SBE (2.2).

We cook up a definition which will allow us to rigorously perform the computations of the Itô trick in a general setting.

Definition 1 (Controlled Process) Denote with $\mathcal{Q}$ the space of pairs $(u, \mathcal{A})_{0 \leqslant t \leqslant T}$ of generalized stochastic processes with continuous paths in $\mathscr{S}'$ such that

1. the law of u_t is the white noise μ for all $t \in [0, T]$;
2. For any test function $\varphi \in \mathscr{S}$ the process $t \mapsto \mathcal{A}_t(\varphi)$ is a.s. of zero quadratic variation, $\mathcal{A}_0(\varphi) = 0$ and the pair $(u(\varphi), \mathcal{A}(\varphi))_{0 \leqslant t \leqslant T}$ satisfies the equation

$$u_t(\varphi) = u_0(\varphi) + \int_0^t u_s(\Delta\varphi)\mathrm{d}s + \mathcal{A}_t(\varphi) + \partial_x M_t(\varphi) \qquad (2.18)$$

where $(M_t(\varphi))_{0 \leqslant t \leqslant T}$ is a martingale with respect to the filtration generated by $(u, \mathcal{A})_{0 \leqslant t \leqslant T}$ with quadratic variation $[M(\varphi)]_t = 2t\|\varphi\|_{L^2(\mathbb{T})}^2$;

3. the reversed processes $\hat{u}_t = u_{T-t}$, $\hat{\mathcal{A}}_t = -(\mathcal{A}_T - \mathcal{A}_{T-t})$ satisfies the same equation with respect to its own filtration (the backward filtration of u).

When $\mathcal{A} = 0$ the process $(X, 0) \in \mathcal{Q}$ is the stationary Ornstein–Uhlenbeck (OU) process with invariant measure μ. It is the unique solution of the SPDE

$$\mathrm{d}X = \Delta X \mathrm{d}t + \partial_x \mathrm{d}M$$

with initial condition $u_0 \sim \mu$. Allowing $\mathcal{A} \neq 0$ has the intuitive meaning of considering perturbation of the OU process with zero quadratic variation antisymmetric drifts. In this sense we say that a couple $(u, \mathcal{A}) \in \mathcal{Q}$ is a process *controlled* by the Ornstein–Uhlenbeck process.

Controlled processes allow the definition of some interesting non-linear functionals. We do not insists on generality here, all we need is that the Burgers drift makes sense as a space-time distribution:

Theorem 2 *Assume that* $(u, \mathcal{A}) \in \mathcal{Q}$ *and for any* $N \geqslant 1$ *and* $0 \leqslant t \leqslant T$. *Let*

$$\mathcal{B}_t^N(\varphi) = \int_0^t \partial_x \left(\Pi_0^N u_s \right)^2 (\varphi) \mathrm{d}s$$

then $(\mathcal{B}_t^N)_{N \geqslant 1}$ *converges in probability in* $C^{1/2+} H^{-1-} \cap C^{0+} H^{0-}$, *we denote the limit by*

$$\int_0^t \partial_x u_s^2(\varphi) \mathrm{d}s.$$

Proof The proof can be obtained as a particular case of the Boltzmann–Gibbs principle proved in Proposition 1 below and valid for all controlled processes. We leave the reader to explicit the details.   □

We are ready to give a rigorous meaning to the SBE (2.2) in the class of controlled processes:

Definition 2 A controlled process $(u, \mathcal{A}) \in \mathcal{Q}$ is an energy solution of SBE iff

$$\mathcal{A}_t(\varphi) = \int_0^t \partial_x u_s^2(\varphi) \mathrm{d}s.$$

almost surely for all $t \in [0, T]$ and $\varphi \in \mathscr{S}$.

2.6.1 Regularization by Noise for Controlled Processes

In this section u will always denote a generic controlled process in $\mathcal{Q}$, not a solution to SBE. Controlled processes have regularization properties coming from the fast decorrelation in time of the OU process associated to the Laplacian Δ. As showed above in the particular case of the OU process itself and of finite dimensional approximations to SBE we are able to exploit an Itô formula to replace time-averages of some functional with a sum of forward and backward martingales whose quadratic variation is controlled in terms of a more regular (or smaller) functional.

For any test function $\varphi \in \mathscr{S}$ the processes $(u_t(\varphi))_{t \in [0,T]}$ and $(\hat{u}_t(\varphi))_{t \in [0,T]}$ are Dirichlet processes: the sum of a martingale and a zero quadratic variation process. This is compatible with the regularity of our solutions and there is no clue that solutions of SBE are distributional semimartingales. Dirichlet processes can be handled via the stochastic calculus by regularization developed by Russo and Vallois [22]. In this approach the Itô formula holds also for Dirichlet processes and if $X = (X^i)_{i=1,\ldots,k}$ is an $\mathbb{R}^k$ valued Dirichlet process and g is a $C^2(\mathbb{R}^k; \mathbb{R})$ function then we have

$$g(X_t) = g(X_0) + \sum_{i=1}^{k} \int_0^t \partial_i g(X_s) \mathrm{d}^- X_s^i + \frac{1}{2} \sum_{i,j=1}^{k} \int_0^t \partial_{i,j}^2 g(X_s) \mathrm{d}^- [X^i, X^j]_s$$

(2.19)

where d^- denotes the forward integral and $[X, X]$ the quadratic covariation of the vector process X. Decomposing $X = M + N$ as the sum of a martingale M and a zero quadratic variation process N we have $[X, X] = [M, M]$ and

$$g(X_t) = g(X_0) + \sum_{i=1}^{k} \int_0^t \partial_i g(X_s) \mathrm{d}^- M_s^i + \sum_{i=1}^{k} \int_0^t \partial_i g(X_s) \mathrm{d}^- N_s^i$$

$$+ \sum_{i,j=1}^{k} \frac{1}{2} \int_0^t \partial_{i,j}^2 g(X_s) \mathrm{d}^- [M^i, M^j]_s$$

where now $\mathrm{d}^- M$ coincide with the usual Itô integral and $[M, M]$ is the usual quadratic variation of the martingale M. The integral $\int_0^t \partial_i g(X_s) \mathrm{d}^- N_s^i$ is well-defined due to the fact that all the other terms in this formula are well defined. The case the function g depends explicitly on time can be handled by the above formula by considering time as an additional (0-th) component of the process X and using the fact that $[X^i, X^0] = 0$ for all $i = 1, \ldots, k$. In the computations which follows we will only need to apply the Itô formula to smooth functions.

For any smooth cylinder function $h : [0, T] \times \Pi_0^N H_0^0 \to \mathbb{R}$ the Itô formula for the finite quadratic variation process $(u_t^N = \Pi_0^N u_t)_t$ gives

$$h(t, u_t^N) = h(0, u_0^N) + \int_0^t (\partial_s + L_0^N) h(s, u_s^N) \mathrm{d}s + \int_0^t Dh(s, u_s^N) \mathrm{d}\mathscr{I}^N \mathcal{A}_s + M_t^+$$

where

$$L_0^N h(s, x) = \sum_{0 < |k| \le N} |k|^2 (-x_k D_k h(s, x) + D_k D_{-k} h(s, x))$$

is the restriction of the operator L_0 to $\Pi^N H_0^0$. The martingale part, denoted M^+, has quadratic variation given by $[M^+]_t = \int_0^t \mathcal{E}^N(h(s, \cdot))(u_s^N) \mathrm{d}s$, where

$$\mathcal{E}^N(\varphi)(x) = \sum_{0 < |k| \le N} |k|^2 |D_k \varphi(x)|^2 = \left\| \mathscr{I}_0^N D\varphi \right\|_{H_0^1}^2 ,$$

is the density of the Dirichlet form associated to the generator L_0^N. The Itô formula on the backward process reads

$$h(T - t, u_{T-t}^N) = h(T, u_T^N) + \int_0^t (-\partial_s + L_0^N) h(T - s, u_{T-s}^N) ds$$

$$- \int_0^t Dh(T - s, u_{T-s}^N) d\mathscr{I}^N \mathcal{A}_{T-s} + M_t^-$$

with $[M^-]_t = \int_0^t \mathcal{E}^N(h(T - s, \cdot))(u_{T-s}^N) ds$. Adding the two Itô formulas we have the key equality

$$\int_0^t 2L_0^N h(s, u_s^N) ds = -M_t^+ + M_{T-t}^- - M_T^-. \tag{2.20}$$

which allows us to represent the time integral of $2L_0^N h$ as a sum of martingales which allows better control since their quadratic variation depends only on $\mathcal{E}^N(h)$. From this we can prove easily (see [11]) the following lemma.

Lemma 6 (Itô Trick) *Let* $h : [0, T] \times \Pi_N H \to \mathbb{R}$ *be a cylinder function. Then for any* $p \geq 1$,

$$\left\| \sup_{t \in [0,T]} \left| \int_0^t L_0 h(s, u_s) ds \right| \right\|_{L^p(\mathbb{P}_\mu)} \lesssim_p T^{1/2} \sup_{s \in [0,T]} \| \mathcal{E}(h(s, \cdot)) \|_{L^{p/2}(\mu)}^{1/2} \tag{2.21}$$

where $\mathcal{E}(\varphi)(x) = \frac{1}{2} \sum_{|k|>0} |k|^2 |D_k \varphi(x)|^2$. *In the particular case* $h(s, x) = e^{a(T-s)} \tilde{h}(x)$ *for some* $a \in \mathbb{R}$ *we have the improved estimate*

$$\left\| \int_0^T e^{a(T-s)} L_0 \tilde{h}(u_s) ds \right\|_{L^p(\mathbb{P}_\mu)} \lesssim_p \left(\frac{1 - e^{2aT}}{2a} \right)^{1/2} \| \mathcal{E}(\tilde{h}) \|_{L^{p/2}(\mu)}^{1/2}. \tag{2.22}$$

If we take $p = 2$ in Lemma 6 we have

$$\mathbb{E} \left[\sup_{t \in [0,T]} \left| \int_0^t L_0 h(u_s) ds \right|^2 \right] \lesssim T \| h \|_1^2$$

where

$$\| h \|_1^2 = \mathbb{E} \mathcal{E}(h(u_0)) = \mathbb{E} \| Dh(u_0) \|_{H^1}^2$$

is the Dirichlet form of the OU process.

We denote by $\mathcal{H}^1$ the completion of the space of smooth cylinder functions with respect to the norm $\| * \|_1$ and denote with $\mathcal{H}^{-1}$ the dual of $\mathcal{H}^1$ with norm

$$\|f\|_{-1} = \sup_{g \in L^2(\mu)} (2\langle f, g \rangle - \|g\|_1^2)$$

(see [20, Chap. 2] for details).

In order to exploit the Itô trick of Lemma 6 we need to be able to solve the Poisson equation $L_0 h_f = f$ for given function f. Sometimes we can do it explicitly and estimate directly the Dirichlet energy of the solution. In other situations howevever it is preferable to have a bound involving directly f without need to find h_f explicitly. (In order not to bother us with domain considerations the reader can think of cylindrical and smooth f, h since this will be enough for our purposes.) Fix an arbitrary h and consider the decomposition

$$f = L_0 h + L_0 h - f$$

then

$$\left\| \sup_{t \in [0,T]} \left| \int_0^t L_0 h(u_s) ds \right| \right\|_{L^2(\mathbb{P})} \leqslant C T^{1/2} \|h\|_1$$

and

$$\left\| \sup_{t \in [0,T]} \left| \int_0^t (L_0 h - f)(u_s) ds \right| \right\|_{L^2(\mathbb{P})} \leqslant \int_0^T \|(L_0 h - f)(u_s)\|_{L^2(\mathbb{P})} ds$$

$$\leqslant T \|L_0 h - f\|_{L^2(\mu)}$$

so

$$\left\| \sup_{t \in [0,T]} \left| \int_0^t f(u_s) ds \right| \right\|_{L^2(\mathbb{P})} \leqslant C T^{1/2} \|h\|_1 + T \|L_0 h - f\|_{L^2(\mu)}$$

Now note that, for fixed $\lambda > 0$ a solution h_λ to $(\lambda - L_0) h_\lambda = -f$ is given explicitly by

$$h_\lambda(x) = - \int_0^\infty e^{-\lambda s} (P_s^{OU} f)(x) ds$$

where P^{OU} is the contraction semigroup generated by L_0 on $L^2(\mu)$. Then

$$\langle h_\lambda, -f \rangle = \langle h_\lambda, (\lambda - L_0) h_\lambda, \rangle = \lambda \|h_\lambda\|_{L^2}^2 + \|h_\lambda\|_1^2$$

and if $f \in \mathscr{H}^{-1}$ we have

$$\lambda \|h_\lambda\|_{L^2}^2 + \|h_\lambda\|_1^2 = |\langle h_\lambda, f \rangle| \leq \|f\|_{-1} \|h_\lambda\|_1$$

which implies $\|h_\lambda\|_1 \leq \|f\|_{-1}$ and $\|h_\lambda\|_{L^2} \leq \lambda^{-1/2} \|f\|_{-1}$. So noting that $L_0 h_\lambda - f = \lambda h_\lambda$ we have

$$CT^{1/2} \|h_\lambda\|_1 + T \|L_0 h_\lambda - f\|_{L^2(\mu)} = CT^{1/2} \|h_\lambda\|_1 + T \|\lambda h_\lambda\|_{L^2(\mu)}$$
$$\leq CT^{1/2} \|f\|_{-1} + T \lambda^{1/2} \|f\|_{-1}.$$

Taking $\lambda \to 0$ we end up with the following lemma which extends the Kipnis–Varadhan lemma (cfr. Lemma 2.4 in [20]) to controlled processes:

Lemma 7 *Assume that $f \in \mathscr{H}^{-1}$ then for every controlled process u we have*

$$\left\| \sup_{t \in [0,T]} \left| \int_0^t f(u_s) ds \right| \right\|_{L^2(\mathbb{P})} \leq CT^{1/2} \|f\|_{-1}.$$

with some constant C which does not depends on u.

2.7 Boltzmann–Gibbs Principle

In the theory of interacting particle systems the phenomenon that local quantities of the microscopic fields can be replaced in time averages by simple functionals of the conserved quantities is called *Boltzmann–Gibbs principle*. In this section we investigate a similar phenomenon in order to control quantities of the form

$$\int_0^t \partial_x F(\varepsilon^{1/2}(u_s^\varepsilon)(x)) ds \tag{2.23}$$

as $N \to +\infty$ where $\varepsilon = \pi/N$ and $u_s^\varepsilon = \Pi_0^N v_s^\varepsilon$ where v^ε is a controlled process which could depend on ε. Note that we have $\mathbb{E}[(\varepsilon^{1/2} u_s^\varepsilon(x))^2] = 1$ for all N, and therefore the Gaussian random variables $(\varepsilon^{1/2} u_s^\varepsilon(x))_N$ stay bounded in L^2 for fixed (s, x), but for large N there will be wild fluctuations in (s, x). We will show that the quantity in (2.23) can be replaced by simpler expressions that are constant, linear, or quadratic in u^ε.

2.7.1 A First Computation

In the following $G \in C(\mathbb{R}, \mathbb{R})$ denotes a generic continuous function. A first interesting computation is to consider the random field $x \mapsto G(\varepsilon^{1/2} \Pi_0^N \eta(x))$ and to derive its chaos expansion in the variables $(\eta_k)_k$ where $\eta_k = \langle \eta, e_{-k} \rangle$ are the Fourier coordinates of η (a space white noise as usual). To do so consider the standard Gaussian random variable

$$\eta^N(x) := \varepsilon^{1/2} \Pi_0^N \eta(x) = \varepsilon^{1/2} \sum_{0 < |k| \leqslant N} e_k(x) \eta_k,$$

and observe that the chaos expansion in $L^2(\mathrm{Law}(\eta^N(x)))$ yields

$$G(\eta^N(x)) = \sum_{n \geqslant 0} c_n(G) H_n(\eta^N(x)),$$

where H_n is the n-th Hermite polynomial and

$$c_n(G) = \frac{1}{n!} \mathbb{E}[G(\eta^N(x)) H_n(\eta^N(x))] = \frac{1}{n!} \int_{\mathbb{R}} G(x) H_n(x) \gamma(x) \mathrm{d}x,$$

where γ is the standard Gaussian density. Since $H_n(x) = (-1)^n e^{x^2/2} \partial_x^n e^{-x^2/2}$, we get

$$c_n(G) = \frac{1}{n!} \int_{\mathbb{R}} G(x)(-1)^n \partial_x^n \gamma(x) \mathrm{d}x = \frac{\psi_G^{(n)}(0)}{n!},$$

where $\psi_G(\lambda) = \mathbb{E}[G(\lambda + \eta^N(x))]$.

Our next aim is to relate the Hermite polynomials of $\eta^N(x)$ with the Wick powers of the family $(\eta_k)_k$. To do so we observe that, on one hand the monomials $H_n(\eta^N(x))$ are the coefficients of the powers of λ in $\exp(\lambda \eta^N(x) - \lambda^2/2)$, and on the other hand

$$\sum_n \frac{\lambda^n}{n!} H_n(\eta^N(x)) = \exp(\lambda \eta^N(x) - \lambda^2/2)$$

$$= \exp\left(\lambda \varepsilon^{1/2} \sum_{0 < |k| \leqslant N} e_k(x) \eta_k - \frac{1}{4\pi} \sum_{0 < |k| \leqslant N} (\lambda \varepsilon^{1/2})^2 \right).$$

Writing $[\![\cdot]\!]_n$ for the projection onto the n-th homogeneous chaos generated by η, we have

$$\exp\left(\sum_{0<|k|\leqslant N}\mu_k\eta_k - \frac{1}{2}\sum_{0<|k|\leqslant N}\mu_k\mu_{-k}\right) = \sum_{n\geqslant 0}\sum_{k_1\cdots k_n}\frac{\mu_{k_1}\cdots\mu_{k_n}}{n!}[\![\eta_{k_1}\cdots\eta_{k_n}]\!]_n,$$

where the sum on the right hand side and all the following sums in $k_1\ldots k_n$ are over $0 < |k_1|, \ldots, |k_n| \leqslant N$. Setting $\mu_k = \varepsilon^{1/2}\lambda e_k(x)$ and identifying the coefficients for different powers of λ, we get

$$H_n(\varepsilon^{1/2}\Pi_0^N\eta(x)) = \varepsilon^{n/2}\sum_{k_1\cdots k_n}\frac{e^{i(k_1+\cdots+k_n)x}}{(2\pi)^{n/2}}[\![\eta_{k_1}\cdots\eta_{k_n}]\!]_n,$$

which can also be obtained by writing $H_n(\varepsilon^{1/2}\Pi_0^N\eta(x)) = [\![(\varepsilon^{1/2}\Pi_0^N\eta(x))^n]\!]_n$ and expanding the power $(\cdot)^n$ inside the projection. We can thus represent the function $G(\eta^N(x))$ as

$$G(\eta^N(x)) = \sum_{n\geqslant 0}c_n(G)H_n(\varepsilon^{1/2}\Pi_0^N\eta(x))$$

$$= \sum_{n\geqslant 0}c_n(G)\varepsilon^{n/2}\sum_{k_1,\ldots,k_n}\frac{e^{i(k_1+\cdots+k_n)x}}{(2\pi)^{n/2}}[\![\eta_{k_1}\cdots\eta_{k_n}]\!]_n.$$

If $\varphi \in C^\infty(\mathbb{T})$ is a test function, we get

$$\langle G(\eta^N), \varphi\rangle = \sum_{n\geqslant 0}c_n(G)\varepsilon^{n/2}\sum_{k_1,\ldots,k_n}\frac{\hat{\varphi}(-k_1-\cdots-k_n)}{(2\pi)^{(n-1)/2}}[\![\eta_{k_1}\cdots\eta_{k_n}]\!]_n. \qquad (2.24)$$

So in particular the q-th Littlewood-Paley block (see [12] for the definition of Littlewood-Paley blocks) of $G(\eta^N)$ is given by

$$\Delta_q G(\eta^N)(x) = \sum_{n\geqslant 0}c_n(G)\varepsilon^{n/2}\sum_{k_1,\ldots,k_n}\theta_q(k_1+\cdots+k_n)\frac{e^{i(k_1+\cdots+k_n)x}}{(2\pi)^{(n-1)/2}}[\![\eta_{k_1}\cdots\eta_{k_n}]\!]_n,$$

where $(\theta_q)_{q\geqslant-1}$ is a dyadic partition of unity, and

$$\mathbb{E}[|\Delta_q(G(\eta^N) - c_0(G))(x)|^2] \leqslant \sum_{n\geqslant 1}c_n(G)^2\frac{z_n\varepsilon^n}{(2\pi)^{n-1}}\sum_{k_1,\ldots,k_n}\theta_q(k_1+\cdots+k_n)^2$$

$$\lesssim \sum_{n\geqslant 1}c_n(G)^2 z_n\frac{\varepsilon^n(2N)^{n-1}}{(2\pi)^{n-1}}(2^q\wedge N)$$

$$\lesssim \varepsilon\sum_{n\geqslant 1}c_n(G)^2 z_n(2^q\wedge N),$$

where $z_n = \max_{k_1 \ldots k_n} \mathbb{E}[|[\![\eta_{k_1} \cdots \eta_{k_n}]\!]_n|^2] \leqslant n!$ is a combinatorial factor. We thus obtain

$$\mathbb{E}[\|\Delta_q(G(\eta^N) - c_0(G))\|^2_{L^2(\mathbb{T})}] \lesssim \min\{\varepsilon 2^q, 1\},$$

uniformly in N, and then

$$\mathbb{E}\left[\left|\int_s^t \Delta_q(G(\varepsilon^{1/2} u_r^\varepsilon(x)) - c_0(G))dr\right|^2\right] \leqslant |t-s| \int_s^t \mathbb{E}[|\Delta_q(G(\varepsilon^{1/2} u_r^\varepsilon(x))$$

$$- c_0(G))|^2]dr$$

$$\lesssim |t-s|^2 \min\{\varepsilon 2^q, 1\},$$

where in the last step we used that $\varepsilon^{1/2} u_r^\varepsilon$ has the same distribution as η^N, which easily implies the following first result.

Lemma 8 *Assume that* $\mathbb{E}[|G(U)|^2] < \infty$ *for a standard normal variable* U, *and let* $c_0(G) = \mathbb{E}[G(U)]$. *Then*

$$\lim_{N \to \infty} \int_0^t G(\varepsilon^{1/2} u_s^\varepsilon(x))ds = c_0(G)t,$$

where the convergence is in $C([0, T], H^{0-})$. *If* $c_0(G) = 0$, *then*

$$\varepsilon^{-1/2} \int_0^t G(\varepsilon^{1/2} u_s^\varepsilon(x))ds$$

is bounded in $C([0, T], H^{-1/2-})$.

To analyse the case where $c_0(G) = 0$ we need a more refined argument which is provided by the Itô trick for controlled paths.

Let us write $\mathscr{L}_0^\varepsilon$ for the generator of the mollified Ornstein–Uhlenbeck process

$$\partial_t X^\varepsilon = \Delta X^\varepsilon + \partial_x \Pi_0^N \xi.$$

For $\Psi \in \text{dom}\left(\mathscr{L}_0^\varepsilon\right)$ and $T > 0$, $p \geqslant 1$ we have:

$$\mathbb{E}\left[\sup_{t \in [0,T]} \left|\int_0^t \mathscr{L}_0^\varepsilon \Psi(u_s^\varepsilon)ds\right|^p\right] \lesssim T^{p/2} \mathbb{E}[\mathcal{E}^\varepsilon(\Psi)^{p/2}].$$

To apply this Itô trick we need to solve the Poisson equation. In our setting this can be done efficiently by using the chaos expansion (2.24). Recall that we wrote $\eta_k = \langle \eta, e_k \rangle$ for the Fourier coefficients of a truncated spatial white noise $\Pi_0^N \eta$ (which therefore has law μ^ε), and that $[\![\cdot]\!]_n$ denotes the projection onto the n-th chaos. We need to compute $\mathscr{L}_0^\varepsilon [\![\eta_{k_1} \dots \eta_{k_n}]\!]_n$, as these are the random variables appearing in a general chaos expansion. Let us start by considering $\varphi \in Y_N = \Pi_0^N L^2(\mathbb{T}, \mathbb{R})$ with $\|\varphi\|_{L^2} = 1$ for which we have $[\![\langle \eta, \varphi \rangle^n]\!]_n = H_n(\langle \eta, \varphi \rangle)$, where H_n is the n-th Hermite polynomial. Itô's formula gives

$$dH_n(\langle X_t^\varepsilon, \varphi \rangle) = H_n'(\langle X_t^\varepsilon, \varphi \rangle) \langle X_t^\varepsilon, \Delta\varphi \rangle dt + H_n''(\langle X_t^\varepsilon, \varphi \rangle) \langle \Pi_0^N \partial_x \varphi, \Pi_0^N \partial_x \varphi \rangle dt + dM_t,$$

with a square integrable martingale M. The Hermite polynomials satisfy $H_n' = n H_{n-1}$, so we get

$$H_n'(\langle X_t^\varepsilon, \varphi \rangle) \langle X_t^\varepsilon, \Delta\varphi \rangle + H_n''(\langle X_t^\varepsilon, \varphi \rangle) \langle \Pi_0^N \partial_x \varphi, \Pi_0^N \partial_x \varphi \rangle$$
$$= n H_{n-1}(\langle X_t^\varepsilon, \varphi \rangle) H_1(\langle X_t^\varepsilon, \Delta\varphi \rangle) - n(n-1) H_{n-2}(\langle X_t^\varepsilon, \varphi \rangle) \langle \Pi_0^N \varphi, \Pi_0^N \Delta\varphi \rangle.$$

The projection onto the n-th chaos of the first term is explicitly given by

$$[\![H_{n-1}(\langle X_t^\varepsilon, \varphi \rangle) H_1(\langle X_t^\varepsilon, \Delta\varphi \rangle)]\!]_n = [\![[\![\langle X_t^\varepsilon, \varphi \rangle^{n-1}]\!]_{n-1} [\![\langle X_t^\varepsilon, \Delta\varphi \rangle]\!]_1]\!]_n$$
$$= [\![\langle X_t^\varepsilon, \varphi \rangle^{n-1}]\!]_{n-1} [\![\langle X_t^\varepsilon, \Delta\varphi \rangle]\!]_1$$
$$- (n-1) [\![\langle X_t^\varepsilon, \varphi \rangle^{n-2}]\!]_{n-2} \langle \Pi_0^N \varphi, \Pi_0^N \Delta\varphi \rangle,$$

which is obtained by contracting $\langle X_t^\varepsilon, \Delta\varphi \rangle$ with each of the $n-1$ variables $\langle X_t^\varepsilon, \varphi \rangle$ inside the projector $[\![\cdot]\!]_{n-1}$. Therefore, we have

$$dH_n(\langle X_t^\varepsilon, \varphi \rangle) = n [\![H_{n-1}(\langle X_t^\varepsilon, \varphi \rangle) H_1(\langle X_t^\varepsilon, \Delta\varphi \rangle)]\!]_n dt + dM_t$$
$$= n [\![\langle X_t^\varepsilon, \varphi \rangle^{n-1} \langle X_t^\varepsilon, \Delta\varphi \rangle]\!]_n dt + dM_t,$$

which shows that

$$\mathscr{L}_0^\varepsilon [\![\langle \eta, \varphi \rangle^n]\!]_n = n [\![\langle \eta, \varphi \rangle^{n-1} \langle \eta, \Delta\varphi \rangle]\!]_n.$$

So far we assumed $\|\varphi\|_{L^2} = 1$, but actually this last formula is invariant under scaling so it extends to all $\varphi \in \Pi_0^N L^2(\mathbb{T}, \mathbb{R})$, and then to $\varphi \in \Pi_0^N L^2(\mathbb{T}, \mathbb{C})$, and for general products we obtain by polarization

$$\mathscr{L}_0^\varepsilon [\![\langle \eta, \varphi_1 \rangle \dots \langle \eta, \varphi_n \rangle]\!]_n = \sum_{k=1}^{n} [\![\langle \eta, \varphi_1 \rangle \dots \text{\textit{text}} \dots \langle \eta, \varphi_n \rangle \langle \eta, \Delta\varphi_k \rangle]\!]_n.$$

So finally we deduce that

$$\mathscr{L}_0^\varepsilon [\![\eta_{k_1} \cdots \eta_{k_n}]\!] = -(k_1^2 + \cdots + k_n^2) [\![\eta_{k_1} \cdots \eta_{k_n}]\!], \tag{2.25}$$

for all $0 < |k_1|, \ldots, |k_n| \leqslant N$. Combining that formula with (2.24), we obtain the following lemma.

Lemma 9 *Consider a function of the form* $\Phi(\eta) = \langle G(\varepsilon^{1/2} \Pi_0^N \eta), \varphi \rangle$ *and assume that* $\mathbb{E}[G(U)] = 0$, *where* U *is a standard normal variable, or that* $\hat{\varphi}(0) = 0$. *Then the solution* Ψ *to the Poisson equation* $\mathscr{L}_0^\varepsilon \Psi = \Phi$ *is explicitly given by*

$$\Psi(\eta) = -\sum_{n \geqslant 1} c_n(G) \varepsilon^{n/2} \sum_{k_1 \cdots k_n} \frac{\hat{\varphi}(-k_1 - \cdots - k_n)}{(2\pi)^{(n-1)/2}} \frac{[\![\eta_{k_1} \cdots \eta_{k_n}]\!]_n}{(k_1^2 + \cdots + k_n^2)},$$

where the sum is over all $0 < |k_1|, \ldots, |k_n| \leqslant N$.

Remark 2 Incidentally note that the solution can be represented as

$$\Psi(\eta) = -\int_0^\infty dt \sum_{n \geqslant 1} c_n(G) \varepsilon^{n/2} \sum_{k_1 \cdots k_n} e^{-(k_1^2 + \cdots + k_n^2)t} \frac{e^{i(k_1 + \cdots + k_n)x}}{(2\pi)^{n/2}} [\![\eta_{k_1} \cdots \eta_{k_n}]\!]_n$$

$$= -\int_0^\infty dt \, G(\varepsilon^{1/2}(e^{\Delta t} \Pi_0^N \eta)(x)).$$

To apply the Itô trick we need also to compute $\mathcal{E}(\Psi) = \sum_k k^2 D_{-k} \Psi D_k \Psi$ for the solution Ψ of the Poisson equation. For that purpose consider again $\varphi \in Y_N$ with $\|\varphi\|_{L^2} = 1$ and $H_n(\langle \eta, \varphi \rangle) = [\![\langle \eta, \varphi \rangle^n]\!]_n$, for which we have

$$D_k H_n(\langle \eta, \varphi \rangle) = H_n'(\langle \eta, \varphi \rangle) \langle e_k, \varphi \rangle = n H_{n-1}(\langle \eta, \varphi \rangle) \langle e_k, \varphi \rangle$$

$$= n [\![\langle \eta, \varphi \rangle^{n-1}]\!]_{n-1} \langle e_k, \varphi \rangle,$$

so by polarization

$$D_k [\![\eta_{k_1} \cdots \eta_{k_n}]\!]_n = \sum_j \mathbf{1}_{k_j = k} [\![\eta_{k_1} \cdots \widehat{\eta_{k_j}} \cdots \eta_{k_n}]\!]_{n-1}, \qquad (2.26)$$

where $\widehat{\eta_{k_j}}$ denotes that this particular factor is removed.

To prove the Boltzmann–Gibbs principle we need one more auxiliary result.

Lemma 10 *For all* $M \leqslant N$, $\ell \in \mathbb{Z}$ *and* $0 \leqslant s < t < \infty$ *we have the estimate*

$$\mathbb{E}\left[\left|\int_s^t \langle \partial_x (\Pi_0^M u_r^\varepsilon)^2, e_{-\ell} \rangle dr\right|^2\right] \lesssim \ell^2 |t - s|^2 M.$$

Proof We simply bound

$$\mathbb{E}\left[\left|\int_s^t \langle \partial_x (\Pi_0^M u_r^\varepsilon)^2, e_{-\ell} \rangle dr\right|^2\right] \leqslant |t - s| \int_s^t \mathbb{E}[|\langle \partial_x (\Pi_0^M u_r^\varepsilon)^2, e_{-\ell} \rangle|^2] dr,$$

and since we can replace $(\Pi_0^M u_r^\varepsilon)^2$ by $(\Pi_0^M u_r^\varepsilon)^2 - \mathbb{E}[(\Pi_0^M u_r^\varepsilon)^2]$, the integrand is given by

$$\mathbb{E}[|\langle \partial_x (\Pi_0^M u_r^\varepsilon)^2, e_{-\ell}\rangle|^2] = \ell^2 \int_{\mathbb{T}} dx \int_{\mathbb{T}} dx' \mathbb{E}[[\![(\Pi_0^M u_r^\varepsilon(x))^2]\!]_2 [\![(\Pi_0^M u_r^\varepsilon(x'))^2]\!]_2]$$

$$\lesssim \ell^2 \int_{\mathbb{T}} dx \int_{\mathbb{T}} dx' |\mathbb{E}[\Pi_0^M u_r^\varepsilon(x) \Pi_0^M u_r^\varepsilon(x')]|^2.$$

The expectation on the right hand side can be explicitly computed as

$$|\mathbb{E}[\Pi_0^M u_r^\varepsilon(x) \Pi_0^M u_r^\varepsilon(x')]| = \left| \sum_{0 < |k| \leqslant M} e^{ik(x-x')} \right|$$

$$= \left| \frac{\cos(M(x-x')) - \cos((M+1)(x-x'))}{1 - \cos(x-x')} - 1 \right|$$

$$\leqslant \min\{2M, C|x - x'|^{-1}\},$$

for some constant $C < +\infty$, for which

$$\int_{\mathbb{T}} dx \int_{\mathbb{T}} dx' \min\{2M, C|x - x'|^{-1}\}^2 dx \lesssim 2M,$$

and therefore the claim follows. $\qquad\square$

Proposition 1 (Boltzmann–Gibbs Principle) *Let $G, G' \in L^2(v)$, where v denotes the law of a standard normal variable. Then for all $\ell \in \mathbb{Z}$ and $0 \leqslant s < t \leqslant s+1$ and all $\kappa > 0$*

$$\mathbb{E}\left[\left| \int_s^t \langle \varepsilon^{-1} \partial_x \Pi_0^N G(\varepsilon^{1/2} u_r^\varepsilon) - \varepsilon^{-1/2} c_1(G) \partial_x \Pi_0^N u_r^\varepsilon, e_{-\ell}\rangle dr \right|^2 \right]$$

$$\lesssim |t - s|^{3/2 - \kappa} \ell^2 \int_{\mathbb{R}} |G'(x)|^2 v(dx)$$

uniformly in $N \in \mathbb{N}$, and for all $M \leqslant N/2$

$$\mathbb{E}\left[\left| \int_s^t \langle \varepsilon^{-1} \partial_x \Pi_0^N G(\varepsilon^{1/2} u_r^\varepsilon) - \varepsilon^{-1/2} c_1(G) \partial_x \Pi_0^N u_r^\varepsilon - c_2(G) \partial_x (\Pi_0^M u_r^\varepsilon)^2, e_{-\ell}\rangle dr \right|^2 \right]$$

$$\lesssim |t - s| \ell^2 (M^{-1} + \varepsilon \log^2 N) \int_{\mathbb{R}} |G'(x)|^2 v(dx).$$

Proof We start by showing the second bound. Note that $\partial_x(\Pi_0^M \eta)^2 = \Pi_0^N \partial_x(\Pi_0^M \eta)^2$ for $M \leqslant N/2$ and that by Lemma 9 the solution Ψ to

$$\mathscr{L}_0^\varepsilon \Psi(\eta) = -\varepsilon^{-1}\langle G(\varepsilon^{1/2}\Pi_0^N \eta) - c_1(G)\varepsilon^{1/2}\Pi_0^N \eta - c_2(G)(\varepsilon^{1/2}\Pi_0^M \eta)^2, \partial_x \Pi_0^N e_{-\ell}\rangle$$

is given by

$$\Psi(\eta) = c_2(G) \sum_{k_1,k_2} \mathbf{1}_{|k_1|\vee|k_2|>M}\mathbf{1}_{0<|\ell|\leqslant N}(i\ell)\frac{\mathbf{1}_{k_1+k_2=\ell}}{(2\pi)^{1/2}}\frac{[\![\eta_{k_1}\eta_{k_2}]\!]_2}{(k_1^2+k_2^2)}$$

$$+ \sum_{n\geqslant 3} c_n(G)\varepsilon^{n/2-1} \sum_{k_1\cdots k_n} \mathbf{1}_{0<|\ell|\leqslant N}(i\ell)\frac{\mathbf{1}_{k_1+\cdots+k_n=\ell}}{(2\pi)^{(n-1)/2}}\frac{[\![\eta_{k_1}\cdots\eta_{k_n}]\!]_n}{(k_1^2+\cdots+k_n^2)},$$

where it is understood that all sums in k_i are over $0 < |k_i| \leqslant N$. Therefore (2.26) yields for $0 < |\ell| \leqslant N$

$$D_k\Psi(\eta) = c_2(G)2 \sum_{k_1} \mathbf{1}_{|k|\vee|k_1|>M}i\ell\frac{\mathbf{1}_{k+k_1=\ell}}{(2\pi)^{1/2}}\frac{[\![\eta_{k_1}]\!]_1}{(k^2+k_1^2)}$$

$$+ \sum_{n\geqslant 2} c_{n+1}(G)\varepsilon^{(n-1)/2}(n+1)$$

$$\times \sum_{k_1\cdots k_n} i\ell\frac{\mathbf{1}_{k+k_1+\cdots+k_n=\ell}}{(2\pi)^{n/2}}\frac{[\![\eta_{k_1}\cdots\eta_{k_n}]\!]_n}{(k^2+k_1^2+\cdots+k_n^2)}.$$

Applying the Itô trick we get

$$\mathbb{E}\left[\left|\int_s^t \langle\varepsilon^{-1}\partial_x\Pi_0^N G(\varepsilon^{1/2}u_r^\varepsilon)\right.\right.$$

$$\left.\left.- \varepsilon^{-1/2}c_1(G)\partial_x\Pi_0^N u_r^\varepsilon - c_2(G)\partial_x(\Pi_0^M u_r^\varepsilon)^2, e_{-\ell}\rangle dr\right|^2\right]$$

$$\lesssim |t-s| \sum_{0<|k|\leqslant N} k^2\mathbb{E}[|D_k\Psi|^2]$$

$$= |t-s| \sum_{0<|k|\leqslant N} k^2 c_2(G)^2 2^2 \ell^2 \sum_{k_1} \mathbf{1}_{|k|\vee|k_1|>M}\frac{\mathbf{1}_{k+k_1=\ell}}{2\pi}\frac{\mathbb{E}[|[\![\eta_{k_1}]\!]_1|^2]}{(k^2+k_1^2)^2}$$

$$+ |t-s| \sum_{0<|k|\leqslant N} k^2 \sum_{n\geqslant 2} c_{n+1}(G)^2\varepsilon^{(n+1)-2}(n+1)^2\ell^2$$

$$\times \sum_{k_1\cdots k_n} \frac{\mathbf{1}_{k+k_1+\cdots+k_n=\ell}}{(2\pi)^n}\frac{\mathbb{E}[|[\![\eta_{k_1}\cdots\eta_{k_n}]\!]_n|^2]}{(k^2+k_1^2+\cdots+k_n^2)^2}$$

$$= |t-s| \sum_{n\geqslant 1} A_n,$$

where the (A_n) are implicitly defined by the equation. Now $\mathbb{E}[|[\![\eta_{k_1} \cdots \eta_{k_n}]\!]_n|^2] \leqslant n!$ for all $k_1, \ldots, k_n$, so that

$$A_1 \lesssim \sum_{0 < |k|, |k_1| \leqslant N} k^2 c_2(G)^2 \ell^2 \mathbf{1}_{k + k_1 = \ell} \frac{\mathbf{1}_{|k| \vee |k_1| > M}}{(k^2 + k_1^2)^2}$$

$$\leqslant \sum_{0 < |k|, |k_1| \leqslant N} c_2(G)^2 \ell^2 \mathbf{1}_{k + k_1 = \ell} \frac{\mathbf{1}_{|k| \vee |k_1| > M}}{k^2 + k_1^2}$$

$$\lesssim c_2(G)^2 \ell^2 \sum_{0 < |k| < \infty} \frac{\mathbf{1}_{\ell \neq k} \mathbf{1}_{|k| \vee |\ell - k| > M}}{k^2 + (\ell - k)^2}$$

$$\leqslant c_2(G)^2 \ell^2 \sum_{0 < |k| < \infty} \left(\frac{\mathbf{1}_{\ell \neq k}}{M^2 + (\ell - k)^2} + \frac{\mathbf{1}_{\ell \neq k}}{k^2 + M^2} \right)$$

$$\lesssim c_2(G)^2 \ell^2 M^{-1},$$

while for $n > 1$

$$A_n = \sum_{0 < |k| \leqslant N} k^2 c_{n+1}(G)^2 \varepsilon^{n-1} (n+1)^2 \ell^2$$

$$\times \sum_{k_1 \cdots k_n} \frac{\mathbf{1}_{k + k_1 + \cdots + k_n = \ell}}{(2\pi)^n} \frac{\mathbb{E}[|[\![\eta_{k_1} \cdots \eta_{k_n}]\!]_n|^2]}{(k^2 + k_1^2 + \cdots + k_n^2)^2}$$

$$\leqslant \frac{\varepsilon^{n-1}}{(2\pi)^n} \ell^2 (n+1)^2 c_{n+1}(G)^2 n! \sum_{0 < |k|, |k_1|, \ldots, |k_n| \leqslant N} k^2 \frac{\mathbf{1}_{k + k_1 + \cdots + k_n = \ell}}{(k^2 + k_1^2 + \cdots + k_n^2)^2}$$

$$\leqslant \frac{\varepsilon^{n-1}}{(2\pi)^n} \ell^2 (n+1)^2 c_{n+1}(G)^2 n! \sum_{0 < |k_1|, \ldots, |k_n| \leqslant N} \frac{1}{k_1^2 + \cdots + k_n^2}$$

$$\leqslant \frac{\varepsilon^{n-1}}{(2\pi)^n} \ell^2 (n+1)^2 c_{n+1}(G)^2 n! \sum_{0 < |k_1|, \ldots, |k_n| \leqslant N} \frac{1}{k_1^2 + k_2^2}$$

$$= \frac{\varepsilon^{n-1}}{(2\pi)^n} \ell^2 (n+1)^2 c_{n+1}(G)^2 n! (2N)^{n-2} \sum_{0 < |k_1|, |k_2| \leqslant N} \frac{1}{k_1^2 + k_2^2}$$

$$\lesssim \varepsilon \ell^2 (n+1)^2 c_{n+1}(G)^2 n! \log^2 N.$$

The sum over n is bounded by

$$\sum_{n=2}^\infty c_{n+1}(G)^2 n! (n+1)^2 = \sum_{n=1}^\infty n c_n(G)^2 n! \lesssim \int_\mathbb{R} |G'(x)|^2 \nu(\mathrm{d}x),$$

so that overall we get

$$\mathbb{E}\left[\left|\int_s^t \langle \varepsilon^{-1}\partial_x \Pi_0^N G(\varepsilon^{1/2}u_r^\varepsilon) - \varepsilon^{-1/2}c_1(G)\partial_x \Pi_0^N u_r^\varepsilon - c_2(G)\partial_x(\Pi_0^M u_r^\varepsilon)^2, e_{-\ell}\rangle dr\right|^2\right] \tag{2.27}$$

$$\lesssim |t-s|\ell^2(M^{-1} + \varepsilon \log^2 N)\int_\mathbb{R}|G'(x)|^2 \nu(dx), \tag{2.28}$$

which is our second claimed bound.

To get the first bound, we take $M \simeq |t-s|^{-1/2}$ in (2.27) (which requires $N > |t-s|^{-1/2}$), and combine this with Lemma 10 to obtain

$$\mathbb{E}\left[\left|\int_s^t \langle \varepsilon^{-1}\partial_x \Pi_0^N G(\varepsilon^{1/2}u_r^\varepsilon) - \varepsilon^{-1/2}c_1(G)\partial_x \Pi_0^N u_r^\varepsilon, e_{-\ell}\rangle dr\right|^2\right]$$

$$\lesssim |t-s|\ell^2(M^{-1} + \varepsilon \log^2 N + |t-s|M)\int_\mathbb{R}|G'(x)|^2\nu(dx)$$

$$\lesssim |t-s|^{3/2-\kappa}\ell^2 \int_\mathbb{R}|G'(x)|^2\nu(dx).$$

If $N \leqslant |t-s|^{-1/2}$ we use another estimate: as in the proof of Lemma 10 we have

$$\mathbb{E}\left[\left|\int_s^t \langle \varepsilon^{-1}\partial_x \Pi_0^N G(\varepsilon^{1/2}u_r^\varepsilon) - \varepsilon^{-1/2}c_1(G)\partial_x \Pi_0^N u_r^\varepsilon, e_{-\ell}\rangle dr\right|^2\right]$$

$$\leqslant |t-s|^2 \mathbb{E}[|\langle \varepsilon^{-1}\partial_x \Pi_0^N G(\varepsilon^{1/2}u_0^\varepsilon) - \varepsilon^{-1/2}c_1(G)\partial_x \Pi_0^N u_0^\varepsilon, e_{-\ell}\rangle|^2]$$

$$\lesssim |t-s|^2 \sum_{n\geqslant 2}\ell^2\varepsilon^{-2}c_n(G)^2 \int_\mathbb{T} dx \int_\mathbb{T} dx' \mathbb{E}[H_n(\varepsilon^{1/2}u_0^\varepsilon(x))H_n(\varepsilon^{1/2}u_0^\varepsilon(x'))],$$

$$\lesssim |t-s|^2 \sum_{n\geqslant 2}\ell^2\varepsilon^{-2}n!c_n(G)^2 \int_\mathbb{T} dx \int_\mathbb{T} dx' |\mathbb{E}[\varepsilon^{1/2}u_0^\varepsilon(x)\varepsilon^{1/2}u_0^\varepsilon(x')]^n|$$

$$\lesssim |t-s|^2 \sum_{n\geqslant 2}\ell^2\varepsilon^{n-2}n!c_n(G)^2 \int_\mathbb{T} dx \int_\mathbb{T} dx' \min\{2N, C|x-x'|^{-1}\}^n$$

$$\lesssim |t-s|^2 \sum_{n\geqslant 2}\ell^2\varepsilon^{n-2}n!c_n(G)^2(2N)^{n-1} \lesssim |t-s|^2 \sum_{n\geqslant 2}\ell^2\varepsilon^{-1}c_n(G)^2 n!$$

$$\lesssim \ell^2|t-s|^{3/2}\int_\mathbb{R}|G'(x)|^2\nu(dx),$$

where in the last step we used that $|t-s|^{-1/2}N^{-1} \geqslant 1$. $\square$

2.8 The Hairer–Quastel Invariance Principle

In this section we will study convergence to SBE for the large scale limit of a class of (non-singular) SPDEs of the form

$$\partial_t v = \Delta v + \varepsilon^{1/2} \partial_x F(v) + \partial_x \chi^\varepsilon \qquad (2.29)$$

on $[0, \infty) \times \mathbb{T}_\varepsilon$ with $\mathbb{T}_\varepsilon = \mathbb{R}/(2\pi\varepsilon^{-1}\mathbb{Z})$, where χ^ε is a Gaussian noise that is white in time and spatially smooth. The Hairer–Quastel universality result [16] states that there exist constants $c_1, c_2 \in \mathbb{R}$ such that the rescaled process $\varepsilon^{-1/2} v_{t\varepsilon^{-2}}((x - c_1\varepsilon^{-1/2}t)\varepsilon^{-1})$ converges (for small time) to the solution u of the stochastic Burgers equation (as defined in the context of Regularity Structures)

$$\partial_t u = \Delta u + c_2 \partial_x u^2 + \partial_x \xi,$$

where ξ is a space-time white noise. Here we give an alternative proof of this result, based on the concept of energy solutions and in a stationary situation. The proof turns out to be simpler in this language, on the other hand our method only applies at stationarity and moreover we need an explicit control of the invariant measure which will force us to formulate a bit differently the initial problem.

Let us state the result more precisely. We modify (2.29) such that after rescaling $\tilde{u}_t^\varepsilon(x) = \varepsilon^{-1/2} v_{t\varepsilon^{-2}}(x\varepsilon^{-1})$ we have

$$\partial_t \tilde{u}^\varepsilon = \Delta \tilde{u}^\varepsilon + \varepsilon^{-1} \partial_x \Pi_0^N F(\varepsilon^{1/2} \tilde{u}^\varepsilon) + \partial_x \Pi_0^N \tilde{\xi}, \qquad \tilde{u}_0^\varepsilon = \Pi_0^N \eta, \qquad (2.30)$$

where $\tilde{\xi}$ is a space-time white noise on $[0, \infty) \times \mathbb{T}$ (where $\mathbb{T} = \mathbb{T}_1$) with variance 2, η is a space white noise which is independent of $\tilde{\xi}$, Π_0^N denotes the projection onto the Fourier modes $0 < |k| \leqslant N$, and we always link N and ε via

$$N = \pi/\varepsilon.$$

Theorem 3 *Let F be almost everywhere differentiable and assume that for all $\varepsilon > 0$ there is a unique solution $\tilde{u}^\varepsilon$ to (2.30) which does not blow up before $T > 0$. Assume also that $F, F' \in L^2(\nu)$ where ν is the standard normal distribution. Then $u_t^\varepsilon(x) := \tilde{u}_t^\varepsilon(x - \varepsilon^{-1/2} c_1(F)t)$, $(t, x) \in [0, T] \times \mathbb{T}$, converges in distribution to the unique stationary energy solution u of*

$$\partial_t u = \Delta u + c_2(F) \partial_x u^2 + \partial_x \xi,$$

where ξ is a space-time white noise with variance 2 and for $U \sim \nu$ and $k \geqslant 0$ and H_k the k-th Hermite polynomial

$$c_k(F) = \frac{1}{k!} \mathbb{E}[F(U) H_k(U)].$$

Remark 3 If F is even, then $c_1(F) = 0$ while $c_2(F) = 0$ if F is odd.

Note that we introduced a second regularization in (2.30) compared to (2.29) which acts on $F(\varepsilon^{1/2}u^\varepsilon)$. The reason is that we need to keep track of the invariant measure and this second regularization allows us to write it down explicitly. For simplicity here we only consider the mollification operator Π_0^N, but it is possible to extend everything to more general operators $\rho(\varepsilon|\partial_x|)u = \mathscr{F}^{-1}(\rho(\varepsilon\cdot)\mathscr{F}u)$, where ρ is an even, compactly supported, bounded function which is continuous in a neighborhood of 0 and satisfies $\rho(0) = 1$. We should then modify the equation as

$$\partial_t \tilde{u}^\varepsilon = \Delta\tilde{u}^\varepsilon + \varepsilon^{-1}\partial_x \rho(\varepsilon|\partial_x|)\rho(\varepsilon|\partial_x|)F(\varepsilon^{1/2}\tilde{u}^\varepsilon) + \partial_x \rho(\varepsilon|\partial_x|)\tilde{\xi}, \quad \tilde{u}_0^\varepsilon = \rho(\varepsilon|\partial_x|)\eta,$$

to keep control of the invariant measure, see [6].

While this result only applies in the stationary state, we have more freedom in choosing the nonlinearity F than [16] who require it to be an even polynomial. Also, the energy solution method extends without great difficulty to the modified equation on $[0, T] \times \mathbb{R}$.

Let us start by making some basic observations concerning the solution to (2.30).

Galilean Transformation Recall that $\tilde{u}^\varepsilon$ solves

$$\partial_t \tilde{u}^\varepsilon = \Delta\tilde{u}^\varepsilon + \varepsilon^{-1}\partial_x \Pi_0^N F(\varepsilon^{1/2}\tilde{u}^\varepsilon) + \partial_x \Pi_0^N \xi,$$

and that $u_t^\varepsilon(x) = \tilde{u}_t^\varepsilon(x - \varepsilon^{-1/2}c_1(F)t)$. We define the modified test function $\tilde{\varphi}_t(x) = \varphi(x + \varepsilon^{-1/2}c_1(F)t)$ and then $\langle u_t^\varepsilon, \varphi \rangle = \langle \tilde{u}_t^\varepsilon, \tilde{\varphi}_t \rangle$. The Itô–Wentzell formula gives

$$\begin{aligned}
\mathrm{d}\langle u_t^\varepsilon, \varphi \rangle &= \langle \mathrm{d}\tilde{u}_t^\varepsilon, \tilde{\varphi}_t \rangle + \langle \tilde{u}_t^\varepsilon, \partial_t\tilde{\varphi}_t \rangle \mathrm{d}t \\
&= \langle \Delta\tilde{u}_t^\varepsilon, \tilde{\varphi}_t \rangle \mathrm{d}t + \langle \varepsilon^{-1}\partial_x \Pi_0^N F(\varepsilon^{1/2}\tilde{u}^\varepsilon), \tilde{\varphi}_t \rangle \mathrm{d}t + \langle \mathrm{d}\partial_x \tilde{M}_t^\varepsilon, \tilde{\varphi}_t \rangle \\
&\quad + \langle \varepsilon^{-1/2}c_1(F)\tilde{u}_t^\varepsilon, \partial_x\tilde{\varphi}_t \rangle \mathrm{d}t,
\end{aligned}$$

where $\tilde{M}_t^\varepsilon(x) = \int_0^t \Pi_0^N \tilde{\xi}(s, x)\mathrm{d}s$. Integrating the last term on the right hand side by parts, we get

$$\begin{aligned}
\mathrm{d}\langle u_t^\varepsilon, \varphi \rangle &= \langle \Delta u_t^\varepsilon, \varphi \rangle \mathrm{d}t + \langle \varepsilon^{-1}\partial_x \Pi_0^N F(\varepsilon^{1/2}u^\varepsilon), \varphi \rangle \mathrm{d}t - \varepsilon^{-1/2}c_1(F)\langle \partial_x u_t^\varepsilon, \varphi \rangle \mathrm{d}t \\
&\quad + \langle \mathrm{d}\partial_x \tilde{M}_t^\varepsilon, \tilde{\varphi}_t \rangle.
\end{aligned}$$

The martingale term has quadratic variation

$$\mathrm{d}[\langle \partial_x \tilde{M}^\varepsilon, \tilde{\varphi}_t \rangle]_t = \mathrm{d}[\langle \tilde{M}^\varepsilon, \partial_x\tilde{\varphi}_t \rangle]_t = 2\|\Pi_0^N \partial_x\tilde{\varphi}_t\|_{L^2}^2 \mathrm{d}t = 2\|\Pi_0^N \partial_x\varphi\|_{L^2}^2 \mathrm{d}t,$$

which means that the process $\langle M_t^\varepsilon, \varphi \rangle := \langle \tilde{M}_t^\varepsilon, \tilde{\varphi}_t \rangle$ is of the form $M_t^\varepsilon = \int_0^t \Pi_0^N \xi(s, x) ds$ for a new space-time white noise $\tilde{\xi}$ with variance 2. In conclusion, u^ε solves

$$\partial_t u^\varepsilon = \Delta u^\varepsilon + \varepsilon^{-1} \partial_x \Pi_0^N (F(\varepsilon^{1/2} u^\varepsilon) - c_1(F)\varepsilon^{1/2} u^\varepsilon) + \partial_x \Pi_0^N \xi, \qquad u_0^\varepsilon = \Pi_0^N \eta, \tag{2.31}$$

so in other words by performing the change of variables $u_t^\varepsilon(x) = \tilde{u}_t^\varepsilon(x - \varepsilon^{-1/2} c_1(F)t)$ we replaced the function F by $\tilde{F}(x) = F(x) - c_1(F)x$, and now it suffices to study Eq. (2.31).

Invariant Measure Note that (2.31) actually is an SDE in the finite dimensional space $Y_N = \Pi_0^N L^2(\mathbb{T}, \mathbb{R}) \simeq \mathbb{R}^{2N}$, so that we can apply Echeverria's criterion [4] to show the stationarity of a given distribution. The natural candidate is $\mu^\varepsilon = \mathrm{law}(\Pi_0^N \eta)$, where η is a space white noise, since we know that the dynamics of the regularized Ornstein–Uhlenbeck process

$$\partial_t X^\varepsilon = \Delta X^\varepsilon + \partial_x \Pi_0^N \xi$$

are invariant and even reversible under μ^ε and that for models in the KPZ universality class the asymmetric version often has the same invariant measure as the symmetric one. Let us write

$$B_F^\varepsilon(u) = \varepsilon^{-1} \partial_x \Pi_0^N (F(\varepsilon^{1/2} u) - c_1(F)\varepsilon^{1/2} u) =: \varepsilon^{-1} \partial_x \Pi_0^N \tilde{F}(\varepsilon^{1/2} u),$$

where $\tilde{F} = F - c_1(F)x$.

Lemma 11 *The vector field $B_F^\varepsilon : Y_N \to Y_N$ leaves the Gaussian measure μ^ε invariant. More precisely, if D denotes the gradient with respect to the Fourier monomials $(e_k)_{0 < |k| \leqslant N}$ on Y_N, then*

$$\int_{Y_N} (B_F^\varepsilon(u) \cdot D\Phi(u)) \Psi(u) \mu^\varepsilon(du) = -\int_{Y_N} \Phi(u) B_F^\varepsilon(u) \cdot D\Psi(u) \mu^\varepsilon(du)$$

for all $\Phi, \Psi \in L^2(\mu^\varepsilon)$ with $B_F^\varepsilon \cdot D\Phi$, $B_F^\varepsilon \cdot D\Psi \in L^2(\mu^\varepsilon)$.

Proof In this proof it is more convenient to work with the orthonormal basis

$$\left\{ \frac{1}{\sqrt{\pi}} \sin(k\cdot), \frac{1}{\sqrt{\pi}} \cos(k\cdot), 0 < k \leqslant N \right\}$$

of Y_N, rather than with Fourier monomials. We write $(\varphi_k)_{k=1,\dots,2N}$ for an enumeration of these trigonometric functions. Then $B_F^\varepsilon \cdot D$ can also be expressed in terms of the (φ_k), and we have

$$\Phi(u) = f(\langle u, \varphi_1 \rangle, \dots, \langle u, \varphi_{2N} \rangle), \qquad \Psi(u) = g(\langle u, \varphi_1 \rangle, \dots, \langle u, \varphi_{2N} \rangle)$$

for some $f, g : \mathbb{R}^{2N} \to \mathbb{R}$. We assume that f and g are continuously differentiable, with polynomial growth of the first order derivatives. The general case then follows by an approximation argument (note that Hermite polynomials of linear combinations of $(\langle u, \varphi_k \rangle)_k$ form an orthogonal basis of $L^2(\mu^\varepsilon)$). Identifying Y_N with $\mathbb{R}^{2N}$, we can write $\mu^\varepsilon(du) = \gamma_{2N}(u)du$, where γ_{2N} is the density of a $2N$-dimensional standard normal variable. Integrating by parts we therefore have

$$\int_{Y_N} (B_F^\varepsilon(u) \cdot D\Phi(u))\Psi(u)\mu^\varepsilon(du) = -\int_{Y_N} (B_F^\varepsilon(u) \cdot D\Psi(u))\Phi(u)\mu^\varepsilon(du)$$

(2.32)

$$-\int_{Y_N} \sum_{k=1}^{2N} (\langle \partial_{\langle u, \varphi_k \rangle} B_F^\varepsilon(u), \varphi_k \rangle$$

$$- \langle B_F^\varepsilon(u), \varphi_k \rangle \langle u, \varphi_k \rangle) \Psi(u)\Phi(u)\mu^\varepsilon(du)$$

(2.33)

and it suffices to show that the zero order differential operator terms on the right hand side vanish. For the first one of them we have

$$\sum_{k=1}^{2N} \langle \partial_{\langle u, \varphi_k \rangle} B_F^\varepsilon(u), \varphi_k \rangle = \sum_{k=1}^{2N} \langle \partial_{\langle u, \varphi_k \rangle} \varepsilon^{-1} \partial_x \Pi_0^N \tilde{F}(\varepsilon^{1/2} u), \varphi_k \rangle$$

$$= \sum_{k=1}^{2N} \langle \varepsilon^{-1/2} \partial_x (\Pi_0^N \tilde{F}'(\varepsilon^{1/2} u)\varphi_k), \varphi_k \rangle$$

$$= -\sum_{k=1}^{2N} \langle \varepsilon^{-1/2} \Pi_0^N \tilde{F}'(\varepsilon^{1/2} u)\varphi_k, \partial_x \varphi_k \rangle$$

$$= -\frac{\varepsilon^{-1/2}}{2} \langle \Pi_0^N \tilde{F}'(\varepsilon^{1/2} u), \partial_x \sum_{k=1}^{2N} \varphi_k^2 \rangle,$$

and since $\sin(mx)^2 + \cos(mx)^2 = 1$ the sum of the squares of the φ_k does not depend on x so its derivative is 0. For the remaining term in (2.32) we get μ^ε-almost surely

$$\sum_{k=1}^{2N} \langle B_F^\varepsilon(u), \varphi_k \rangle \langle u, \varphi_k \rangle = \langle B_F^\varepsilon(u), u \rangle = \langle \varepsilon^{-1} \partial_x \Pi_0^N \tilde{F}(\varepsilon^{1/2} u), u \rangle$$

$$= \varepsilon^{-1} \langle \partial_x \tilde{F}(\varepsilon^{1/2} u), \Pi_0^N u \rangle = -\varepsilon^{-1} \langle \tilde{F}(\varepsilon^{1/2} u), \partial_x \Pi_0^N u \rangle.$$

Now observe that there exists G with $G' = \tilde{F}$, and that under μ^ε we have $u = \Pi_0^N u$ almost surely, which yields

$$-\varepsilon^{-1}\langle \tilde{F}(\varepsilon^{1/2}u), \partial_x \Pi_0^N u\rangle = -\varepsilon^{-1}\langle G'(\varepsilon^{1/2}\Pi_0^N u), \partial_x \Pi_0^N u\rangle$$
$$= -\varepsilon^{-3/2}\langle \partial_x G(\varepsilon \Pi_0^N u), 1\rangle = 0,$$

and therefore the proof is complete. □

The previous lemma, together with the reversibility of the Ornstein–Uhlenbeck dynamics under μ^ε, implies that the Itô SDE (2.31) has μ^ε as invariant measure and that for $T > 0$ the time reversed process $\hat{u}_t^\varepsilon = \hat{u}_{T-t}^\varepsilon$ solves

$$\partial_t \hat{u}^\varepsilon = \Delta \hat{u}^\varepsilon - \varepsilon^{-1}\partial_x \tilde{F}(\varepsilon^{1/2}\Pi_0^N \hat{u}^\varepsilon) + \partial_x \Pi_0^N \hat{\xi} \qquad (2.34)$$

with a time-reversed space-time white noise $\hat{\xi}$.

2.8.1 The Invariance Principle

We now have all the tools to prove the convergence of (u^ε) to an energy solution of the stochastic Burgers equation. We proceed in two steps. First we establish the tightness of (u^ε), and in a second step we show that every weak limit is an energy solution. Using the uniqueness of energy solutions, we therefore obtain the convergence of (u^ε).

Tightness Let (u^ε) solve (2.31) and write $\tilde{F}(x) = F(x) - c_1(F)x$. To prove the tightness of (u^ε) it suffices to show that for all $\ell \in \mathbb{Z}$ the complex-valued process $\langle u^\varepsilon, e_{-\ell}\rangle$ is tight and satisfies a polynomial bound in ℓ, uniformly in ε. We decompose $\langle u_t^\varepsilon, e_{-\ell}\rangle$ as

$$\langle u_t^\varepsilon, e_{-\ell}\rangle = \langle u_0^\varepsilon, e_{-\ell}\rangle + \int_0^t \langle u_s^\varepsilon, \Delta e_{-\ell}\rangle ds - \int_0^t \langle \varepsilon^{-1}\Pi_0^N \tilde{F}(\varepsilon^{1/2}u_s^\varepsilon), \partial_x e_{-\ell}\rangle ds$$

$$(2.35)$$

$$- \int_0^t \langle \Pi_0^N \xi_s, \partial_x e_{-\ell}\rangle ds \qquad (2.36)$$

$$=: \langle u_0^\varepsilon, e_{-\ell}\rangle + \langle S_t^\varepsilon, e_{-\ell}\rangle + \langle A_t^\varepsilon, e_{-\ell}\rangle + \langle M_t^\varepsilon, e_{-\ell}\rangle, \qquad (2.37)$$

where S^ε, A^ε, M^ε stand for symmetric, antisymmetric and martingale part, respectively, and we show tightness for each term on the right hand side separately. The convergence of $\langle u_t^\varepsilon, e_{-\ell}\rangle$ at a fixed time (in particular $t = 0$) follows from the fact that the law of u_t^ε is that of μ^ε for all t, and (μ^ε) obviously converges to the law of the white noise as $\varepsilon \to 0$. The linear term is tight because

$$\mathbb{E}\left[\left|\int_s^t \langle u_r^\varepsilon, \Delta e_\ell \rangle \mathrm{d}r\right|^p\right] \leqslant |t-s|^{p-1}\int_s^t \mathbb{E}[|\langle u_r^\varepsilon, \ell^2 e_\ell\rangle|^p]\mathrm{d}r$$

$$\lesssim |t-s|^{p-1}\int_s^t \mathbb{E}[|\langle u_r^\varepsilon, \ell^2 e_\ell\rangle|^2]^{p/2}\mathrm{d}r = |t-s|^p|\ell|^{2p}.$$

For all $\varepsilon > 0$ the martingale term is a mollified space-time white noise, so its convergence is immediate.

Only the nonlinear contribution to the dynamics is nontrivial to control. Here we use the Boltzmann–Gibbs principle stated in Proposition 1 to get

$$\mathbb{E}\left[\left|\int_s^t \langle \varepsilon^{-1}\Pi_0^N \tilde{F}(\varepsilon^{1/2}u_r^\varepsilon), \partial_x e_{-\ell}\rangle \mathrm{d}r\right|^2\right] \lesssim |t-s|^{3/2-\kappa}\ell^2 \int_{\mathbb{R}} |F'(x)|^2 v(\mathrm{d}x).$$

This bound gives readily tightness in $C([0, T], \mathbb{C})$ and also that any limit point has zero quadratic variation.

Similarly we have for the time reversed process $\hat{u}_t^\varepsilon = u_{T-t}^\varepsilon$

$$\langle \hat{u}_t^\varepsilon, e_{-\ell}\rangle = \langle \hat{u}_0^\varepsilon, e_{-\ell}\rangle + \int_0^t \langle \hat{u}_s^\varepsilon, \Delta e_{-\ell}\rangle \mathrm{d}s + \int_0^t \langle \varepsilon^{-1}\Pi_0^N \tilde{F}(\varepsilon^{1/2}\hat{u}_s^\varepsilon), \partial_x e_{-\ell}\rangle \mathrm{d}s$$

(2.38)

$$-\int_0^t \langle \Pi_0^N \hat{\xi}_s, \partial_x e_{-\ell}\rangle \mathrm{d}s$$

(2.39)

$$=: \langle \hat{u}_0^\varepsilon, e_{-\ell}\rangle + \langle \hat{S}_t^\varepsilon, e_{-\ell}\rangle + \langle \hat{A}_t^\varepsilon, e_{-\ell}\rangle + \langle \hat{M}_t^\varepsilon, e_{-\ell}\rangle,$$

(2.40)

and the same arguments as before show that each term on the right hand side is tight in $C([0, T], \mathbb{C})$, satisfies a uniform polynomial bound, and that any limit point of $\langle \hat{A}^\varepsilon, e_{-\ell}\rangle$ has zero quadratic variation. Since we have suitable moment bounds for each term, we actually get the joint tightness:

Lemma 12 *Consider the decomposition (2.35), (2.38). Then the tuple*

$$(u_0^\varepsilon, \hat{u}_0^\varepsilon, S^\varepsilon, \hat{S}^\varepsilon, A^\varepsilon, \hat{A}^\varepsilon, M^\varepsilon, \hat{M}^\varepsilon)$$

is tight in $(\mathscr{S}')^2 \times C\left([0, T], \mathscr{S}'\right)^6$. For every weak limit $(u_0, \hat{u}_0, S, \hat{S}, A, \hat{A}, M, \hat{M})$ and any $\varphi \in C^\infty(\mathbb{T})$ the processes $\langle A, \varphi\rangle$ and $\langle \hat{A}, \varphi\rangle$ have zero quadratic variation and satisfy $\hat{A}_t = -(A_T - A_{T-t})$. Moreover, $u_t = u_0 + S_t + A_t + M_t$, $t \in [0, T]$, is for every fixed time a spatial white noise.

Theorem 4 *Let $(u, \mathcal{A})$ be as in Lemma 12. Then $(u, \mathcal{A}) \in \mathcal{Q}$ and u is an energy solution to*

$$\partial_t u = \Delta u + c_2(F)\partial_x u^2 + \partial_x \xi.$$

Proof The tuple $(u_0^\varepsilon, \hat{u}_0^\varepsilon, S^\varepsilon, \hat{S}^\varepsilon, A^\varepsilon, \hat{A}^\varepsilon, M^\varepsilon, \hat{M}^\varepsilon)$ converges along a subsequence $\varepsilon_n \to 0$, but to simplify notation we still denote this subsequence by the same symbol. Since $(u_0^\varepsilon, S^\varepsilon, A^\varepsilon, M^\varepsilon)$ converges jointly and for every fixed ε the process u^ε solves (2.30), we get for $\varphi \in C^\infty(\mathbb{T})$

$$\langle u_t, \varphi \rangle = \langle u_0, \varphi \rangle + \langle S_t, \varphi \rangle + \langle \mathcal{A}_t, \varphi \rangle + \langle M_t, \varphi \rangle,$$

and since $\langle S_t^\varepsilon, \varphi \rangle = \int_0^t \langle u_s^\varepsilon, \Delta\varphi \rangle ds$ also $\langle S_t, \varphi \rangle = \int_0^t \langle u_s, \Delta\varphi \rangle ds$. The same argument works for the backward process, so that $(u, \mathcal{A}) \in \mathcal{Q}$. It remains to show that $\mathcal{A} = c_2(F)\partial_x u^2$, which follows from the Boltzmann–Gibbs principle, Proposition 1. For all $\varepsilon > 0$ and $M \leqslant N/2 = \pi/(2\varepsilon)$,

$$\mathbb{E}\left[\left| \int_s^t \langle A_r^\varepsilon - c_2(F)\partial_x (\Pi_0^M u_r^\varepsilon)^2, e_{-\ell} \rangle dr \right|^2 \right]$$

$$\lesssim |t - s|\ell^2 (M^{-1} + \varepsilon \log^2 N) \int_{\mathbb{R}} |F'(x)|^2 \nu(dx),$$

so by Fatou's lemma

$$\mathbb{E}\left[\left| \int_s^t \langle \mathcal{A} - c_2(F)\partial_x (\Pi_0^M u_r)^2, e_{-\ell} \rangle dr \right|^2 \right]$$

$$\leqslant \liminf_{\varepsilon \to 0} \mathbb{E}\left[\left| \int_s^t \langle A_r^\varepsilon - c_2(F)\partial_x (\Pi_0^M u_r^\varepsilon)^2, e_{-\ell} \rangle dr \right|^2 \right]$$

$$\lesssim |t - s|\ell^2 M^{-1} \int_{\mathbb{R}} |F'(x)|^2 \nu(dx).$$

It now suffices to send $M \to \infty$.   □

2.9 Uniqueness of Energy Solutions

Now we prove uniqueness of energy solutions.

2.9.1 Mapping to the SHE

The strategy to prove uniqueness is to perform the Hopf–Cole transformation at the level of controlled processes and check that after the transformation we get an Itô solution to the SHE which we already know it is unique. In order to implement this idea we need first to regularise the energy solution, apply Itô formula to the Hopf–Cole transformation. The critical step is to control the error term in the Itô formula and show that it goes to zero. This is done via the Itô trick, in the form of the Kipnis–Varadhan lemma for energy solutions and a careful analysis of the related estimation. The control of this remainder is very similar to some computation in the paper of Funaki–Quastel [6].

Fix $L \geqslant 1$ and consider the process $\phi_t^L(x) = \exp(Iu_t^L(x))$ where $u_t^L = \mathscr{I}_0^L u_t$ and $I = \partial_x^{-1}$ is the antiderivative defined on $L_0^2(\mathbb{T})$ by $\partial_x I \varphi = \varphi$ for all $\varphi \in L_0^2(\mathbb{T})$. Let $I^* = -I$ the adjoint of I and $\rho_x^L = \mathscr{I}_0^L \delta_x$ is a smooth test function of zero mean. Now $I\rho_x^L(y) = (\delta^L * \Theta)(y - x)$ where $\Theta : \mathbb{T} \to R$ is the odd periodic function given by $\Theta(y) = \frac{1}{2}\operatorname{sgn}(y) - y/(2\pi)$ for $y \in (-\pi, \pi]$ with $\Theta(0) = 0$.

The process $(\phi_t^L(x))_{t,x}$ is smooth in x and of finite quadratic variation in t. Note that $\phi_t^L(x) = \exp(Iu_t^L(x)) = \exp(u_t(I^*\rho_x^L))$ is not a cylindrical function but by a direct approximation procedure we can pretend it is and since $(u_t, \mathcal{A})$ is controlled we have

$$d(Iu_t^L)(x) = du_t(I^*\rho_x^L) = u_t(\Delta I^*\rho_x^L)dt + d\mathcal{A}_t(I^*\rho_x^L) - dM_t(\partial_x I^*\rho_x^L).$$

The Itô formula for Dirichlet processes (2.19) gives

$$
\begin{aligned}
d\phi_t^L(x) &= \phi_t^L(x)(u_t(\Delta I^*\rho_x^L)dt \\
&\quad + d\mathcal{A}_t(I^*\rho_x^L) - dM_t(\partial_x I^*\rho_x^L)) + \phi_t(x)\langle \partial_x I^*\rho_x^L, \partial_x I^*\rho_x^L\rangle dt \\
&= \phi_t^L(x)(u_t(\Delta I^*\rho_x^L)dt + d\mathcal{A}_t(I^*\rho_x^L) + dM_t(\rho_x^L)) + \phi_t(x)\langle \rho_x^L, \rho_x^L\rangle dt
\end{aligned}
$$

Moreover since $\rho_x^L(y) = \rho_0^L(y - x)$ we have $\Delta(I^*\rho_x^L) = \Delta_x(I^*\rho_x^L)$ and

$$
\begin{aligned}
\phi_t^L(x)u_t(\Delta I^*\rho_x^L) &= \phi_t^L(x)\Delta_x u_t(I^*\rho_x^L) = \Delta_x\phi_t^L(x) - \phi_t^L(x)(\partial_x u_t(I^*\rho_x^L))^2 \\
&= \Delta_x\phi_t^L(x) - \phi_t^L(x)(u_t(\partial_x I^*\rho_x^L))^2 \\
&= \Delta_x\phi_t^L(x) - \phi_t^L(x)(u_t(\rho_x^L))^2.
\end{aligned}
$$

Being ϕ_t^L is smooth in space we can rewrite this as an approximate stochastic heat equation (SHE)

$$d\phi_t^L(x) = \Delta_x\phi_t^L(x)dt + \phi_t^L(x)(K^L - Q_t^L)dt + \phi_t^L(x)dM_t(\rho_x^L) + dR_t^L(x)$$

where the remainder R^L is given by

$$R_t^L(x) = \int_0^t \phi_s^L(x)(\mathrm{d}\mathcal{A}_s(I^*\rho_x^L) - \mathscr{I}_0(u_s(\rho_*^L))^2(x)\mathrm{d}s - K^L\mathrm{d}s)$$

and where we introduced a constant K^L and the one-dimensional process

$$Q_t^L = \frac{1}{2\pi}\int_{\mathbb{T}}[(u_t(\rho_x^L))^2 - \langle\rho_x^L, \rho_x^L\rangle]\mathrm{d}x$$

By definition of energy solution of SBE we have also

$$\mathcal{A}_t(I^*\rho_x^L) = -\lim_{N\to 0}\int_0^t\left(\mathscr{I}_0^N u_s\right)^2(\partial_x I^*\rho_x^L)\mathrm{d}s = \lim_{N\to 0}\int_0^t\left(\mathscr{I}_0^N u_s\right)^2(\rho_x^L)\mathrm{d}s$$

$$(2.41)$$

Thanks to Eq. (2.41) we can define for any $N \geqslant 1$ the process

$$R_t^{L,N}(x) = \int_0^t\phi_s^L(x)(\left(\mathscr{I}_0^N u_s\right)^2(\rho_x^L) - \mathscr{I}_0\left((\mathscr{I}_0^L u_s)^2\right)(x) - K^{L,N})\mathrm{d}s$$

$$= \int_0^t\phi_s^L(x)(\mathscr{I}_0^L\left(\mathscr{I}_0^N u_s\right)^2(x) - \mathscr{I}_0\left((\mathscr{I}_0^L u_s)^2\right)(x) - K^{L,N})\mathrm{d}s$$

and observe that $\lim_{N\to\infty} R^{L,N} = R^L$ if we choose the sequence of constants $K^{L,N}$ so that $K^{L,N} \to K^L$. In Sect. 2.9.2 below we prove the following lemma which is the key for the convergence

Lemma 13 *There exists a choice of $K^{L,N}$ such that for any $\varphi \in \mathscr{S}$ we have* $\mathbb{E}[R_t^{L,N}(\varphi)] = 0$ *and*

$$\lim_{L\to\infty}\lim_{N\to\infty}\mathbb{E}[\sup_{0\leqslant t\leqslant T}(R_t^{L,N}(\varphi))^2] = 0$$

where $R_t^{L,N}(\varphi) = \langle R_t^{L,N}, \varphi\rangle_{L^2(\mathbb{T})}$. Moreover $K^{L,N} \to K^L$ as $N \to +\infty$ and $K^L \to K = -1/12$ as $L \to +\infty$.

We are now ready to prove the uniqueness result. We have

$$R_t^L(\varphi) = \phi_t^L(\varphi) - \phi_0^L(\varphi) - \int_0^t\phi_s^L(\Delta\varphi)\mathrm{d}s$$

$$- \int_0^t\phi_s^L(\varphi)(K^L - Q_s^L)\mathrm{d}s - \int_0^t\int_{\mathbb{T}}\phi_s^L(x)\varphi(x)\mathrm{d}M_s(\rho_x^L)\mathrm{d}x$$

We already know that the family $(\phi_t^L(x))_{t,x}$ convergences in the space of continuous paths on $[0, T] \times \mathbb{T}$ to $\phi_t(x) = \exp((Iu_t)(x))$. It is not difficult also to show that the process $(Q_s^L)_s$ converges in $L^2([0, T])$ to a limit which we call Q. The quadratic variation of the martingale part converges also, so we have that, for fixed $\varphi \in \mathscr{S}$

$$R_t(\varphi) = \lim_{L \to \infty} R_t^L(\varphi)$$

$$= \phi_t(\varphi) - \phi_0(\varphi) - \int_0^t \phi_s(\Delta\varphi)ds$$

$$- \int_0^t \phi_s(\varphi)(K - Q_s)ds - \int_0^t \int_{\mathbb{T}} \phi_s(x)\varphi(x)d\,(\mathscr{I}_0 M_s)\,(x)dx$$

as a continuous process in $t \in [0, T]$. But from Lemma 13 we have also by Fatou

$$\mathbb{E}[\sup_{0 \leqslant t \leqslant T} (R_t(\varphi))^2] \leqslant \liminf_{L \to \infty} \mathbb{E}[\sup_{0 \leqslant t \leqslant T} (R_t^L(\varphi))^2]$$

$$\leqslant \liminf_{L \to \infty} \liminf_{N \to \infty} \mathbb{E}[\sup_{0 \leqslant t \leqslant T} (R_t^{L,N}(\varphi))^2] = 0$$

so we conclude that ϕ satisfy

$$\phi_t(\varphi) - \phi_0(\varphi) - \int_0^t \phi_s(\Delta\varphi)ds - \int_0^t \phi_s(\varphi)(K - Q_s)ds$$

$$- \int_0^t \int_{\mathbb{T}} \phi_s(x)\varphi(x)d\,(\mathscr{I}_0 M_s)\,(x)dx = 0$$

almost surely for all $\varphi \in \mathscr{S}$ and $t \in [0, T]$. Note also that the process

$$\tilde{\phi}_t(x) = \phi_t(x)\exp\left(M_t(1) - \frac{t}{2} + Kt - \int_0^t Q_s ds\right)$$

$$= \exp\left((Iu_t)(x) + M_t(1) - \frac{t}{2} + Kt - \int_0^t Q_s ds\right)$$

satisfies

$$\tilde{\phi}_t(\varphi) - \tilde{\phi}_0(\varphi) - \int_0^t \tilde{\phi}_s(\Delta\varphi)ds - \int_0^t \int_{\mathbb{T}} \tilde{\phi}_s(x)\varphi(x)dM_s(x)dx = 0.$$

However we also know that there exists a unique strictly positive solution ψ for this equation which is adapted to the filtration generated by M, so we must have $\tilde{\phi}_t(x) = \psi_t(x)$ and in particular

$$(Iu_t)(x) + M_t(1) - \frac{t}{2} + Kt - \int_0^t Q_s ds = \log \tilde{\phi}_t(x) = \log \psi_t(x)$$

By differentiating this equality (in the sense of distributions) and projecting away the constants we get

$$u_t(x) = \partial_x \log \psi_t(x)$$

which shows that the only energy solution of the SBE is the Cole–Hopf solution obtained from solving the SHE associated to the noise M.

2.9.2 Convergence of the Remainder

Let

$$r^{L,N}(u_s, x) = \phi_s^L(x)(\mathscr{I}_0^L\left(\mathscr{I}_0^N u_s\right)^2(x) - \mathscr{I}_0(\mathscr{I}_0^L u_s)^2(x) - K^{L,N})$$

so that $R_t^{L,N}(x) = \int_0^t r^{L,N}(u_s, x)\mathrm{d}s$. Using Lemma 7 we can estimate

$$\mathbb{E}[\sup_{0 \leqslant t \leqslant T}(R_t^{L,N}(\varphi))^2] \lesssim T\|r^{L,N}(\cdot, \varphi)\|_{-1}^2$$

where

$$\|r^{L,N}(\cdot, \varphi)\|_{-1}^2 = \sup_{\Phi \in L^2(\mu)} 2\mathbb{E}[r^{L,N}(u_0, \varphi)\Phi(u_0)] - \|\Phi\|_1^2$$

with $\|\Phi\|_1^2 = \mathbb{E}\|D\Phi\|_{H^1}^2$ in terms of the Malliavin derivative D associated to the measure μ. We prove below that we can choose $K^{L,N}$ so that $\mathbb{E}[r^{L,N}(u_t, x)] = \mathbb{E}[r^{L,N}(u_0, x)] = 0$ for all $x \in \mathbb{T}$. This is necessary in order for $\|r^{L,N}(\cdot, \varphi)\|_{-1}$ to be finite for all φ. At this point everything boils down to control

$$\mathbb{E}[r^{L,N}(u_0, \varphi)\Phi(u_0)]$$

and show that it goes to zero as $N \to +\infty$ and $L \to +\infty$ where φ is a smooth test function and $\Phi \in L^2(\mu)$ is such that $\|\Phi\|_1 < +\infty$.

For any $n \geqslant 1$ we denote by

$$W_n(h) = \int_{\mathbb{T}^n} h(z_1, \ldots, z_n)W(\mathrm{d}z_1 \cdots \mathrm{d}z_n)$$

the n-th chaos of the white noise u_0. It is also possible to check that when Ψ is in the domain of the Malliavin derivative D we can partially integrate by parts in the expression $\mathbb{E}[W_n(h)\Psi]$ to obtain the equality

$$\mathbb{E}[W_n(h)\Psi] = \int_{\mathbb{T}} \mathbb{E}[W_{n-1}(h(z_1, \cdot, \ldots, \cdot))D_{z_1}\Psi]dz_1$$

this will be used repeatedly below.

Observe that the random variable $(\mathscr{I}_0^L \left(\mathscr{I}_0^N u_0\right)^2(x) - \mathscr{I}_0\left((\mathscr{I}_0^L u_0)^2\right)(x))$ is an element of the second chaos of u_0. Let us compute its kernel:

$$\mathscr{I}_0^L \left(\mathscr{I}_0^N u_0\right)^2(x) = \int_{\mathbb{T}^2} \left[\int_{\mathbb{T}} dy \rho_x^L(y)\rho_y^N(z_1)\rho_y^N(z_2)\right] W(dz_1 dz_2)$$

and

$$\mathscr{I}_0\left((\mathscr{I}_0^L u_0)^2\right)(x) = \int_{\mathbb{T}^2} \left[\int_{\mathbb{T}} dy \rho_x^\infty(y)\rho_y^L(z_1)\rho_y^L(z_2)\right] W(dz_1 dz_2)$$

where

$$\rho_x^\infty(y) = \lim_{K \to \infty} \rho_y^K(y) = \delta_x(y) - \frac{1}{2\pi}$$

is the distributional kernel of $\mathscr{I}_0$. Let

$$g_x^{L,N}(z_1, z_2) = \int_{\mathbb{T}} dy (\rho_x^L(y)\rho_y^N(z_1)\rho_y^N(z_2) - \rho_x^\infty(y)\rho_y^L(z_1)\rho_y^L(z_2))$$

so that

$$(\mathscr{I}_0^L \left(\mathscr{I}_0^N u_0\right)^2(x) - \mathscr{I}_0\left((\mathscr{I}_0^L u_0)^2\right)(x)) = W_2(g_x^{L,N})$$

$$= \int_{\mathbb{T}^2} g_x^{L,N}(z_1, z_2)W(dz_1 dz_2).$$

We let also

$$W_1(g_x^{L,N}(z_1, \cdot)) = \int_{\mathbb{T}} g_x^{L,N}(z_1, z_2)W(dz_2).$$

Consider now

$$\mathbb{E}[r^\varepsilon(u_0, \varphi)\Phi(u_0)] = \int_{\mathbb{T}} \varphi(x)\mathbb{E}[(W_2(g_x^{L,N}) - K^{L,N})\phi_0^L(x)\Phi(u_0)]dx$$

Partially integrating by parts W_2 we have

$$\mathbb{E}[W_2(g_x^{L,N})\phi_0^L(x)\Phi(u_0)] = \int_{\mathbb{T}} \mathbb{E}[W_1(g_x^{L,N}(z_1, \cdot))D_{z_1}[\phi_0^L(x)\Phi(u_0)]]dz_1$$

$$= \int_{\mathbb{T}} \mathbb{E}[W_1(g_x^{L,N}(z_1, \cdot))\phi_0^L(x)D_{z_1}\Phi(u_0)]dz_1$$

$$+ \int_{\mathbb{T}} \mathbb{E}[W_1(g_x^{L,N}(z_1, \cdot))D_{z_1}\phi_0^L(x)\Phi(u_0)]dz_1$$

The second term can be again integrated by parts to obtain

$$\int_{\mathbb{T}} \mathbb{E}[W_1(g_x^{L,N}(z_1, \cdot))D_{z_1}\phi_0^L(x)\Phi(u_0)]dz_1$$

$$= \int_{\mathbb{T}^2} g_x^{L,N}(z_1, z_2)\mathbb{E}[D_{z_1,z_2}^2\phi_0^L(x)\Phi(u_0)]dz_1dz_2$$

$$+ \int_{\mathbb{T}^2} g_x^{L,N}(z_1, z_2)\mathbb{E}[D_{z_1}\phi_0^L(x)D_{z_2}\Phi(u_0)]dz_1dz_2.$$

while the first can be written as

$$\int_{\mathbb{T}} \mathbb{E}[W_1(g_x^{L,N}(z_1, \cdot))\phi_0^L(x)D_{z_1}\Phi(u_0)]dz_1$$

$$= \int_{\mathbb{T}} \mathbb{E}[(W_1(g_x^{L,N}(z_1, \cdot)) \diamond \phi_0^L(x))D_{z_1}\Phi(u_0)]dz_1$$

$$+ \int_{\mathbb{T}^2} g_x^{L,N}(z_1, z_2)\mathbb{E}[D_{z_2}\phi_0^L(x)D_{z_1}\Phi(u_0)]dz_1dz_2.$$

where

$$(W_1(g_x^{L,N}(z_1, \cdot)) \diamond \phi_0^L(x)) = W_1(g_x^{L,N}(z_1, \cdot))\phi_0^L(x) - \int_{\mathbb{T}} g_x^{L,N}(z_1, z_2)D_{z_2}\phi_0^L(x)dz_2$$

is a (partial) Wick contraction of the two terms. In the end we have the decomposition

$$\mathbb{E}[r^{L,N}(u_0, \varphi)\Phi(u_0)] = A^{L,N} + B^{L,N} + C^{L,N}$$

where

$$A^{L,N} = \int_{\mathbb{T}} \varphi(x) \int_{\mathbb{T}} \mathbb{E}[(W_1(g_x^{L,N}(z_1, \cdot)) \diamond \phi_0^L(x))D_{z_1}\Phi(u_0)]dz_1dx$$

$$B^{L,N} = \int_{\mathbb{T}} \varphi(x) \left[\int_{\mathbb{T}^2} g_x^{L,N}(z_1, z_2)\mathbb{E}\left[(D_{z_1,z_2}^2\phi_0^L(x))\Phi(u_0) \right]dz_1dz_2 \right.$$

$$\left. - K^{L,N}\mathbb{E}[\phi_0^L(x)\Phi(u_0)] \right]dx$$

and

$$C^{L,N} = 2 \int_{\mathbb{T}} \varphi(x) \int_{\mathbb{T}^2} g_x^{L,N}(z_1, z_2) \mathbb{E}[(D_{z_1}\phi_0^L(x))D_{z_2}\Phi(u_0)]dz_1dz_2dx.$$

So we can bound these three terms independently. In order to proceed, observe that

$$D_{z_1}\phi_0^L(x) = \phi_0^L(x)(I^*\rho_x^L)(z_1)$$

and

$$D_{z_1,z_2}^2\phi_0^L(x) = \phi_0^L(x)(I^*\rho_x^L)(z_1)(I^*\rho_x^L)(z_2)$$

so

$$B^{L,N} = \int_{\mathbb{T}} \varphi(x)\left[\int_{\mathbb{T}^2} g_x^{L,N}(z_1, z_2)(I^*\rho_x^L)(z_1)(I^*\rho_x^L)(z_2)dz_1dz_2 - K^{L,N}\right]$$
$$\times \mathbb{E}[\phi_0^L(x)\Phi(u_0)]dx$$

and

$$C^{L,N} = 2 \int_{\mathbb{T}} \varphi(x) \int_{\mathbb{T}^2} g_x^{L,N}(z_1, z_2)(I^*\rho_x^L)(z_1)\mathbb{E}[\phi_0^L(x)D_{z_2}\Phi(u_0)]dz_1dz_2dx.$$

Let

$$K^{L,N} = \int_{\mathbb{T}^2} g_x^{L,N}(z_1, z_2)(I^*\rho_x^L)(z_1)(I^*\rho_x^L)(z_2)dz_1dz_2$$

to have $B^{L,N} = 0$. From the expression of $A^{L,N}$ and $C^{L,N}$ we deduce that with this choice

$$\mathbb{E}[r^{L,N}(u_0, \varphi)] = 0$$

as required. By Cauchy–Schwarz we have

$$(A^{L,N})^2 \leqslant \mathbb{E}\left\|\int_{\mathbb{T}} \varphi(x)(W_1(g_x^{L,N}(z_1, \cdot)) \diamond \phi_0^L(x))dx\right\|_{H_{z_1}^{-1}}^2 \mathbb{E}\|D_{z_1}\Phi(u_0)\|_{H_{z_1}^1}^2$$
$$= (A_1^{L,N})^2\|\Phi\|_1^2$$

where

$$(A_1^{L,N})^2 = \mathbb{E}\left\|\int_{\mathbb{T}} \varphi(x)W_1(g_x^{L,N}(z_1,\cdot)) \diamond \phi_0^L(x)dx\right\|_{H_{z_1}^{-1}}^2$$

$$= \int_{\mathbb{T}^2} \varphi(x)\varphi(x')\mathbb{E}$$

$$\times \left[\langle \phi_0^L(x) \diamond W_1(g_x^{L,N}(z_1,\cdot)), \phi_0^L(x') \diamond W_1(g_{x'}^{L,N}(z_1,\cdot))\rangle_{H_{z_1}^{-1}}\right]dxdx'$$

and

$$(C^{L,N})^2 \lesssim \mathbb{E}\left\|\int_{\mathbb{T}} \varphi(x)\phi_0^L(x)\int_{\mathbb{T}} g_x^{L,N}(z_1,z_2)(I^*\rho_x^L)(z_1)dz_1dx\right\|_{H_{z_2}^{-1}}^2$$

$$\times \mathbb{E}\|D_{z_2}\Phi(u_0)\|_{H_{z_2}^1}^2$$

$$\lesssim (C_1^{L,N})^2\|\Phi\|_1^2$$

with

$$(C_1^{L,N})^2 = \mathbb{E}\left\|\int_{\mathbb{T}} \varphi(x)\phi_0^L(x)\int_{\mathbb{T}} g_x^{L,N}(z_1,z_2)(I^*\rho_x^L)(z_1)dz_1dx\right\|_{H_{z_2}^{-1}}^2$$

So we have proven

Lemma 14 *We have*

$$|\mathbb{E}[r^{L,N}(u_0,\varphi)\Phi(u_0)]| \lesssim \|\Phi\|_1(A_1^{L,N} + C_1^{L,N})$$

and in particular

$$\|r^{L,N}(u_0,\varphi)\|_{-1} \lesssim A_1^{L,N} + C_1^{L,N}.$$

Then it remains to bound each of these constants to show that they vanish in the limit $N, L \to +\infty$ and to study the limit of $K^{L,N}$. This is the aim of the following lemmas which unfortunately have to be obtained by tedious and careful explicit computations.

Lemma 15 *We have* $\lim_{N\to+\infty} K^{L,N} = K^L$ *and* $\lim_{L\to+\infty} K^L = K = -1/12$.

Proof Since

$$\int_{\mathbb{T}} (I^*\rho_x^L)(z_1)\rho_y^N(z_1)dz_1 = \left\langle I^*\mathscr{I}_0^L\delta_x, \mathscr{I}_0^N\delta_y\right\rangle_{L^2}$$

$$= \left\langle \delta_x, I\mathscr{I}_0^L\mathscr{I}_0^N\delta_y\right\rangle_{L^2} = \left(I\mathscr{I}_0^L\mathscr{I}_0^N\delta_y\right)(x)$$

we have

$$K^{L,N} = \int_{\mathbb{T}^2} g_x^{L,N}(z_1, z_2)(I^* \rho_x^L)(z_1)(I^* \rho_x^L)(z_2)dz_1 dz_2$$

$$= \int_{\mathbb{T}} dy(\rho_x^L(y) \left(I \mathscr{I}_0^{L \wedge N} \delta_y\right)(x)^2 - \rho_x^\infty(y) \left(I \mathscr{I}_0^L \delta_y\right)(x)^2).$$

Taking $N \to \infty$ we get

$$K^{L,N} \to K^L = \int_{\mathbb{T}} dy(\rho_x^L(y) \left(I \mathscr{I}_0^L \delta_y\right)(x)^2 - \rho_x^\infty(y) \left(I \mathscr{I}_0^L \delta_y\right)(x)^2).$$

Now note that the integral of the first term in the expression of K^L vanishes, indeed:

$$\int_{\mathbb{T}} dy \rho_x^L(y) \left(I \mathscr{I}_0^L \delta_y\right)(x)^2 = \int_{\mathbb{T}} dy \rho_x^L(y) \left(I \mathscr{I}_0^L \delta_x\right)(y)^2$$

$$= \int_{\mathbb{T}} dy \left(\partial_y I \mathscr{I}_0^L \delta_x\right)(y) \left(I \mathscr{I}_0^L \delta_x\right)(y)^2$$

$$= \frac{1}{3} \int_{\mathbb{T}} dy \partial_y \left(I \mathscr{I}_0^L \delta_x\right)(y)^3 = 0.$$

Moreover the second term satisfies

$$\int_{\mathbb{T}} dy \rho_x^\infty(y) \left(I \mathscr{I}_0^L \delta_y\right)(x)^2 = \left(I \mathscr{I}_0^L \delta_x\right)(x)^2 - \frac{1}{2\pi} \int_{\mathbb{T}} dy \left(I \mathscr{I}_0^L \delta_y\right)(x)^2$$

since $\rho_x^\infty = \delta_x - 1/2\pi$. By symmetry also the $\left(I \mathscr{I}_0^L \delta_x\right)(x)^2$ contribution vanishes, since

$$\left(I \mathscr{I}_0^L \delta_x\right)(x) = \left(I^* \mathscr{I}_0^L \delta_x\right)(x) = -\left(I \mathscr{I}_0^L \delta_x\right)(x) = 0.$$

So we ends up with

$$K^L = -\frac{1}{2\pi} \int_{\mathbb{T}} dy \left(I \mathscr{I}_0^L \delta_x\right)(y)^2 = -\frac{1}{2\pi} \left\langle I \mathscr{I}_0^L \delta_x, I \mathscr{I}_0^L \delta_x \right\rangle$$

$$= -\frac{1}{4\pi^2} \sum_{0 < |k|} \frac{|\hat{\rho}^L(k)|^2}{|k|^2}$$

since

$$\mathscr{F} \left(I \mathscr{I}_0^L \delta_x\right)(k) = \frac{\hat{\rho}^L(k)}{ik} e_k(x)^*.$$

And as $L \to \infty$ we obtain $\hat{\rho}^L(k) \to 1$ for all $k \neq 0$ so

$$K^L \to -\frac{1}{2\pi^2} \sum_{n=1}^{\infty} \frac{1}{n^2} = -\frac{1}{12} = K.$$

$\square$

Lemma 16 *We have* $A_1^{L,N} \to 0$ *as* $N \to \infty$ *and* $L \to \infty$.

Proof

$$(A_1^{L,N})^2 = \mathbb{E} \left\| \int_{\mathbb{T}} \varphi(x) W_1(g_x^{L,N}(z_1, \cdot)) \diamond \phi_0^L(x) dx \right\|_{H_{z_1}^{-1}}^2$$

$$= \int_{\mathbb{T}^2} \varphi(x)\varphi(x')\mathbb{E}$$

$$\times \left[\langle \phi_0^L(x) \diamond W_1(g_x^{L,N}(z_1, \cdot)), \phi_0^L(x') \diamond W_1(g_{x'}^{L,N}(z_1, \cdot)) \rangle_{H_{z_1}^{-1}} \right] dx dx'$$

Integrating by parts the W_1 terms and taking into account the cancellations due to the partial $\diamond$ contractions we get

$$\mathbb{E} \left[\langle \phi_0^L(x) \diamond W_1(g_x^{L,N}(z_1, \cdot)), \phi_0^L(x') \diamond W_1(g_{x'}^{L,N}(z_1, \cdot)) \rangle_{H_{z_1}^{-1}} \right]$$

$$= \mathbb{E}[\phi_0^L(x)\phi_0^L(x')] \int_{\mathbb{T}^2} \langle g_x^{L,N}(z_1, z_2), g_{x'}^{L,N}(z_1, z_2) \rangle_{H_{z_1}^{-1}} dz_2$$

$$+ \int_{\mathbb{T}^2} \mathbb{E}[(D_{z_3}\phi_0^L(x))(D_{z_2}\phi_0^L(x'))] \langle g_x^{L,N}(z_1, z_2), g_{x'}^{L,N}(z_1, z_3) \rangle_{H_{z_1}^{-1}} dz_2 dz_3$$

The second term can be written as

$$\mathbb{E}[\phi_0^L(x)\phi_0^L(x')] \int_{\mathbb{T}^2} (I^* \rho_x^L)(z_3)(I^* \rho_{x'}^L)(z_2) \langle g_x^{L,N}(z_1, z_2), g_{x'}^{L,N}(z_1, z_3) \rangle_{H_{z_1}^{-1}} dz_2 dz_3$$

so letting

$$V^L(x, x') = \varphi(x)\varphi(x')\mathbb{E}[\phi_0^L(x)\phi_0^L(x')]$$

we have

$$(A_1^{L,N})^2 = \int_{\mathbb{T}^2} V^L(x, x') \int_{\mathbb{T}^2} (I^* \rho_x^L)(z_3)(I^* \rho_{x'}^L)(z_2)$$

$$\times \langle g_x^{L,N}(z_1, z_2), g_{x'}^{L,N}(z_1, z_3) \rangle_{H_{z_1}^{-1}} dz_2 dz_3 dx dx'$$

$$+ \int_{\mathbb{T}^2} V^L(x, x') \int_{\mathbb{T}^2} \langle g_x^{L,N}(z_1, z_2), g_{x'}^{L,N}(z_1, z_2) \rangle_{H_{z_1}^{-1}} dz_2 dx dx'$$

$$= A_{1,1} + A_{1,2}$$

Let us consider $A_{1,1}$ first :

$$A_{1,1} = \int_{\mathbb{T}^4} V^L(x, x') \langle G_{x,x'}, G_{x',x} \rangle_{H^{-1}} dx dx'$$

where

$$G_{x,x'}(z_1) = \int_{\mathbb{T}} (I^* \rho^L_{x'})(z_2) g^{L,N}_x(z_1, z_2) dz_2.$$

Now observe that

$$G_{x,x'}(z_1) = \int_{\mathbb{T}} dy \rho^L_x(y) \rho^N_y(z_1)(\mathscr{I}^N_0 I^* \rho^L_{x'})(y) - \int_{\mathbb{T}} dy \rho^\infty_x(y) \rho^L_y(z_1)(I^* \rho^L_{x'})(y)$$

so the H^{-1} scalar product in $A_{1,1}$ can be expanded as

$$\langle G_{x,x'}, G_{x',x} \rangle_{H^{-1}} = \int_{\mathbb{T}} dy \rho^L_x(y)(\mathscr{I}^N_0 I^* \rho^L_{x'})(y)$$

$$\times \int_{\mathbb{T}} dy' \rho^L_{x'}(y')(\mathscr{I}^N_0 I^* \rho^L_x)(y') \langle \rho^N_y(z_1), \rho^N_{y'}(z_1) \rangle_{H^{-1}_{z_1}}$$

$$- \int_{\mathbb{T}} dy \rho^\infty_x(y)(I^* \rho^L_{x'})(y)$$

$$\times \int_{\mathbb{T}} dy' \rho^L_{x'}(y')(\mathscr{I}^N_0 I^* \rho^L_x)(y') \langle \rho^L_y(z_1), \rho^N_{y'}(z_1) \rangle_{H^{-1}_{z_1}}$$

$$- \int_{\mathbb{T}} dy \rho^L_x(y)(\mathscr{I}^N_0 I^* \rho^L_{x'})(y)$$

$$\times \int_{\mathbb{T}} dy' \rho^\infty_{x'}(y')(I^* \rho^L_x)(y') \langle \rho^N_y(z_1), \rho^L_{y'}(z_1) \rangle_{H^{-1}_{z_1}}$$

$$+ \int_{\mathbb{T}} dy \rho^\infty_x(y)(I^* \rho^L_{x'})(y)$$

$$\times \int_{\mathbb{T}} dy' \rho^\infty_{x'}(y')(I^* \rho^L_x)(y') \langle \rho^L_y(z_1), \rho^L_{y'}(z_1) \rangle_{H^{-1}_{z_1}}$$

Taking $N \to \infty$ we get

$$\langle G_{x,x'}, G_{x',x} \rangle_{H^{-1}} \to \int_{\mathbb{T}} dy \rho^L_x(y)(I^* \rho^L_{x'})(y)$$

$$\times \int_{\mathbb{T}} dy' \rho^L_{x'}(y')(I^* \rho^L_x)(y') \langle \rho^\infty_y(z_1), \rho^\infty_{y'}(z_1) \rangle_{H^{-1}_{z_1}}$$

$$- \int_{\mathbb{T}} dy \rho^\infty_x(y)(I^* \rho^L_{x'})(y)$$

$$\times \int_{\mathbb{T}} \mathrm{d}y' \rho_{x'}^L(y')(I^*\rho_x^L)(y')\langle \rho_y^L(z_1), \rho_{y'}^\infty(z_1)\rangle_{H_{z_1}^{-1}}$$

$$- \int_{\mathbb{T}} \mathrm{d}y \rho_x^L(y)(I^*\rho_{x'}^L)(y)$$

$$\times \int_{\mathbb{T}} \mathrm{d}y' \rho_{x'}^\infty(y')(I^*\rho_x^L)(y')\langle \rho_y^\infty(z_1), \rho_{y'}^L(z_1)\rangle_{H_{z_1}^{-1}}$$

$$+ \int_{\mathbb{T}} \mathrm{d}y \rho_x^\infty(y)(I^*\rho_{x'}^L)(y)$$

$$\times \int_{\mathbb{T}} \mathrm{d}y' \rho_{x'}^\infty(y')(I^*\rho_x^L)(y')\langle \rho_y^L(z_1), \rho_{y'}^L(z_1)\rangle_{H_{z_1}^{-1}}$$

$$= \left\langle I\mathscr{I}_0(\rho_x^L(I^*\rho_{x'}^L)), I\mathscr{I}_0(\rho_{x'}^L(I^*\rho_x^L))\right\rangle$$

$$- \left\langle I\mathscr{I}_0^L(\rho_x^\infty(I^*\rho_{x'}^L)), I\mathscr{I}_0(\rho_{x'}^L(I^*\rho_x^L))\right\rangle$$

$$- \left\langle I\mathscr{I}_0(\rho_x^L(I^*\rho_{x'}^L)), I\mathscr{I}_0^L(\rho_{x'}^\infty(I^*\rho_x^L))\right\rangle$$

$$+ \left\langle I\mathscr{I}_0^L(\rho_x^\infty(I^*\rho_{x'}^L)), I\mathscr{I}_0^L(\rho_{x'}^\infty(I^*\rho_x^L))\right\rangle.$$

The point is to show that, as $L \to +\infty$,

$$\left\langle I\mathscr{I}_0(\rho_x^L(I^*\rho_{x'}^L)), I\mathscr{I}_0(\rho_{x'}^L(I^*\rho_x^L))\right\rangle \to \left\langle I\mathscr{I}_0(\rho_x^\infty(I^*\rho_{x'}^\infty)), I\mathscr{I}_0(\rho_{x'}^\infty(I^*\rho_x^\infty))\right\rangle$$

and the same limit for the other three quantities. Writing things in Fourier space we have

$$J_L(x - x') = \left\langle I\mathscr{I}_0(\rho_x^L(I^*\rho_{x'}^L)), I\mathscr{I}_0(\rho_{x'}^L(I^*\rho_x^L))\right\rangle$$

$$= \frac{1}{2\pi} \sum_{\substack{0 < |k_1|, |k_2|, |k_3|, |k_4| \\ k_1 + k_2 + k_3 + k_4 = 0}} \frac{\mathbb{I}_{k_1+k_2\neq 0}}{(k_1 + k_2)^2} e^{ik_1 x} \frac{e^{ik_2 x'}}{-ik_2} e^{ik_3 x'} \frac{e^{ik_4 x}}{-ik_4}$$

$$\times \hat{\rho}^L(k_1)\hat{\rho}^L(k_2)\hat{\rho}^L(k_3)\hat{\rho}^L(k_4)$$

$$= \frac{1}{2\pi} \sum_{\substack{0 < |k_1|, |k_2|, |k_3|, |k_4| \\ k_1 + k_2 + k_3 + k_4 = 0}} \frac{\mathbb{I}_{k_1+k_2\neq 0}}{(k_1 + k_2)^2} \frac{e^{i(k_2+k_3)(x'-x)}}{-ik_2} \frac{1}{-ik_4}$$

$$\times \hat{\rho}^L(k_1)\hat{\rho}^L(k_2)\hat{\rho}^L(k_3)\hat{\rho}^L(k_4)$$

Now taking Littlewood–Paley blocks $(\Delta_q)_{q \geqslant -1}$ of this quantity we get

$$\Delta_q J_L(x) = \frac{1}{2\pi} \sum_{\substack{0 < |k_1|, |k_2|, |k_3|, |k_4| \\ k_1 + k_2 + k_3 + k_4 = 0}} \frac{\mathbb{I}_{k_1+k_2\neq 0}}{(k_1+k_2)^2} \frac{e^{i(k_2+k_3)(x'-x)}}{-ik_2} \frac{\theta_q(k_2+k_3)}{-ik_4}$$

$$\times \hat{\rho}^L(k_1)\hat{\rho}^L(k_2)\hat{\rho}^L(k_3)\hat{\rho}^L(k_4).$$

Where θ_q is the Fourier multiplier associated to Δ_q by $\Delta_q = \theta_q(D)$. Taking into account that $\sup_k |\hat{\rho}^L(k)| < +\infty$ if we show that the quantity

$$S_q = \frac{1}{2\pi} \sum_{\substack{0 < |k_1|, |k_2|, |k_3|, |k_4| \\ k_1 + k_2 + k_3 + k_4 = 0}} \mathbb{I}_{k_1+k_2\neq 0} \frac{\theta_q(k_2+k_3)}{(k_1+k_2)^2|k_2||k_3|}$$

is finite then we can conclude by dominated convergence. Note that

$$S_q \lesssim 2^{q\varepsilon} \sum_{0 < |k_1|, |k_2|, |k_3|} \frac{\mathbb{I}_{k_1+k_2\neq 0, k_2+k_3\neq 0}}{(k_1+k_2)^2|k_2||k_3||k_2+k_3|^\varepsilon}$$

for any $\varepsilon > 0$ since

$$|\theta_q(k)| = |\theta(2^{-q}k)| \lesssim (1+|2^{-q}k|)^{-\varepsilon} \lesssim 2^{q\varepsilon}|k|^{-\varepsilon}$$

due to the fact that the function θ is supported in a ball of finite radius. We can now perform the sum over k_3 and get

$$S_q \lesssim 2^{q\varepsilon} \sum_{0 < |k_1|, |k_2|} \frac{\mathbb{I}_{k_1+k_2\neq 0}}{(k_1+k_2)^2|k_2|^{1+\varepsilon}}$$

since

$$\sum_{0 < |k_3|} \frac{\mathbb{I}_{k_2+k_3\neq 0}}{|k_3||k_2+k_3|^\varepsilon} \lesssim \frac{1}{|k_2|^\varepsilon}$$

uniformly in $k_2 \neq 0$ and L. Now performing the other two sums we show simply that

$$S_q \lesssim 2^{q\varepsilon} \sum_{0 < |k_2|} \frac{1}{|k_2|^{1+\varepsilon}} \lesssim 2^{q\varepsilon}$$

where we used the fact that

$$\sum_{0<|k_1|} \frac{\mathbb{I}_{k_1+k_2\neq 0}}{(k_1+k_2)^2} \leqslant \sum_{k_1\in\mathbb{Z}} \frac{\mathbb{I}_{k_1+k_2\neq 0}}{(k_1+k_2)^2} = \sum_{k_1\in\mathbb{Z}} \frac{\mathbb{I}_{k_1\neq 0}}{(k_1)^2} \lesssim 1.$$

By dominated convergence we can conclude that

$$\left\langle I\mathscr{I}_0(\rho_x^L(I^*\rho_{x'}^L)), I\mathscr{I}_0(\rho_{x'}^L(I^*\rho_x^L)) \right\rangle \to \left\langle I\mathscr{I}_0(\rho_x^\infty(I^*\rho_{x'}^\infty)), I\mathscr{I}_0(\rho_{x'}^\infty(I^*\rho_x^\infty)) \right\rangle$$

as distributions of arbitrarily small negative order of the variable $x - x'$. This is ok since we test this quantity against the function V^L which belongs to $C^{1/2-}(\mathbb{T}^2)$ uniformly in L. The other terms can be handled similarly and thus we conclude that $A_{1,1} \to 0$.

Let us turn now to $A_{1,2}$. The relevant computation is the following:

$$\int_{\mathbb{T}^2} \langle g_x^{L,N}(z_1,z_2), g_{x'}^{L,N}(z_1,z_2) \rangle_{H_{z_1}^{-1}} dz_2$$

$$= \int_{\mathbb{T}} dy \rho_x^L(y) \int_{\mathbb{T}} dy' \rho_{x'}^L(y') \langle \rho_y^N(z_1), \rho_{y'}^N(z_1) \rangle_{H_{z_1}^{-1}} \langle \rho_y^N(z_2), \rho_{y'}^N(z_2) \rangle_{L_{z_2}^2}$$

$$- \int_{\mathbb{T}} dy \rho_x^L(y) \int_{\mathbb{T}} dy' \rho_{x'}^\infty(y') \langle \rho_y^N(z_1), \rho_{y'}^L(z_1) \rangle_{H_{z_1}^{-1}} \langle \rho_y^N(z_2), \rho_{y'}^L(z_2) \rangle_{L_{z_2}^2}$$

$$- \int_{\mathbb{T}} dy \rho_x^\infty(y) \int_{\mathbb{T}} dy' \rho_{x'}^L(y') \langle \rho_y^L(z_1), \rho_{y'}^N(z_1) \rangle_{H_{z_1}^{-1}} \langle \rho_y^L(z_2), \rho_{y'}^N(z_2) \rangle_{L_{z_2}^2}$$

$$+ \int_{\mathbb{T}} dy \rho_x^\infty(y) \int_{\mathbb{T}} dy' \rho_{x'}^\infty(y') \langle \rho_y^L(z_1), \rho_{y'}^L(z_1) \rangle_{H_{z_1}^{-1}} \langle \rho_y^L(z_2), \rho_{y'}^L(z_2) \rangle_{L_{z_2}^2}$$

Now consider one of the terms:

$$\int_{\mathbb{T}} dy \rho_x^L(y) \int_{\mathbb{T}} dy' \rho_{x'}^L(y') \langle \rho_y^N(z_1), \rho_{y'}^N(z_1) \rangle_{H_{z_1}^{-1}} \langle \rho_y^N(z_2), \rho_{y'}^N(z_2) \rangle_{L_{z_2}^2}$$

$$= \int_{\mathbb{T}} dy \rho_x^L(y) \int_{\mathbb{T}} dy' \rho_{x'}^L(y') \rho_y^N(y')(I^*I\rho_y^N)(y')$$

since

$$\langle \rho_y^N(z_1), \rho_{y'}^N(z_1) \rangle_{H_{z_1}^{-1}} = \langle I\rho_y^N, I\rho_{y'}^N \rangle_{L^2} = \langle I^*I\rho_y^N, \delta_{y'} \rangle_{L^2} = (I^*I\rho_y^N)(y')$$

As $N \to \infty$ this quantity becomes

$$\to \int_{\mathbb{T}} dy \rho_x^L(y) \int_{\mathbb{T}} dy' \rho_{x'}^L(y') \rho_y^\infty(y') (I^* I \rho_y^\infty)(y')$$

$$= \int_{\mathbb{T}} dy \rho_x^L(y) \rho_{x'}^L(y) (I^* I \rho_y^\infty)(y) - \frac{1}{2\pi} \int_{\mathbb{T}} dy \rho_x^L(y) \int_{\mathbb{T}} dy' \rho_{x'}^L(y') (I^* I \rho_y^\infty)(y')$$

$$= (I^* I \rho_0^\infty)(0) \int_{\mathbb{T}} dy \rho_x^L(y) \rho_{x'}^L(y) - \frac{1}{2\pi} (I^* I \rho_x^L)(x')$$

$$= (I^* I \rho_0^\infty)(0) \rho_x^L(x') - \frac{1}{2\pi} (I^* I \rho_x^L)(x')$$

since $(I^* I \rho_y^\infty)(y) = (I^* I \rho_0^\infty)(0)$. As $L \to \infty$ we obtain

$$= (I^* I \rho_0^\infty)(0) \rho_x^\infty(x') - \frac{1}{2\pi} (I^* I \rho_x^\infty)(x')$$

For the other terms the discussion of the convergence is similar and the limit is the same so we obtain that

$$\int_{\mathbb{T}^2} \langle g_x^{L,N}(z_1, z_2), g_{x'}^{L,N}(z_1, z_2) \rangle_{H_{z_1}^{-1}} dz_2 \to 0$$

as a distribution in the x, x' variables and we conclude that $A_{1,2} \to 0$.   □

Lemma 17 *We have $C_1^{L,N} \to 0$ as $N \to \infty$ and $L \to \infty$.*

Proof Let

$$G_x(z_2) = \int_{\mathbb{T}} g_x^{L,N}(z_1, z_2)(I^* \rho_x^L)(z_1) dz_1 = G_{x,x}(z_2)$$

and recall that

$$(C_1^{L,N})^2 = \mathbb{E} \left\| \int_{\mathbb{T}} \varphi(x) \phi_0^L(x) G_x(z_2) dx \right\|_{H_{z_2}^{-1}}^2 = \int_{\mathbb{T}^2} V^L(x, x') \langle G_x, G_{x'} \rangle_{H^{-1}} dx dx'$$

In order to estimate the scalar product can take $f \in H^1$ and consider $\int G_{x,x}(z) f(z) dz$ instead. Recall that

$$\int G_{x,x}(z) f(z) dz = \int_{\mathbb{T}} dy \rho_x^L(y) \mathscr{I}_0^N f(y) (\mathscr{I}_0^N I^* \rho_x^L)(y)$$

$$- \int_{\mathbb{T}} dy \rho_x^\infty(y) \mathscr{I}_0^L f(y) (I^* \rho_x^L)(y)$$

replace $\rho_x^\infty(y)$ by $\delta_x(y) - 1/2\pi$ and $\rho_x^L(y)$ by $\delta_x^L(y) - 1/2\pi$ to get

$$= \int_{\mathbb{T}} dy \, \delta_x^L(y) \left(\mathscr{I}_0^N f(y) \right) (\mathscr{I}_0^N I^* \rho_x^L)(y) - \left(\mathscr{I}_0^L f(x) \right) (I^* \rho_x^L)(x)$$

$$- \frac{1}{2\pi} \int_{\mathbb{T}} dy \, \mathscr{I}_0^N f(y) (\mathscr{I}_0^N I^* \rho_x^L)(y) + \frac{1}{2\pi} \int_{\mathbb{T}} dy \, \mathscr{I}_0^L f(y) (I^* \rho_x^L)(y)$$

$$= \int_{\mathbb{T}} dy \, \delta_x^L(y) \left(\mathscr{I}_0^N f(y) - \mathscr{I}_0^N f(x) \right) (\mathscr{I}_0^N I^* \rho_x^L)(y)$$

$$+ \left(\mathscr{I}_0^N f(x) \right) (I^* \left(\delta^L * \mathscr{I}_0^N \rho_x^L \right))(x) - \left(\mathscr{I}_0^L f(x) \right) (I^* \rho_x^L)(x)$$

$$- \frac{1}{2\pi} \int_{\mathbb{T}} dy \, \mathscr{I}_0^N f(y) (\mathscr{I}_0^N I^* \rho_x^L)(y) + \frac{1}{2\pi} \int_{\mathbb{T}} dy \, \mathscr{I}_0^L f(y) (I^* \rho_x^L)(y).$$

We note that, by symmetry,

$$\left(\mathscr{I}_0^N f(x) \right) (I^* \left(\delta^L * \mathscr{I}_0^N \rho_x^L \right))(x) = \left(\mathscr{I}_0^L f(x) \right) (I^* \rho_x^L)(x) = 0$$

and moreover that in this expression we have that

$$\sup_x \left| \left\langle \mathscr{I}_0^L f, I^* \rho_x^L \right\rangle - \langle f, \Theta_x \rangle \right| + \sup_x \left| \left\langle \mathscr{I}_0^N f, \mathscr{I}_0^N I^* \rho_x^L \right\rangle - \langle f, \Theta_x \rangle \right|$$

$$\lesssim \|f\|_{H^1} o((L \wedge N)^{-1})$$

so finally it remains to bound

$$H = \int_{\mathbb{T}} dy \, \delta_x^L(y) \left(\mathscr{I}_0^N f(y) - \mathscr{I}_0^N f(x) \right) (\mathscr{I}_0^N I^* \rho_x^L)(y)$$

However using $H^1 \subseteq C^{1/2}$ compactly and that $\|\mathscr{I}_0^N I^* \rho_x^L\|_\infty < +\infty$ uniformly in x, L, N we have

$$|H| \lesssim \left\| \mathscr{I}_0^N f \right\|_{H^1} \int_{\mathbb{T}} dy \, \delta_x^L(y) |y - x|^{1/2} \lesssim \|f\|_{H^1} L^{-1/2} \to 0.$$

In conclusion

$$\sup_{x \in \mathbb{T}} \left| \int G_{x,x'}(z) f(z) dz \right| \lesssim \|f\|_{H^1} o((L \wedge N)^{-1})$$

which implies

$$\sup_{x \in \mathbb{T}} \langle G_{x,x}, G_{x,x} \rangle_{H^{-1}} = o((L \wedge N)^{-1}).$$

and thus we can conclude that $C_1^{L,N} \to 0$. $\qquad\square$

References

1. S. Assing, A pregenerator for Burgers equation forced by conservative noise. Commun. Math. Phys. **225**(3), 611–632 (2002)
2. L. Bertini, G. Giacomin, Stochastic Burgers and KPZ equations from particle systems. Commun. Math. Phys. **183**(3), 571–607 (1997)
3. J. Diehl, M. Gubinelli, N. Perkowski, The Kardar–Parisi–Zhang equation as scaling limit of weakly asymmetric interacting Brownian motions. Commun. Math. Phys. **354**(2), 549–589 (2017)
4. P. Echeverría, A criterion for invariant measures of Markov processes. Zeitschrift für Wahrscheinlichkeitstheorie und Verwandte Gebiete **61**(1), 1–16 (1982)
5. T. Franco, P. Gonçalves, G.M. Schütz, Scaling limits for the exclusion process with a slow site. Stoch. Process. Appl. **126**(3), 800–831 (2016)
6. T. Funaki, J. Quastel, KPZ equation, its renormalization and invariant measures. Stoch. Part. Differ. Equ. Anal. Comput. **3**(2), 159–220 (2015)
7. P. Gonçalves, Derivation of the stochastic Burgers equation from the WASEP, in *From Particle Systems to Partial Differential Equations. II*. Springer Proceedings in Mathematics and Statistics, vol. 129 (Springer, Cham, 2015), pp. 209–229
8. P. Goncalves, M. Jara, Universality of KPZ equation (2010). arXiv:1003.4478
9. P. Gonçalves, M. Jara, Nonlinear fluctuations of weakly asymmetric interacting particle systems. Arch. Ration. Mech. Anal. **212**(2), 597–644 (2014)
10. P. Gonçalves, M. Jara, S. Sethuraman, A stochastic Burgers equation from a class of microscopic interactions. Ann. Probab. **43**(1), 286–338 (2015)
11. M. Gubinelli, M. Jara, Regularization by noise and stochastic burgers equations. Stoch. Part. Differ. Equ. Anal. Comput. **1**(2), 325–350 (2013)
12. M. Gubinelli, N. Perkowski, *Lectures on Singular Stochastic PDEs*. Ensaios Matemáticos [Mathematical Surveys], vol. 29 (Sociedade Brasileira de Matemática, Rio de Janeiro, 2015)
13. M. Gubinelli, N. Perkowski, The Hairer-Quastel universality result at stationarity, in *Stochastic Analysis on Large Scale Interacting Systems*. RIMS Kôkyûroku Bessatsu, vol. B59 (Research Institute for Mathematical Sciences, Kyoto, 2016), pp. 101–115
14. M. Gubinelli, N. Perkowski, Energy solutions of KPZ are unique. J. Am. Math. Soc. **31**(2), 427–471 (2018)
15. M. Gubinelli, N. Perkowski, The infinitesimal generator of the stochastic Burgers equation (2018). arXiv:1810.12014 [math]
16. M. Hairer, J. Quastel, A class of growth models rescaling to KPZ. Forum Math. Pi **6**, e3, 112 (2018)
17. S. Janson, *Gaussian Hilbert Spaces*. Cambridge Tracts in Mathematics, vol. 129 (Cambridge University Press, Cambridge, 1997)
18. I. Karatzas, S.E. Shreve, *Brownian Motion and Stochastic Calculus* (Springer, New York, 1998)
19. M. Kardar, G. Parisi, Y.-C. Zhang, Dynamic scaling of growing interfaces. Phys. Rev. Lett. **56**(9), 889–892 (1986)
20. T. Komorowski, C. Landim, S. Olla, *Fluctuations in Markov Processes: Time Symmetry and Martingale Approximation*, 2012 edn. (Springer, Heidelberg, 2012)
21. J. Quastel, *Introduction to KPZ*. Lecture Notes in Mathematics (Arizona School of Analysis and Mathematical Physics, Tucson, 2012)
22. F. Russo, P. Vallois, Elements of stochastic calculus via regularization, in *Séminaire de Probabilités XL*, ed. by C. Donati-Martin, M. Émery, A. Rouault, C. Stricker. Lecture Notes in Mathematics, vol. 1899 (Springer, Berlin, 2007), pp. 147–185
23. F. Russo, P. Vallois, J. Wolf, A generalized class of Lyons-Zheng processes. Bernoulli **7**(2), 363–379 (2001)

Chapter 3
Pathwise Solutions for Fully Nonlinear First- and Second-Order Partial Differential Equations with Multiplicative Rough Time Dependence

Panagiotis E. Souganidis

Abstract The notes are an overview of the theory of pathwise weak solutions of two classes of scalar fully nonlinear first- and second-order degenerate parabolic partial differential equations with multiplicative rough time dependence, a special case being Brownian. These are Hamilton-Jacobi, Hamilton-Jacobi-Isaacs-Bellman and quasilinear divergence form equations including multidimensional scalar conservation laws. If the time dependence is "regular", the weak solutions are respectively the viscosity and entropy/kinetic solutions. The main results are the well-posedness and qualitative properties of the solutions. Some concrete applications are also discussed.

3.1 Introduction

I present an overview of the theory of pathwise weak solutions of two classes of scalar fully nonlinear first- and second-order degenerate parabolic (stochastic) partial differential equations (spde for short) with multiplicative rough time dependence, a special case being Brownian. These are Hamilton-Jacobi, Hamilton-Jacobi-Isaacs-Bellman and quasilinear divergence form partial differential equations (pde for short) including multidimensional scalar conservation laws. If the time dependence is "regular", the weak solutions are respectively the viscosity and entropy/kinetic solutions. The main results are the well-posedness and qualitative properties of the solutions. Some concrete applications are also discussed both to motivate as well as to show the scope of the theory. Most of the results presented here are part of the ongoing development of the theory in collaboration with Lions [71, 72, 74–79]. The results about quasilinear divergence form equations are based on joint work with Lions et al. [38–41, 65–67].

P. E. Souganidis (✉)
Department of Mathematics, University of Chicago, Chicago, IL, USA
e-mail: souganidis@math.uchicago.edu

© Springer Nature Switzerland AG 2019
F. Flandoli et al. (eds.), *Singular Random Dynamics*, Lecture Notes
in Mathematics 2253, https://doi.org/10.1007/978-3-030-29545-5_3

Problems of the type discussed here arise in several applied contexts and models for a wide variety of phenomena and applications including mean field games, turbulence, phase transitions and front propagation with random velocity, nucleations in physics, macroscopic limits of particle systems, pathwise stochastic control theory, stochastic optimization with partial observations, stochastic selection, etc.

The general classes of evolution equations considered in these notes are

$$\begin{cases} du = F(D_u^2, Du, u, x, t) \\ \quad + \sum_{i=1}^m H^i(Du, u, x, t) \cdot dB_i \text{ in } Q_T := \mathbb{R}^d \times (0, T], \end{cases} \tag{3.1}$$

and

$$du + \sum_{i=1}^d \partial_{x_i}(A^i(u, x, t)) \cdot dB_i - \text{div}(A(u, x, t)Du)dt = 0 \text{ in } Q_T, \tag{3.2}$$

with initial condition

$$u(\cdot, 0) = u_0 \text{ on } \mathbb{R}^d. \tag{3.3}$$

Here $F = F(X, p, u, x, t)$, $H^1 = H^1(p, u, x, t), \ldots, H^m = H^m(p, u, x, t)$, $A^1 = A^1(u, x, t), \ldots, A^d = A^d(u, x, t)$ and $A = A(u, x, t)$ are (at least) continuous functions of their arguments (exact assumptions will be shown later), F and A are respectively degenerate elliptic in X and monotone in u, $B := (B_1, \ldots, B_m)$ and $B = (B_1, \ldots, B_d)$ are, for example, continuous geometric rough in time and "$\cdot$" simply denotes the way B acts on the H^i and A^i. When B is a Brownian path, "$\cdot$" becomes the usual Stratonovich differential "$\circ$", something justified by the fact that the pathwise solutions may be obtained as the limit of solutions of equations with smooth signals. The B_i's can be taken to be approximations of "colored white noise." For simplicity, below we assume that any spatial dependence on the signal B_i is part of H^i and the A^i. Finally, $Q_\infty := \mathbb{R}^d \times (0, \infty)$.

When B is either smooth or has bounded variation, then "d" is the regular time derivative and (3.1) and (3.2) are "regular" equations, which have been studied using respectively the viscosity and entropy/kinetic theories. When the driving signals are regular ("non rough"), I refer to the equations as "deterministic" or "non-rough". If the signals are "rough", the equations will be called "rough" or "stochastic" when the path is Brownian.

The theory presented in these notes is a pathwise one and simply treats B as the time derivative of a continuous function. When the H^i's and A^i's are respectively independent of (u, x) and x, the general qualitative theory does not need any other assumption but continuity. When there is spatial dependence, then it is necessary to argue differently.

There is a vast literature for linear and quasilinear versions of (3.1) as well as work for some versions of (3.2). Listing all the references is not possible in this introduction. Some connections are made in main the body of the notes.

3.1.1 Organization of the Notes

Concrete examples where (3.1) and (3.2) arise are presented in Sect. 3.2. Section 3.3 discusses the main difficulties and explains why the Stratonovich formulation is more appropriate. Sections 3.4 to 3.14 are devoted to the pathwise solutions of Hamilton-Jacobi and Hamilton-Jacobi-Isaacs equations. In Sect. 3.4, I present new results about nonlinear equations with linear rough path dependence, I introduce the system of characteristics, and I discuss a short time classical result about stochastic Hamilton-Jacobi equations in the smooth regime. Section 3.5 is about fully nonlinear equations with semilinear rough path dependence. Section 3.6 is about formulae or the lack thereof for Hamilton-Jacobi equations with time dependence. Section 3.7 discusses the simplest possible nonlinear pde with rough time signals as the limit of regular approximations. Section 3.8 is about pathwise solutions of nonlinear first-order pde with nonsmooth Hamiltonians and rough signals. In Sect. 3.9, I present new results about the qualitative properties of the pathwise solutions. Section 3.10 is devoted to the well-posedness theory of the pathwise solutions with spatially depended H^i's. Section 3.11 is about Perron's method, while Sect. 3.12 discusses the convergence of approximation schemes with error estimates. In Sect. 3.13 I present new results about the homogenization of pathwise solutions. Section 3.14 is about the asymptotics of stochastically perturbed reaction-diffusion equations. The results about quasilinear divergence form equations including multi-dimensional stochastic conservation laws are presented in Section 3.15. Finally, Appendix summarizes few basic things from the classical theory of viscosity solutions that are used in the notes.

3.2 Motivation and Some Examples

A discussion follows about a number of results that have been or may be solved using the theory presented in here. In several places, to keep the discussion simple, the presentation is informal.

3.2.1 Motion of Interfaces

An important question in pde and geometry as well as applications like phase transitions is the understanding of the long time behavior of solutions of reaction-diffusion equations and the properties of the developing interfaces, which separate the regions where the solutions approach the different equilibria of the equation.

A classical and well studied problem in this context is the asymptotic behavior of the solution u^ε to the so called Allen-Cahn equation

$$u_t^\varepsilon - \Delta u^\varepsilon + \frac{1}{\varepsilon^2} W'(u^\varepsilon) = 0 \text{ in } Q_T,$$

where $W : \mathbb{R} \to \mathbb{R}$ is a double-well potential with wells of equal depth located at, for example, at ± 1. It is well known that as, $\varepsilon \to 0$, $u^\varepsilon \to \pm 1$ inside and outside an interface moving with normal velocity $V = -\kappa$, where κ is the mean curvature. The interface is the zero-level set of the solution of the level-set pde

$$v_t = \left(I - \frac{Dv}{|Dv|} \otimes \frac{Dv}{|Dv|} \right) : D^2 v \text{ in } Q_T, \tag{3.4}$$

where for $A, B \in \mathcal{S}^d$, the space of symmetric $d \times d$ matrices, $A : B := \text{tr}(AB)$ and I is the identity matrix in $\mathbb{R}^d$.

For the applications, however, it is interesting to consider potentials with wells at locations which change with the scale ε and to identify the exact scaling at which something nontrivial comes up. An example of such a problem is

$$u_t^\varepsilon - \Delta u^\varepsilon + \frac{1}{\varepsilon^2}(W'(u^\varepsilon) + \varepsilon c(t)) = 0 \text{ in } Q_T,$$

for some smooth function $c = c(t)$, which leads, as $\varepsilon \to 0$, to an interface moving with normal velocity $V = -\kappa + \alpha c(t)$, where $\alpha \in \mathbb{R}$ is a "universal" constant which is independent of c.

A natural question is what happens if c is irregular and, in particular, if $c = dB$, where B is a Brownian path. Note that such perturbations often appear in the hydrodynamic limit of interacting particle systems. It turns out that in this case the oscillations of the wells due to dB are too strong for the system to stabilize. However, as it was it was shown by Lions and Souganidis [69], if B is replaced by a "mild" approximation B^ε, then the asymptotic interface moves with normal velocity

$$V = -\kappa + \alpha dB,$$

and is characterized as a level set of the solution of the "stochastic" level-set pde

$$dv = \left[\left(I - \frac{Dv}{|Dv|} \otimes \frac{Dv}{|Dv|} \right) : D^2 v \right] dt + \alpha |Dv| \cdot dB \text{ in } Q_T. \tag{3.5}$$

More details including references as well as a sketch of the proof of the result in [69] are presented in Sect. 3.14.

3.2.2 A Stochastic Selection Principle

A classical question in the theory of level set interfacial motions is whether there is "fattening", that is, if there are configurations (initial data) such that the zero level set of the solution v to (3.4) develops interior. For the motion by mean curvature, it is known that, if the initial configuration is two touching balls, then, for positive times, the evolving front is a "surface" that looks like the boundary of either two separated shrinking balls or some connected open set which moves in time, and there are well defined minimal and maximal moving boundaries.

As it is often the case the introduction of stochasticity resolves this ambiguity and provides a definitive selection principle. Indeed, it was proved by Souganidis and Yip [104] without any regularity restrictions on the evolving set (see also Dirr et al. [23] for a short time result), that the zero level sets of the solutions $v^{\pm \varepsilon}$ of the stochastically perturbed level set pde

$$dv^{\pm \varepsilon} = \left[\left(I - \frac{Dv^{\pm \varepsilon}}{|Dv^{\pm \varepsilon}|} \otimes \frac{Dv^{\pm \varepsilon}}{|Dv^{\pm \varepsilon}|} \right) : D^2 v^{\pm \varepsilon} \right] dt \pm \varepsilon |Dv^{\pm \varepsilon}| \circ dB \text{ in } Q_T,$$

with initial data two touching balls, never develop interior and, as $\varepsilon \to 0$, converge in the Hausdorff distance to the maximal interface of the unperturbed problem.

3.2.3 Pathwise Stochastic Control Theory

To keep the notation simple I assume here that $d = 1$. A typical stochastic control problem with finite horizon $T > 0$ consists of

1. a controlled stochastic differential equation (sde for short)

$$\begin{cases} dX_s = b(X_s, \alpha_s)ds + \sqrt{2}\sigma_1(X_s, \alpha_s)dB_{1,s} + \sqrt{2}\sigma_2(X_s) \circ dB_{2,s} \text{ in } (t, T], \\ X_t = x, \end{cases}$$

where $(B_{1,t})_{t \geq 0}$ and $(B_{2,t})_{t \geq 0}$ are two independent Brownian motions with respective filtrations $(\mathcal{F}_t^{B_1})_{t \geq 0}$ and $(\mathcal{F}_t^{B_2})_{t \geq 0}$, $(\alpha_t)_{t \geq 0} \in \mathcal{A}$, the set of admissible $\mathcal{F}_t^{B_1}$-progressively measurable controls with values in A a subset of some $\mathbb{R}^k$, and

2. a pay-off functional, which, to simplify the presentation, here is taken to be

$$J(x, t; \alpha) = E_{x,t}[g(X_T)|\mathcal{F}_T^{B_2}],$$

the goal being to minimize the pay-off over $\mathcal{A}$.

The associated value function, which is defined by

$$u(x, t) = \text{essinf}_{\alpha \in \mathcal{A}} J(x, t; \alpha),$$

has been shown in Lions and Souganidis [71, 75] (see also Buckdahn and Ma [10] for a special case) to be the pathwise solution of the stochastic associated Bellman equation

$$du + \inf_{\alpha \in A} \left[\sigma_1^2(x, \alpha) u_{xx} + b(x, \alpha) u_x \right] dt$$

$$+ \sqrt{2} \sigma_2(x) u_x \circ dB_2 = 0 \text{ in } Q_T \quad u(\cdot, T) = g,$$

which is a special case of (3.1) with F nonlinear and H linear; notice that to be consistent with control theoretic formulation of the problem the equation is written backwards in time.

The aim of the classical stochastic control theory with the stochastic dynamics above, is to minimize over $\mathcal{A}$ the "averaged" payoff

$$\overline{J}(x, t; \alpha) = E_{x,t}[g(X_T)].$$

It is a classical fact that the value function

$$\overline{u}(x, t) = \text{essinf}_{\alpha \in \mathcal{A}} \overline{J}(x, t; \alpha)$$

is the unique viscosity solution of the deterministic Bellman terminal valued problem

$$\begin{cases} \overline{u}_t + \inf_{\alpha \in A} \left[\left(\sigma_1^2(x, \alpha) + \sigma_2^2(x) \right) \overline{u}_{xx} + \left(b(x, \alpha) + \sigma_{2,x} \sigma_2(x) \right) \overline{u}_x \right] = 0 \text{ in } Q_T, \\ u(\cdot, T) = g. \end{cases}$$

3.2.4 Mean Field Games

A typical example of the Lasry-Lions mean field theory [53–55] is the study of the asymptotic behavior, as $L \to \infty$, of the law $\mathcal{L}(X_t^1, \ldots, X_t^L)$ of the solution of the sde

$$dX^i = \sigma \left(X^i, \frac{1}{L-1} \sum_{j \neq i} \delta_{X^j} \right) \circ dB \qquad (i = 1, \ldots, L).$$

Here δ_y is the Dirac mass at y and $\sigma \in C^{0,1}(\mathbb{R}^d \times \mathcal{P}(\mathbb{R}^d); \mathcal{S}^d)$, $\mathcal{P}(X)$ being the set of probability measures on X.

The result (see Lions [57]) is that, as $L \to \infty$, in the sense of measures and for all $t > 0$,

$$\mathcal{L}(X_t^1, \ldots, X_t^L) \to \pi_t \in \mathcal{P}(\mathcal{P}(\mathbb{R}^d)),$$

where the density $(m_t)_{t \geq 0}$ of the evolution in time of $(\pi_t)_{t \geq 0}$, which is defined, for all $U \in C(\mathcal{P}(\mathbb{R}^d))$, by

$$\int U(m) d\pi_t(m) = E[U(m_t)],$$

solves the stochastic conservation law

$$dm + \text{div}_x(\sigma^T(m, x) \circ dB) = 0 \text{ in } Q_T,$$

which is a special case of (3.2). Here σ^T is the transpose of the matrix σ.

3.3   The Main Difficulties and the Choice of Stochastic Calculus

3.3.1   Difficulties

Given that, in general and without rough signals, (3.1) and (3.2) do not have global smooth solutions, it is natural to expect that this is the case in the presence of rough time dependence.

It is also not possible to use directly the standard viscosity and entropy solutions of the "deterministic" theory, since they depend on inequalities satisfied either at some special points or after integration. Consider, for example, (3.2) with $d = 1$ and $A \equiv 0$. An entropy solution must satisfy, in the sense of distributions, the weak entropy inequality $dS(u) + Q(u)_x \cdot dB \leq 0$ for all pairs (S, Q) of convex entropy S and entropy flux Q. The inequality does not make sense if B is a rough path. A similar difficulty arises when dealing with viscosity inequalities.

Moreover, the lack of regularity does not allow to express the solutions in any form involving time integration as is the case for sde, that is to say, for example, that u solves $du = H(Du) \cdot dB$ in Q_T if, for all $x \in \mathbb{R}^d$ and $s, t \in [0, T]$ with $s > 0$,

$$u(x, t) = u(x, s) + \int_s^t H(Du(x, \tau)) \cdot dB(\tau).$$

Another possibility, at least when $m = 1$, is to take advantage of the multiplicative noise to change time and obtain an equation without rough parts. For example, formally, if $du + H(Du) \cdot dB = 0$, the change of time $u(x, t) = U(x, B(t))$ yields that U must be a global smooth solution to the forward-backward time homogeneous Hamilton-Jacobi equation $U_t + H(DU) = 0$ in $\mathbb{R}^d \times (-\infty, \infty)$. It is, of course, well known that such solutions do not exist in general. Behind this difficulty is the basic fact that the nonlinear problems develop shocks which are not reversible, while the changing sign of the rough signals, in some sense, forces the solutions to move forward and backward in time. Note that the time change works in intervals where dB does not change sign. More details about this are given later in the notes.

A natural question is whether it is possible to solve the equations in law. Recall that solving the sde $dX = \sqrt{2}\sigma(X_t)dB$ in law is equivalent to understanding, for all smooth ϕ and $T > 0$, the solutions u of the initial value problem

$$u_t = \sigma\sigma^T : D^2u \text{ in } Q_T \quad u(\cdot, 0) = \phi.$$

For the equations here the state variable must belong to a suitable function space and the corresponding spde is set in infinite dimensions. For example, the infinite dimensional pde describing the law of $du = \sqrt{2}\, H(Du) \circ dB$ is, formally,

$$U_t = D^2U(H(Df), H(Df)).$$

The problem is that the Hessian D^2U is an unbounded operator independently of the choice of the base space. Such pdes are far away from the theory of viscosity solutions in infinite dimensions developed by Crandall and Lions [14, 15].

Solving linear stochastic pde in law is related to the martingale approach which has been used successfully in linear and some quasilinear settings. A partial list of references is Chueshov and Vuillermot [12, 13], Da Prato et al. [18], Gerencsér et al. [37], Huang and Kushner [45], Krylov [48, 49], Krylov and Röckner [50], Rozovskiĭ [94, 95], Pardoux [85–87], Watanabe [105]. The methodology requires some tightness (compactness) which typically follows from estimates on the derivatives of the solutions. In general, the latter are not available for nonlinear problems.

3.3.2   The Choice of Stochastic Calculus: Stratonovich vs Itô

When studying sdes, it is important to decide if they are written in Stratonovich or Itô form, each of which having advantages and disadvantages; for example, more regularity and chain rule for the former and less regularity but no chain rule for the latter.

At first glance, the choice of calculus does not seem to be relevant for the nonlinear problems discussed here due to the lack of regularity. This is, however, not the case. The actual formulation plays an important role in the interpretation, well-posedness, stability and construction of the solutions, which, typically, are obtained as limits of solutions with regular time dependence. The discussion below touches upon some of these issues.

The advantage of the Stratonovich formulation can be seen in the following rather simple example. Consider, for $\lambda \geq 0$, the Itô-form spde

$$du = \lambda u_{xx}dt + \sqrt{2}u_x dB \text{ in } Q_T.$$

The change of variables $u(x,t) = v(x + \sqrt{2}B(t), t)$ yields that v satisfies the (deterministic) pde

$$v_t = (\lambda - 1)v_{xx} \text{ in } Q_T,$$

which is well-posed if and only if $\lambda \geq 1$.

Of course this is not an issue if the spde was in Stratonovich form to begin with. In that case the change of variables yields the equation

$$v_t = \lambda v_{xx} \text{ in } Q_T,$$

which is well posed if and only if $\lambda \geq 0$, as is this case when B is a smooth path.

Consider, for example, a family $(B^\varepsilon)_{\varepsilon > 0}$ of smooth approximations of the Brownian motion B and the solution u^ε of the equation

$$u_t^\varepsilon = u_{xx}^\varepsilon + u_x^\varepsilon \dot{B}^\varepsilon.$$

It is immediate that $u^\varepsilon(x, t) = v(x + B^\varepsilon(t), t)$ with v solving $v_t = v_{xx}$. Letting $\varepsilon \to 0$ then yields that $u^\varepsilon \to u$, which solves

$$du = u_{xx}dt + u_x \circ dB.$$

Another example, where the use of Stratonovich appears to be necessary, is the application to front propagation via the level set pde. One of the important elements of the theory is that the moving interfaces depend only on the initial one and not the particular choice of the initial datum of the pde. This is equivalent to the requirement that the equations are invariant under increasing changes of the unknown.

Consider, for example, the pde

$$u_t + |Du| = 0.$$

Arguing as if the solution u were smooth (the argument can be made rigorous using viscosity solutions), it is straightforward to check that, for nondecreasing ϕ, $\phi(u)$ is also a solution; note that the monotonicity of ϕ is important when dealing with viscosity solutions.

The next example shows that the Itô formulation is the wrong one. Assume that level set pde of the interfacial motion $V = dB$ with B a Brownian motion is

$$du = |Du|dB.$$

If u is a smooth solution and $\phi : \mathbb{R} \to \mathbb{R}$ is smooth and nondecreasing, Itô's formula yields that

$$d\phi(u) = |D\phi(u)|dB + \frac{1}{2}\phi''(u)|Du|^2,$$

which is not the same equation as the one satisfied by u. This is of course not the case if the level set pde was written in the Stratonovich form, which, however, requires a priori additional regularity which is not available here. Indeed, if $du = H(Du) \circ dB$, then, in Itô's form

$$du = H(Du)dB + \frac{1}{2}\Big\langle D^2uDH(Du), DH(Du)\Big\rangle dt.$$

where, for $x, y \in \mathbb{R}^d$, $\langle x, y \rangle$ is the usual inner product. To make, however, sense of this last equation, it is necessary to have information about D^2u which, in general, is not available.

In the context of second- and first-order (deterministic) pde the difficulties due to the lack of regularity are overcome using viscosity solutions. Their definition is based on inequalities which, as mentioned earlier, cannot be expected to make sense in the presence of rough signals.

There is, however, a reformulation of the definition for viscosity solutions, which, at first glance, appears to be more conducive to stochastic calculus.

Indeed, for B smooth, consider again the equation $u_t = H(Du, x)\dot{B}$. The definition of viscosity subsolutions is equivalent to the requirement that, for any smooth $\phi : \mathbb{R}^d \to \mathbb{R}$, the map $t \to \max(u - \phi)$ satisfies, in the viscosity sense, the differential inequality

$$\frac{d}{dt}\sup(u(\cdot, t) - \phi) \leqq \sup_{\bar{x}(t)\in\mathrm{argmax}(u(\cdot,t)-\phi)} (H(D\phi(\bar{x}(t)), \bar{x}(t))\dot{B}).$$

If B is a Brownian motion, then, assuming that there exists a unique maximum point $\bar{x}(t)$ of $u(\cdot, t) - \phi$, the Stratonovich formulation should be

$$\frac{d}{dt}\max(u(\cdot, t) - \phi) \leqq H(D\phi(\bar{x}(t)), \bar{x}(t)) \circ dB,$$

a fact which, however, breaks down due to the lack of regularity in t of the map $t \mapsto \bar{x}(t)$.

If $\dot{B} \in L^1((0, T))$, then the above inequality is meaningful and has been used by Lions and Perthame [61] and Ishii [46] to study viscosity solutions of Hamilton-Jacobi equations with L^1-time dependence.

The regularity concerns can, of course, be relaxed, if the inequality above is required to hold in Itô's sense. This, however, leads to a contradiction to the classical fact that the maximum of two subsolutions is a subsolution.

Recall that, if u and v are actually differentiable with respect to t, then

$$\frac{d}{dt}(\max(u, v)) = \mathbb{1}_{\{u(\cdot,t)>v(\cdot,t)\}}u_t + \mathbb{1}_{\{u(\cdot,t)\leq v(\cdot,t)\}}v_t,$$

where $\mathbb{1}_A$ denotes the characteristic function of the set A.

If

$$u_t = H(Du), \quad v_t = H(Dv) \quad \text{and} \quad H(0) = 0,$$

it follows that

$$\frac{d}{dt}\max(u, v) \leqq H(D(\max(u, v))),$$

and, hence, $\max(u, v)$ is a subsolution.

Checking the same claim in the Itô's formulation yields

$$d\max(u, v) \geqq \mathbb{1}_{\{u(\cdot,t)>v(\cdot,t)\}}du + \mathbb{1}_{\{u(\cdot,t)\leqq v(\cdot,t)\}}dv,$$

which suggests that $\max(u, v)$ is not necessarily a subsolution.

The final justification for considering the Stratonovich vs Itô's formulation when studying, for example, the equation

$$du = H(Du) \cdot dB$$

comes from considering the family of problems

$$u_t^\varepsilon = H(Du^\varepsilon)\dot{B}^\varepsilon,$$

where B^ε are smooth approximations of the Brownian motion B. If u^ε and u are smooth and, as $\varepsilon \to 0$, $u^\varepsilon \to u$ in $C^2(\mathbb{R}^d \times (0, \infty))$, it is not difficult to see that u must solve the equation in the Stratonovich sense.

Note that, under suitable assumptions on the initial datum of the regularized equation and the Hamiltonian, it is possible to show, using arguments from the theory of viscosity solutions, that the solutions u^ε are, uniformly in ε, bounded and Lipschitz continuous in x, and, hence, converge uniformly along subsequences for each t. This observation is the starting point of the theory, since it provides a candidate for a possible solution of (3.1).

3.4 Single Versus Multiple Signals, the Method of Characteristics and Nonlinear pde with Linear Rough Dependence on Time

3.4.1 Single Versus Multiple Signals

The next example illustrates that there is a difference between one single and many signals and indicates the role that rough paths may play in the theory.

Consider two smooth paths B_1 and B_2 and the linear pde

$$u_t = u_x \dot{B}_1 + f(x)\dot{B}_2 \text{ in } Q_T \quad u(\cdot, 0) = u_0. \tag{3.6}$$

It is immediate that $v(x, t) = u(x - B_1(t), t)$ solves

$$v_t = f(x - B_1(t))\dot{B}_2 \text{ in } Q_T \quad v(\cdot, 0) = u_0,$$

and, hence,

$$u(x, t) = v(x + B_1(t), t) = u_0(x + B_1(t)) + \int_0^t f(x + B_1(t) - B_1(s))\dot{B}_2(s)ds .$$

To extend this expression to non smooth paths, it is necessary to deal with integrals of the form

$$\int_a^b g(B_1(s)) \, dB_2(s),$$

which is one of the key ingredients of Lyons's theory of rough paths; see, for example, Qian and Lyons [82], Lyons [81, 83], Lejay and Lyons [56], etc.

3.4.2 Nonlinear pde with Linear Rough Dependence on Time

The calculation above suggests, however, a possible way to study general linear/nonlinear equations with linear rough dependence, that is, equations of the form

$$\begin{cases} du = F(D^2u, Du, x)dt + \langle a(x), Du \rangle \cdot dB_1 + c(x)u \cdot dB_2 \text{ in } Q_T, \\ u(\cdot, 0) = u_0. \end{cases} \tag{3.7}$$

Consider the system

$$\begin{cases} dX = -a(X) \cdot dB_1 & X(0) = x, \\ dP = \langle Da(X), P \rangle \cdot dB_1 + \langle Dc(X), P \rangle U \cdot dB_2 & P(0) = p, \\ dU = c(X)U \cdot dB_2 & U(0) = u, \end{cases} \qquad (3.8)$$

which, in view of the theory of rough paths, has a solution for any initial datum (x, p, u). Of course, a and c must satisfy appropriate conditions. This, however, is not important for the ongoing discussion.

It is immediate that, with initial condition $X(0) = x$, $P(0) = Du_0(x)$, $U(0) = u_0(x)$, (3.8) is the system of characteristic equations of the linear Hamilton-Jacobi equation

$$du = \langle a(x), Du \rangle \cdot dB_1 + c(x)u \cdot dB_2 \quad u(\cdot, 0) = u_0.$$

The next step is to make the ansatz that the solution u of (3.7) has the form

$$u(x, t) = v(X^{-1}(x, t), t), \qquad (3.9)$$

and to find the equation satisfied by v. Note that, due to the linearity, it is immediate that the map $x \to X(x, t)$ is invertible for all t.

Substituting in (3.7), arguing formally (the calculation can be made rigorous using viscosity solutions when B_1 and B_2 are smooth), and rewriting (3.9) as

$$u(\cdot, t) = S(t)v(\cdot, t),$$

where, for any v_0, $S(t)v_0$ is the solution of the linear Hamilton-Jacobi equation with initial datum v_0, yields

$$\begin{aligned} du = d(S(t)v(\cdot, t)) &= dS(t)v(\cdot, t) + S(t)dv(\cdot, t) \\ &= \langle a(x), DS(t)v(\cdot, t) \rangle \cdot dB_1 + c(x)S(t)v(\cdot, t) \cdot dB_2 + S(t)(v_t(\cdot, t)) \\ &= \langle a(x), DS(t)v(\cdot, t) \rangle \cdot dB_1 + c(x)S(t)v(\cdot, t), x) \cdot dB_2 \\ &\quad + F(D^2 S(t)v(\cdot, t), DS(t)v(\cdot, t), S(t)v(\cdot, t), x)dt, \end{aligned}$$

and, hence,

$$S(t)dv(\cdot, t) = F(D^2 S(t)v(\cdot, t), DS(t)v(\cdot, t), S(t)v(\cdot, t), x)dt,$$

and

$$dv = S^{-1}(t)F(D^2S(t)v(\cdot,t), DS(t)v(\cdot,t), S(t)v(\cdot,t), x)dt.$$

Since the last equation does not contain any singular time dependence, it is convenient to replace dv by v_t and to rewrite the last equation as

$$v_t = S^{-1}(t)F(D^2S(t)v(\cdot,t), DS(t)v(\cdot,t), S(t)v(\cdot,t), x). \tag{3.10}$$

This last expression appears to be more complicated than (3.7), but this is only due to the notation.

The point is that (3.10) actually is simpler since the transformation eliminates the troublesome term

$$\langle a(x), Du \rangle \cdot dB_1 + c(x)u \cdot dB_2.$$

The new equation is of the form

$$v_t = \widetilde{F}(D^2v, Dv, v, x, t) \text{ in } Q_T \quad v(\cdot, 0) = u_0,$$

and can be studied using the viscosity theory as long as $\widetilde{F}$ satisfies the appropriate conditions for well-posedness.

The discussion above gives an alternative way to find pathwise solutions to all the equations studied using the martingale method as well as scalar quasilinear equations of divergence form, always with linear rough time dependence. As a matter of fact, a closer look at the existing theories for linear spde yields that the approach described above allows for the treatment of larger class of equations.

3.4.3 Stochastic Characteristics

The analysis in the previous subsection suggests that to handle equations with non-linear rough dependence, it may be useful to look, at least when the Hamiltonians are smooth, at the associated system of characteristics. When the time signals are smooth this is a classical system of $2d + 1$ ode. In the particular case that the rough dependence is Brownian, the stochastic characteristics were used in the work of Kunita [51] on stochastic flows. In what follows, statements are made without any assumptions and the details are left to the reader.

The characteristics of the Hamilton-Jacobi equation

$$du = \sum_{i=1}^{m} H^i(Du, u, x, t) \cdot dB_i \text{ in } Q_T \quad u(\cdot, 0) = u_0, \tag{3.11}$$

are the solutions to the following system of differential equations:

$$
\begin{cases}
dX = -\sum_{i=1}^{m} D_p H^i(P, U, X, t) \cdot dB_i, \\[2mm]
dP = \sum_{i=1}^{m} \left(D_x H^i(P, U, X, t) + D_u H^i(P, U, X, t)P \right) \cdot dB_i, \\[2mm]
dU = \sum_{i=1}^{m} \left(H^i(P, U, x, t) - \langle D_p H^i(P, U, x, t), P \rangle \right) \cdot dB_i, \\[2mm]
X(x, 0) = x, \qquad P(x, 0) = Du_0(x), \qquad U(x, 0) = u_0(x).
\end{cases}
\tag{3.12}
$$

The connection between (3.11) and (3.12) is made through the relationship

$$
U(x, t) = u(X(x, t), t) \quad \text{and} \quad P(x, t) = Du(X(x, t), t).
$$

The method of characteristics works as long as it is possible to invert the map $t \to X(x, t)$. This can always be done in some interval $(-T^*, T^*)$ for small $T^* > 0$, which depends on bounds on H, u_0, their derivatives and the signal, and, in general, is difficult to estimate in a sharp way.

It then follows that

$$
u(x, t) = U(X^{-1}(x, t), t)
$$

is a smooth solution to (3.11) in $\mathbb{R}^d \times (-T^*, T^*)$. The latter means, for all $s, t \in (-T^*, T^*)$ with $s < t$ and $x \in \mathbb{R}^d$,

$$
u(x, t) = u(x, s) + \int_s^t \sum_{i=1}^{m} H^i(Du(x, r), u(x, r), x, r) \cdot dB_i(r).
$$

If $m = 1$, it is possible to express the solutions of (3.12) using in the characteristics of the "non rough" equation

$$
u_t = H(Du, u, x, t) \text{ in } Q_T \quad u(\cdot, 0) = u_0.
$$

Indeed if (X_d, P_d, U_d) is the solution of

$$
\begin{cases}
\dot{X}_d = -D_p H(P_d, U_d, X_d, t), \\[2mm]
\dot{P}_d = D_x H^i(P_d, U_d, X_d, t) + D_u H^i(P_d, U_d, X_d, t)P_d, \\[2mm]
\dot{U}_d = H^i(P_d, U_d, X_d, t) - \langle D_p H(P_d, U_d, X_d, t), P_d \rangle, \\[2mm]
X_d(x, 0) = x, \qquad P_d(x, 0) = Du_0(x), \qquad U_d(x, 0) = u_0(x),
\end{cases}
\tag{3.13}
$$

then

$$X(x, t) = X_d(x, B(t)), \quad P(x, t) = P_d(x, B(t)), \quad \text{and} \quad U(x, t) = U_d(x, B(t)),$$

and the inversion is possible as long as $|B(t)| < T_d^*$, the maximal time for which X_d is invertible.

This simple expression for the solution of (3.12) is not valid for $m \geq 2$ unless the Hamiltonian H satisfies the involution relationship

$$\{H^i, H^j\} := D_x H^i D_p H^j - D_x H^j D_p H^i = 0 \quad \text{for all} \quad i, j = 1, \ldots, m.$$

The latter yields that the solutions of the system of the characteristics commute, that is

$$X(x, t) = X_d^1(\cdot, B_1(t)) \bullet X_d^2(\cdot, B_2(t)) \bullet \cdots \bullet X_d^M(\cdot, B_m(t))(x),$$

where, for $i = 1, \ldots, m$, (X_d^i, P_d^i, U_d^i) is the solution of (3.12) with $H \equiv H^i$ and $B_i(t) = 1$ and $\bullet$ stands for the composition of maps.

For example, if, for all $i = 1, \ldots, m$, the H^i's are independent of x, u and t, then the involution relationship is satisfied, and (3.12) reduces to

$$dX = -\sum_{i=1}^m DH^i(P) \cdot dB_i, \quad dP = 0, \quad dU = \sum_{i=1}^m [H^i(P) - \langle D_p H^i(P), P \rangle] \cdot B_i.$$

and the X-characteristic is given by

$$X(x, t) = x - \sum_{i=1}^m D_x H^i(Du_0(x)) B_i(t).$$

Finally, either for $m = 1$ or for space homogeneous Hamiltonians when $m \geq 2$, it is possible to find X, P and U for any continuous B. Otherwise it is necessary to appeal to the rough path theory.

3.5 Fully Nonlinear Equations with Semilinear Stochastic Dependence

I describe next the work of Lions and Souganidis [77] about fully nonlinear equations with semilinear stochastic dependence.

Consider the initial value problem

$$du = F(D^2u, Du, u)dt + \sum_{i=1}^m H^i(u) \cdot dB_i \text{ in } Q_T \quad u(\cdot, 0) = u_0, \tag{3.14}$$

with $u_0 \in BUC(\mathbb{R}^d)$, $B = (B_1, \ldots, B_m)$ is a C^α geometric rough path with $\alpha \in (1/3, 1/2)$, for example Brownian motion with Stratonovich, $F \in C(S^d \times \mathbb{R}^d)$ degenerate elliptic, that is, for all $(p, u) \in \mathbb{R}^{d+1}$ and $X, Y \in S^d$,

$$\text{if } X \leq Y, \text{ then } F(X, p, u) \leq F(Y, p, u), \tag{3.15}$$

and

$$H = (H^1, \ldots, H^m) \in (C^5(\mathbb{R}))^m. \tag{3.16}$$

When $m = 1$ and B is continuous path, then (3.16) can be replaced by

$$H \in C^{3,1}(\mathbb{R}). \tag{3.17}$$

Although the results presented here also apply to the more general equations like

$$du = F(D^2 u, Du, u, x, t)dt + \sum_{i=1}^m H^i(u, x, t) \cdot dB_i \text{ in } Q_T, \tag{3.18}$$

for simplicity I concentrate on (3.14) and assume that $m = 1$.

For $v \in \mathbb{R}$, consider the differential equation

$$d\Phi = H(\Phi) \cdot dB \text{ in } (0, \infty) \quad \Phi(v, 0) = v. \tag{3.19}$$

It is assumed that

$$\begin{cases} \text{there exists a unique solution } \Phi \in C([0, T]; C^3(\mathbb{R})) \text{ of (3.19)} \\ \text{such that, for all } T > 0, \\ M(T) = \sup_{0 \leq t \leq T} \left[|\Phi(0, t)| + \sum_{i=1}^3 \|D_v^i \Phi(\cdot, t)\|_\infty \right] < \infty. \end{cases} \tag{3.20}$$

Since $m = 1$, it follows that, for all $t > 0$,

$$\Phi(v, t) = \widehat{\Phi}(v, B(t)), \tag{3.21}$$

where $\widehat{\Phi}$ solves the ode

$$\dot{\widehat{\Phi}} = H(\widehat{\Phi}) \text{ in } \mathbb{R} \quad \widehat{\Phi}(v, 0) = v. \tag{3.22}$$

It is then straightforward to obtain (3.20) from the analogous properties of $\hat{\Phi}$.

Define $\widetilde{F} : S^d \times \mathbb{R} \times [0, \infty) \to \mathbb{R}$ by

$$\begin{aligned} &\widetilde{F}(X, p, v, t) \\ &= \tfrac{1}{\Phi'(v,t)} F(\Phi'(v, t)X + \Phi''(v, t)(p \otimes p), \Phi'(v, t,)p, \Phi(v, t)), \end{aligned} \tag{3.23}$$

where, to simplify the presentation, "$\prime$" denotes the partial derivatives of Φ with respect to v.

The following definitions are motivated by the strategy described in Sect. 3.4 which amounts to inverting the characteristics. For (3.14), the latter are the solutions of (3.19), which, in view of the semilinear form, can be inverted globally.

The definition of weak solution of (3.14) is introduced next.

Definition 5.1 Fix $T > 0$. Then $u \in BUC(\overline{Q}_T)$ is a pathwise subsolution (resp. supersolution) of (3.14), if, for all $\phi \in C^2(Q_T)$ and all local maximum (resp. minimum) points $(x_0, t_0) \in Q_T$ of $(x, t) \to u(x, t) - \Phi(\phi(x, t), t)$,

$$\phi_t(x_0, t_0) \leq \widetilde{F}(D^2\phi(x_0, t_0), D\phi(x_0, t_0), u(x_0, t_0), t_0), \qquad (3.24)$$

(resp.

$$\phi_t(x_0, t_0) \geq \widetilde{F}(D^2\phi(x_0, t_0), D\phi(x_0, t_0), u(x_0, t_0), t_0)) . \qquad (3.25)$$

A function $u \in BUC(\overline{Q}_T)$ is a pathwise (viscosity) solution of (3.14), if it is both subsolution and supersolution of (3.14).

Since the characteristics are globally invertible, it is possible to introduce a global change of the unknown without going through test functions. This leads to the next possible definition.

Definition 5.2 Fix $T > 0$. Then $u \in BUC(\overline{Q}_T)$ is a pathwise subsolution (resp. supersolution) of (3.14), if the function $v : \mathbb{R}^d \times [0, T] \to \mathbb{R}$ defined by

$$u(x, t) = \Phi(v(x, t), t) \qquad (3.26)$$

is a viscosity subsolution (resp. supersolution) of

$$v_t = \widetilde{F}(D^2v, Dv, v, t) \text{ in } Q_T \quad v(\cdot, 0) = u_0. \qquad (3.27)$$

A function $u \in BUC(\overline{Q}_T)$ is a pathwise solution of (3.14) if it is both a subsolution and supersolution.

The two definitions are equivalent, and, moreover, for smooth B's, the solutions introduced in Definitions 5.1 and 5.2 coincide with the classical viscosity solution.

In view of the above, the well-posedness of solutions to (3.14) reduces to the study of the analogous questions for (3.27).

After the work described above was announced, Buckdahn and Ma [9, 10] used the map (3.26), which is known as the Doss-Sussman transformation, to study equations similar to (3.14). The work in [9, 10] covers a more restrictive class of F's and well-posedness is proved under the assumption that the transformed initial value problem admits a comparison principle. In [77] there is no such assumption and the comparison is proved directly.

If H is linear in u, the problem is simpler and the details are left to the reader.

For the rest of the section, H is taken to be nonlinear, and, to simplify the presentation, it is also assumed that F is independent of u.

To deal with $\tilde{F}$, it is necessary to assume that

$$F \in C^{0,1}(\mathcal{S}^d \times \mathbb{R}^d), \tag{3.28}$$

and

$$
\begin{cases}
\text{there exists a constant } C > 0 \text{ such that,} \\[4pt]
\text{for almost every } (X, p), \\[4pt]
\quad \text{either} \quad D_X F(X, p) : X + \langle D_p F(X, p), p \rangle - F \leqq C \\[4pt]
\quad \text{or} \quad D_X F(X, p) : X + \langle D_p F(X, p), p \rangle - F \geqq -C.
\end{cases} \tag{3.29}
$$

It is easy to see that any linear F satisfies (3.29). Moreover, (3.28) implies that F can be written as the minmax of linear functions, that is,

$$F(X, p) = \sup_{\alpha \in A} \inf_{\beta \in B} (a_{\alpha,\beta} : X + \langle b_{\alpha,\beta}, p \rangle + h_{\alpha,\beta}),$$

for $A \subset \mathcal{S}^d$ and $B \subset \mathbb{R}^d$ bounded and $a_{\alpha,\beta} \in \mathcal{S}^d$ and $b_{\alpha,\beta} \in \mathbb{R}^d$ such that

$$\sup_{\alpha \in A} \inf_{\beta \in B} [\|a_{\alpha,\beta}\| + |b_{\alpha,\beta}|] < \infty.$$

Since $D_X F(X, p) : X + \langle D_p F(X, P), P \rangle - F$ is formally the derivative, at $\lambda = 1$, of the map $\lambda \to F(\lambda X, \lambda p) - \lambda F(X, P)$, it follows that (3.29) is related to, a uniform in α, β, one sided bound of $\lambda^{-1} h_{\alpha,\beta} - h_{\alpha,\beta}$ in a neighborhood of $\lambda = 1$.

I present next two explanations for the need for an assumption like (3.29). The first is based on considerations from the method of characteristics. The second relies on viscosity solution arguments.

Consider the following first-order versions of (3.14) and (3.27), namely

$$du = F(Du)dt + H(u) \cdot dB, \tag{3.30}$$

and

$$v_t = \tilde{F}(Dv, v, t), \tag{3.31}$$

with

$$\tilde{F}(p, v, t) = \frac{1}{\Phi'(v, t)} F(\Phi'(v, t)p), \tag{3.32}$$

where $d\Phi = H(\Phi) \cdot dB$, and assume that F, H, B and, hence, $\tilde{F}$ are smooth.

The characteristics of the equations in (3.30) and (3.31) are respectively

$$
\begin{cases}
\dot{X} = -DF(P), \\
\dot{P} = H'(U)P\dot{B}, \\
\dot{U} = [F(P) - \langle DF(P), P \rangle] + H(U)\dot{B},
\end{cases}
\tag{3.33}
$$

and

$$
\begin{cases}
\dot{Y} = -D_Q \tilde{F}(Q, V) = -D_P F(\Phi'(V), Q), \\
\dot{Q} = \tilde{F}_V Q = \Phi''(V)(\Phi'(V))^{-2} \\
\qquad Q[D_P F(\Phi'(V)Q), \Phi'(V)Q)) - F(\Phi'(V)Q)], \\
\dot{V} = \tilde{F} - \langle D_Q \tilde{F}(Q, V), Q \rangle = (\Phi'(V))^{-1}[F(\Phi'(V)Q) \\
\qquad - \langle D_P F(\Phi'(V)Q), \Phi'(V)Q \rangle].
\end{cases}
\tag{3.34}
$$

Of course, (3.33) and (3.34) are equivalent after a change of variables. It is, however, clear that some additional hypotheses are needed in order for (3.33), and, hence, (3.34) to have unique solutions. For example, without any additional assumptions, the right hand side of the P-equation in (3.33) may not be Lipschitz continuous in U. On the other hand, the right hand side of the equations for Q and V in (3.34) contain the quantity $\langle D_p F, P \rangle - F$ appearing in (3.29) and an, at least one-sided, Lipschitz condition is necessary to yield existence and uniqueness.

The second explanation is based on the fact that the comparison principle for the pathwise viscosity solutions of (3.14) will follow from the comparison in $BUC(Q_T)$ of viscosity solutions of (3.27). The latter does not follow directly from the existing theory unless something more is assumed; see, for example, Barles [4] and Crandall et al. [17].

This "additional" assumption is that for each $R > 0$, there exists $C_R > 0$ such that, for all $X \in \mathcal{S}^d$, $p \in \mathbb{R}^d$, $v \in [-R, R]$ and $t \in [0, T]$,

$$
\frac{\partial \tilde{F}}{\partial v}(X, p, v, t) \leqq C_R .
\tag{3.35}
$$

A straightforward calculation, using (3.29), yields that, for all X, p, v and t,

$$
\frac{\partial \tilde{F}}{\partial v} = \frac{\Phi''}{(\Phi')^2}[D_X F : (\Phi' X + \Phi'' p \otimes p) + \langle D_p F, \Phi' p \rangle - F]
$$
$$
+ \Phi'(\frac{\Phi''}{\Phi'})' D_X F : p \otimes p;
\tag{3.36}
$$

note that to keep the formula simple, the explicit dependence of F and its derivatives on $\Phi' X + \Phi'' p \otimes p$ is omitted.

It is immediate that $\dfrac{\partial \tilde{F}}{\partial v}$ cannot satisfy (3.35) without an extra assumption on F and control on the size of p. If a bound on p is not available, it is necessary to know that $\Phi'(\Phi''(\Phi')^{-1})' \geq 0$.

The last point that needs explanation is that (3.36) is nonlocal, in the sense that it depends on v through Φ, while (3.29) is a local one, that is Φ plays no role whatsoever. This can be taken care of in the proof by working in uniformly small time intervals, using the local time behavior of Φ and then iterating in time.

The comparison result is stated next.

Theorem 5.1 *Assume* (3.15), (3.17), (3.20), (3.28) *and* (3.29). *For each $T > 0$ and any geometric rough path B in C^α with $\alpha \in (1/3, 1/2]$, there exists a constant $C = C(F, H, B, T) > 0$ such that, if $\overline{v} \in BUC(\overline{Q}_T)$ and $\underline{v} \in BUC(\overline{Q}_T)$ are respectively a subsolution and a supersolution of* (3.14), *then, for all $t \in [0, T]$,*

$$\sup_{x \in \mathbb{R}^d} (\overline{v}(x, t) - \underline{v}(\cdot, t))_+ \leq C \sup_{x \in \mathbb{R}^d} (\overline{v}(\cdot, 0) - \underline{v}(\cdot, 0))_+.$$

Proof To simplify the presentation, it is assumed that F is smooth. The actual proof follows by writing finite differences instead of taking derivatives and using regularizations.

Since $\Phi(v, 0) = v$, (3.20) yields that, for fixed $\delta > 0$, it is possible to choose $h > 0$ so small that

$$\sup_{0 \leq t \leq h} \left[|\Phi(v, t) - v| + |\Phi'(v, t) - 1| + |\Phi''(v, t)| + |\Phi'''(v, t))| \right] \leq \delta. \qquad (3.37)$$

Next consider the new change of variables

$$v = \phi(z) = z + \delta \psi(z) \quad \text{with } \phi' > 0.$$

If v is a subsolution (resp. supersolution) of (3.27), then z is a subsolution (resp. supersolution) of

$$z_t = \tilde{\tilde{F}}(D^2 z, Dz, z), \qquad (3.38)$$

with

$$\tilde{\tilde{F}}(X, p, z) = \frac{1}{\Phi'(\phi(z), t)\phi'(z)} F\big(\Phi'(\phi(z), t)[\phi'(z)X + \phi''(z)(p \otimes p)]$$

$$+ \Phi''(\phi(z), t)(\phi'(z))^2(p \otimes p), \Phi'(\phi(z), t)\phi'(z)p\big). \qquad (3.39)$$

The comparison result follows from the classical theory of viscosity solutions, if there exists $C = C_R > 0$ where $R = \max(\|\bar{v}\|, \|\underline{v}\|)$, such that, for all X, p and z,

$$\frac{\partial}{\partial z}\tilde{\tilde{F}}(X, p, z) \leq C. \tag{3.40}$$

A straightforward calculation yields

$$\frac{\partial}{\partial z}\tilde{\tilde{F}}(X, p, z) = -\frac{(\Phi'\phi')'}{(\Phi'\phi')^2}F + \frac{1}{(\Phi'\phi')}\Big[\langle D_X F, [(\Phi'\phi')'X$$
$$+ [(\Phi'\phi'')' + (\Phi''(\phi')^2)'](p \otimes p)\rangle\big] + \langle D_p F, (\Phi'\phi')'p)\rangle\Big]$$
$$= \frac{(\Phi'\phi')'}{(\Phi'\phi')^2}\big[-F + \langle D_X F, (\Phi'\phi'X + (\Phi'\phi'' + \Phi''(\phi')^2)(p \otimes p)))\rangle$$
$$+ \langle D_p F, \Phi'\phi'p\rangle\big]$$
$$+ \langle D_X F, \Big[\frac{(\Phi'\phi'' + \Phi''(\phi')^2)'}{\Phi'\phi'} - \frac{(\Phi'\phi'' + \Phi''(\phi')^2)(\Phi'\phi')''}{(\Phi'\phi')^2}\Big]$$
$$\times (p \otimes p)\rangle,$$

where, to simplify the notation, the arguments of F, $D_p F$, $D_X^2 F$, Φ', Φ'', ϕ' and ϕ'' are omitted.

In view of (3.15) and (3.28), to obtain (3.35) it suffices to choose ϕ so that

$$\frac{(\Phi'\phi'' + \Phi''(\phi')^2)'}{\Phi'\phi'} - \frac{(\Phi'\phi'' + \Phi''(\phi')^2)(\Phi'\phi')''}{(\Phi'\phi')^2} \leq 0 \tag{3.41}$$

and, if the second inequality in (3.29) holds,

$$\frac{(\Phi'\phi')'}{(\Phi'\phi')^2} \leq 0 \tag{3.42}$$

or, if the first inequality in (3.29) holds,

$$\frac{(\Phi'\phi')'}{(\Phi'\phi')^2} \geq 0. \tag{3.43}$$

Assumption (3.37) and the special choice of ϕ yield that (3.41) is satisfied if $\psi''' \leq -1$, and that (3.42) (resp. (3.43)) holds, if $\psi'' \leq -1$ (resp. $\psi'' \geq 1$). It is a simple exercise to find ψ so that (3.41) and either (3.42) or (3.43) hold in its domain of definition.

The classical comparison result for viscosity solutions then yields that, if $\overline{v}(\cdot, 0) \leq \underline{v}(\cdot, 0)$ on $\mathbb{R}^d$, then $\overline{v} \leq \underline{v}$ on $\mathbb{R}^d \times [0, h]$. The same argument then yields the comparison in $[h, 2h]$, etc.

The existence of the pathwise solutions of (3.14) is based on the stability properties of the "approximating" initial value problem

$$u_t^\varepsilon = F(D^2 u^\varepsilon, D u^\varepsilon) + \sum_{i=1}^m H^i(u^\varepsilon)\dot{B}_i^\varepsilon \quad \text{in } Q_T \quad u^\varepsilon(\cdot, 0) = u_0^\varepsilon, \tag{3.44}$$

where $u_0^\varepsilon \in BUC(\mathbb{R}^d)$, and

$$\begin{cases} B^\varepsilon = (B_1^\varepsilon, \ldots, B_m^\varepsilon) \in C^1([0, \infty); \mathbb{R}^m), \\ \text{and, for all } T > 0, \text{ as } \varepsilon \to 0, \ B^\varepsilon \to B \text{ in the rough path metric.} \end{cases} \tag{3.45}$$

Note that, if $m = 1$, the assumption in (3.45) can be reduced to $B^\varepsilon \to B$ uniformly on $[0, T]$.

The existence result is stated next.

Theorem 5.2 *Assume* (3.15), (3.17), (3.20), (3.28) *and* (3.29) *and fix* $T > 0$. *Let* $(\zeta^\varepsilon)_{\varepsilon>0}$ *and* $(\xi^\eta)_{\eta>0}$ *satisfy* (3.45) *and consider the solutions* $u^\varepsilon, v^\eta \in BUC(\overline{Q}_T)$ *of* (3.44) *with initial datum* u_0^ε *and* v_0^η *respectively. If, as* $\varepsilon, \eta \to 0$, $u_0^\varepsilon - v_0^\eta \to 0$ *uniformly on* $\mathbb{R}^d$, *then, as* $\varepsilon, \eta \to 0$, $u^\varepsilon - v^\eta \to 0$ *uniformly on* $\overline{Q}_T$. *In particular, each family* $(u^\varepsilon)_{\varepsilon>0}$ *is Cauchy in* $\overline{Q}_T$. *Hence, it converges uniformly to* $u \in BUC(\overline{Q}_T)$, *which is a pathwise viscosity solution to* (3.14). *Moreover, all approximate families converge to the same limit.*

The proof of Theorem 5.2 follows from the comparison between subsolutions and supersolutions of (3.27) for different approximations $(\zeta^\varepsilon)_{\varepsilon>0}$ and $(\xi^\eta)_{\eta>0}$. Since a similar theorem will be proved later when dealing with nonlinear gradient dependent H, the proof is omitted.

Finally the next result is about the Lipschitz continuity of the solutions. Its proof is based on the comparison estimate obtained in Theorem 5.1 and, hence, it is omitted.

Proposition 5.1 *Fix* T *and assume* (3.15), (3.17), (3.20), (3.28) *and* (3.29) *and let* $u \in BUC(\overline{Q}_T)$ *be the unique pathwise solution to* (3.14) *for* $u_0 \in C^{0,1}(\mathbb{R}^d)$. *Then* $u(\cdot, t) \in C^{0,1}(\mathbb{R}^d)$ *for all* $t \in [0, T]$, *and there exists* $C = C(F, H, B, T) > 0$ *such that, for all* $t \in [0, T]$, $\|Du(\cdot, t)\| \leq C$.

Of course Proposition 5.1 is immediate if F and H do not depend on x. The point is that the claim holds in full generality.

3.6 The Extension Operator for Spatially Homogeneous First-Order Problems

The object here is the study the space homogeneous Hamilton-Jacobi equation

$$du = \sum_{i=1}^{m} H^i(Du) \cdot dB_i \text{ in } Q_\infty \quad u(\cdot, 0) = u_0, \tag{3.46}$$

with $B = (B_1, \ldots, B_m) \in C_0([0, \infty); \mathbb{R}^m) = \{B \in C([0, \infty); \mathbb{R}^m) : B(0) = 0\}$.

The aim is to show that, if $H = (H_1, \ldots, H_m) \in C_{loc}^{1,1}(\mathbb{R}^d; \mathbb{R}^m)$, the solution operator of (3.46) with smooth paths has a unique extension to the set of continuous paths.

The result is stated next.

Theorem 6.1 *Fix $H \in C_{loc}^{1,1}(\mathbb{R}^d; \mathbb{R}^m)$, $u_0 \in BUC(\mathbb{R}^d)$ and $B \in C_0([0, \infty); \mathbb{R}^m)$. There exists a unique $u \in BUC(\overline{Q}_\infty)$ such that, for any families $(B^\varepsilon)_{\varepsilon>0}$ in $C_0([0, \infty); \mathbb{R}^m) \cap C^1([0, \infty); \mathbb{R}^m)$ and $(u_0^\varepsilon)_{\varepsilon>0}$ in $BUC(\mathbb{R}^d)$ which approximate respectively B in $C([0, \infty); \mathbb{R}^m)$ and u_0 in $BUC(\mathbb{R}^d)$, if $u^\varepsilon \in BUC(\overline{Q}_\infty)$ is the unique viscosity solution of $du^\varepsilon = \sum_{i=1}^{m} H^i(Du^\varepsilon) \cdot dB_i^\varepsilon$ in Q_∞ and $u^\varepsilon(\cdot, 0) = u_0^\varepsilon$, then, as $\varepsilon \to 0$, $u^\varepsilon \to u$ uniformly in $\overline{Q}_\infty$.*

This unique limit will be also characterized later as the unique pathwise solution of (3.46).

The claim follows from the next theorem which asserts that, if the family of smooth paths $(B^\varepsilon)_{\varepsilon>0}$ and initial data $(u_0^\varepsilon)_{\varepsilon>0}$ are Cauchy in $C_0([0, \infty); \mathbb{R})$ and $BUC(\mathbb{R}^d)$ respectively, then the solutions $u^\varepsilon \in BUC(\overline{Q}_\infty)$ of

$$du^\varepsilon = \sum_{i=1}^{m} H^i(Du^\varepsilon) \cdot dB_i^\varepsilon \text{ in } Q_\infty \quad u^\varepsilon(\cdot, 0) = u_0^\varepsilon, \tag{3.47}$$

form a Cauchy family in $BUC(\overline{Q}_\infty)$.

Theorem 6.2 *Fix $H \in C_{loc}^{1,1}(\mathbb{R}^d; \mathbb{R}^m)$, assume that $\zeta^\varepsilon, \xi^\eta \in C_0([0, \infty); \mathbb{R}^m) \cap C^1([0, \infty); \mathbb{R}^m)$ and $u_0^\varepsilon, v_0^\eta \in BUC(\mathbb{R}^d)$ are such that, as $\varepsilon, \eta \to 0$, $\zeta^\varepsilon - \xi^\eta \to 0$ in $C([0, \infty); \mathbb{R}^m)$ and $u_0^\varepsilon - v_0^\eta \to 0$ in $BUC(\mathbb{R}^d)$. If $u^\varepsilon, v^\eta \in BUC(\overline{Q}_\infty)$ are the viscosity solutions of (3.47) with respective paths and initial condition $(\zeta^\varepsilon, u_0^\varepsilon)$, (ξ^η, v_0^η), then, as $\varepsilon, \eta \to 0$, $u^\varepsilon - v^\eta \to 0$ in $BUC(\overline{Q}_\infty)$.*

Proof A simple density argument implies that it is enough to consider $u_0^\varepsilon, v_0^\eta \in C^{0,1}(\mathbb{R}^d)$. Since H is independent of x, it follows that $u^\varepsilon, v^\eta \in C^{0,1}(\overline{Q}_\infty)$ and, for all $t > 0$, $\max(\|Du^\varepsilon(\cdot, t)\|, \|Dv^\eta(\cdot, t)\|) \leq \max(\|Du_0^\varepsilon\|, \|Dv_0^\eta\|)$. Hence, without any loss of generality, it may be assumed that $H \in C^{1,1}(\mathbb{R}^d)$.

Notice that, for each ε and η, u^ε and v^η are actually also Lipschitz continuous in time. The Lipschitz constants in time, however, depend on $|\dot{\zeta}^\varepsilon|$ and $|\dot{\xi}^\eta|$, and, hence, are not bounded uniformly in ε, η. This is one of the main reasons behind the difficulties here.

To keep the arguments simple, it is also assumed that u_0^ε and v_0^ε, and, hence, u^ε and v^η are periodic in the unit cube $\mathbb{T}^d$. This simplification allows not be concerned about infinity, and, more precisely, the possibility that the suprema below are not achieved. The periodicity can be eliminated as an assumption by introducing appropriate penalizations at infinity that force the sup's to be actually maxima.

Finally, from now on I assume that $m = 1$. This is only done to keep the notation simpler. Since the equation does not depend on the space variable, the extension to $m > 1$ is immediate

The general strategy in the theory of viscosity solutions to show that, as $\varepsilon, \eta \to 0$, $u^\varepsilon - v^\eta \to 0$ in $\overline{Q}_\infty$, is to double the variables and consider the function

$$z(x, y, t) = u^\varepsilon(x, t) - v^\eta(y, t),$$

which satisfies the so-called "doubled" initial value problem

$$z_t = H(D_x z)\dot{\zeta}^\varepsilon - H(-D_y z)\dot{\xi}^\eta \text{ in } \mathbb{R}^d \times \mathbb{R}^d \times (0, \infty),$$

$$z(x, y, 0) = u_0^\varepsilon(x) - v_0^\eta(y). \tag{3.48}$$

The assumptions on u_0^ε and v_0^η imply that, for $\lambda > 0$, there exists $\theta(\lambda) > 0$ such that $\theta(\lambda) \to 0$ as $\lambda \to \infty$ and

$$z_0(x, y) \leq \lambda |x - y|^2 + \theta(\lambda) + \sup(u_0^\varepsilon - v_0^\eta). \tag{3.49}$$

To conclude, it suffices to show that there exists $U^{\varepsilon,\eta,\lambda} : \mathbb{R}^d \times \mathbb{R}^d \times [0, T] \to \mathbb{R}$ such that, as $\varepsilon, \eta \to 0$ and $\lambda \to \infty$,

$$U^{\varepsilon,\eta,\lambda}(x, x, t) \to 0 \text{ uniformly in } \overline{Q}_\infty \text{ and } z \leq U^{\varepsilon,\eta,\lambda} \text{ in } \mathbb{R}^d \times \mathbb{R}^d \times [0, \infty).$$

It would then follow that

$$\lim_{\varepsilon,\eta \to 0} \sup_{(x,t) \in \overline{Q}_\infty} z(x, x, t) = 0,$$

which is one part of the claim. The other direction is proved similarly.

Again, as in the general "non rough" theory, it is natural to try to show that there exists, for some $C > 0$ and $a(\lambda) > 0$ such that $a(\lambda) \to 0$ as $\lambda \to \infty$, a supersolution of (3.48) of the form

$$U^{\varepsilon,\eta,\lambda}(x, y, t) = C\lambda |x - y|^2 + a(\lambda).$$

This is, however, the main difficulty, since both the C and $a(\lambda)$ will depend on $|\dot{\zeta}^{\varepsilon}|$ and $|\dot{\xi}^{\eta}|$, which are not bounded uniformly in ε and η.

The first new idea to circumvent this difficulty is to find sharper upper bounds by considering the solution $\phi^{\lambda,\varepsilon,\eta}(x, y, t)$ of

$$\begin{cases} \phi_t^{\lambda,\varepsilon,\eta} = H(D_x\phi^{\lambda,\varepsilon,\eta})\dot{\zeta}^{\varepsilon} - H(-D_y\phi^{\lambda,\varepsilon,\eta})\dot{\xi}^{\eta}\text{ in } \mathbb{R}^d \times \mathbb{R}^d \times (0, \infty), \\ \phi^{\lambda,\varepsilon,\eta}(x, y, 0) = \lambda|x - y|^2, \end{cases} \tag{3.50}$$

which, in view of the spatial homogeneity of H and the fact that $\phi^{\lambda,\varepsilon,\eta}(\cdot, \cdot, 0)$ depends on $x - y$, is given by

$$\phi^{\lambda,\varepsilon,\eta}(x, y, t) = \Phi^{\lambda,\varepsilon,\eta}(x - y, t),$$

with $\Phi^{\lambda,\varepsilon,\eta}$ solving the initial value problem

$$\Phi_t = H(D\Phi)(\dot{\zeta}^{\varepsilon} - \dot{\xi}^{\eta}) \text{ in } Q_{\infty} \quad \Phi(\cdot, 0) = \lambda|z|^2, \tag{3.51}$$

which is well-posed for each ε, η.

The classical comparison principle for viscosity solutions yields, that for all $x, y \in \mathbb{R}^d, t \geq 0, \lambda > 0$ and ε, η,

$$z(x, y, t) \leq \phi^{\lambda,\varepsilon,\eta}(x, y, t) + \max_{x,y\in\mathbb{R}^d} (z(x, y, 0) - \lambda|x - y|^2),$$

and, hence, for all $x \in \mathbb{R}^d$ and $t \geq 0$,

$$u^{\varepsilon}(x, t) - u^{\eta}(x, t) \leq \phi^{\lambda,\varepsilon,\eta}(x, x, t) + \theta(\lambda) + \sup(u_0^{\varepsilon} - v_0^{\eta}).$$

To conclude, it is necessary to show that there exists $\Theta(\lambda) > 0$ such that $\lim_{\lambda\to\infty} \Theta(\lambda) = 0$ and

$$\overline{\lim_{\varepsilon,\eta\to 0}} \sup_{x\in\mathbb{R}^d} \phi^{\lambda,\varepsilon,\eta}(x, x, t) \leq \Theta(\lambda),$$

a fact that a priori may present a problem since the "usual" viscosity theory yields the existence of $\phi^{\lambda,\varepsilon,\eta}$ but not the desired uniform estimate.

Here comes the second new idea, namely, to use the characteristics to construct a smooth solution $\phi^{\lambda,\varepsilon,\eta}$, at least for a small time, which, of course, depends on ε and η. The aim then will be to show that, as $\varepsilon, \eta \to 0$, the interval of existence becomes of order one.

The characteristics of (3.50) are

$$
\begin{cases}
\dot{X} = -DH(P)\dot{\zeta}^{\varepsilon} \quad \dot{Y} = -DH(Q)\dot{\xi}^{\eta}, \\[1mm]
\dot{P} = 0 \quad \dot{Q} = 0, \\[1mm]
\dot{U} = \big(H(P) - \langle D_p H(P), P\rangle\big)\dot{\zeta}^{\varepsilon} - (H(Q) \\[1mm]
\quad - \langle DH(Q), Q\rangle)\dot{\xi}^{\eta}, \\[1mm]
X(0) = x \, Y(0) = y \, P(0) = Q(0) = 2\lambda(x-y) \, U(0) = \lambda|x-y|^2.
\end{cases}
\tag{3.52}
$$

Note that to keep the equations simpler the system is written for $Q(x,t) = -Dv^{\eta}(Y(t), t)$ instead of $Dv^{\eta}(Y(t), t)$. Similarly I ignore the dependence on λ, ε and η.

The method of characteristics provides a classical solution of (3.50) for some short time $T^*_{\varepsilon,\eta,\lambda}$ as long the map $(x, y) \to (X(t), Y(t))$ is invertible.

The special structure of (3.52) yields that, for all $t \geq 0$,

$$
P(t) = Q(t) = 2\lambda(x - y),
$$

and

$$
(X - Y)(t) = (x - y) - DH(2\lambda(x - y))(\zeta^{\varepsilon}(t) - \xi^{\eta}(t)).
$$

To simplify the notation, let $z = x - y$ and $Z(t) = X(t) - Y(t)$, in which case the last equation can be rewritten as

$$
Z(x, t) = z - DH(2\lambda z)(\zeta^{\varepsilon}(t) - \xi^{\eta}(t)).
$$

Note that $z \to Z(z, t)$ is the position characteristic associated with the simplified initial value problem (3.51), and, in the problem at hand, is the only map that needs to be inverted. Since

$$
D_z Z(z, t) = I + 2\lambda D^2 H(2\lambda z)(\zeta^{\varepsilon}(t) - \xi^{\eta}(t)),
$$

it follows that the map $z \mapsto Z$ is invertible as long as

$$
\sup_{t \in [0,T]} |(\zeta^{\varepsilon}(t) - \xi^{\eta}(t))| \, \|D^2 H\|_\infty < (2\lambda)^{-1}.
\tag{3.53}
$$

This is, of course, possible for any T and λ provided ε and η are small, since, as $\varepsilon, \eta \to 0$, $\zeta^{\varepsilon} - \xi^{\eta} \to 0$ in $C([0, \infty))$.

The above estimates depend on having $H \in C^2$. Since the interval of existence depends only on the $C^{1,1}$ bounds of H, it can be assumed that H has this regularity and then conclude introducing yet another level of approximations.

It now follows that

$$\phi^{\lambda,\varepsilon,\eta}(X(t), Y(t), t) = \lambda |x - y - DH(2\lambda(x - y))(\zeta^\varepsilon(t) - \xi^\eta(t))|^2$$
$$+ [H(2\lambda(x - y)) - \langle DH(2\lambda(x - y)), 2\lambda(x - y)\rangle]$$
$$\times (\zeta^\varepsilon(t) - \xi^\eta(t)).$$

Moreover, it follows from (3.50), that there exists $C > 0$ depending only on $\|H\|_{C^{1,1}}$ such that

$$|\phi^{\lambda,\varepsilon,\eta}(X(t), Y(t), t) - \lambda |x - y|^2| \leq \lambda C \sup_{0 \leq t \leq T} |\zeta^\varepsilon(t) - \xi^\eta(t)|.$$

Returning to the x, y variables, the above estimate gives that, for each fixed $\lambda > 0$ and $T > 0$ and as $\varepsilon, \eta \to 0$,

$$\sup_{\substack{x,y \in \mathbb{R}^d \\ t \in [0,T]}} (\phi^{\lambda,\varepsilon,\eta}(x, y, t) - \lambda |x - y|^2) \to 0.$$

3.6.1 A Summary of the General Strategy

Since the approach and the arguments of the proof above are used several times in the theory and the notes, it is helpful to present a brief summary of the main points.

The conclusion of the theorem is that it is possible to construct, using the classical theory of viscosity solutions, a (unique) $u \in BUC(\overline{Q}_\infty)$, which is the candidate for the solution of (3.46) for any B continuous as long as $H \in C^{1,1}_{\text{loc}}$.

The key technical step in the proof was the fact that, if, as $\varepsilon, \eta \to 0$, $\zeta^\varepsilon - \xi^\eta \to 0$ in $C([0, \infty))$, then, for each $\lambda > 0$ and $T > 0$, as $\varepsilon, \eta \to 0$

$$\sup_{\substack{z \in \mathbb{R}^d \\ t \in [0,T]}} \left(v^\lambda_{\varepsilon,\eta}(z, t) - \lambda |z|\right) \to 0,$$

where $v = v^\lambda_{\varepsilon,\eta}$ is the solution of the initial value problem

$$v_t = H(Dv)(\dot{\zeta}^\varepsilon - \dot{\xi}^\eta) \text{ in } Q_\infty \quad v(z, 0) = \lambda |z|. \tag{3.54}$$

The proof presented earlier used $\lambda |z|^2$ as initial condition in (3.54). It is not hard to see, however, that the same argument will work for initial datum $\lambda |z|$. Indeed it is enough to consider regularizations like $(\delta + |z|^2)^{1/2}$ and to observe that the estimate on $u^\varepsilon(\cdot, t) - v^\eta(\cdot, t)$ is uniform on δ in view of the assumption that $H \in C^{1,1}$. The conclusion for $\lambda |z|$ then follows from the stability properties of viscosity solutions.

The result about the extension can be summarized as follows.

Given sufficiently regular paths $B_n = (B_{1,n}, \ldots, B_{m,n}) : [0, \infty) \to \mathbb{R}^m$, $H = (H^1, \ldots, H^m) \in C(\mathbb{R}^d; \mathbb{R}^m)$ and $\lambda > 0$, let $v_{n,t}^\lambda \in \mathrm{BUC}(\overline{Q}_\infty)$ be the solution of

$$v_{n,t}^\lambda = \sum_{i=1}^m H^i(Dv_n^\lambda)\dot{B}_{n,i} \text{ in } Q_\infty \quad v_n^\lambda(z, 0) = \lambda|z|. \tag{3.55}$$

The following theorem gives a sufficient condition for the existence of the extension.

Theorem 6.3 *If for every $B_n \in C_0([0, \infty); \mathbb{R}^m) \cap C^1([0, \infty); \mathbb{R}^m)$ such that, as $n \to \infty$, $B_n \to 0$ in $C([0, \infty); \mathbb{R}^m)$, and $T > 0$, the solution $v_{n,t}^\lambda$ of (3.55) has the property*

$$\lim_{n \to \infty, \lambda \to \infty} \sup_{(z,t) \in \mathbb{R}^d \times [0,T]} \left(v_n^\lambda(z, t) - \lambda|z| \right) = 0 \tag{3.56}$$

then there is an extension.

3.7 Pathwise Solutions for Equations with Non-smooth Hamiltonians

It is important to extend the class of Hamiltonians for which the solution operator of (3.50) with smooth paths has an extension. The assumption that $H \in C_{\mathrm{loc}}^{1,1}$ is rather restrictive. For example, the typical Hamiltonian $H(p) = |p|$ arising in front propagation does not have this regularity.

The aim of this section is to provide a necessary and sufficient condition on H to have an extension as well as to investigate if it is possible to assume less in H by "increasing" the regularity of the paths, while still covering many cases of interest.

An important question and tool in this direction is to understand/control the cancellations arising from the oscillations of the paths. And for this, it is useful to investigate if there are some formulae for the solutions in the presence of sign changing driving signals.

3.7.1 Formulae for Solutions

The simplest possible formulae for the solutions of

$$u_t = H(Du) \text{ in } Q_T \quad u(\cdot, 0) = u_0, \tag{3.57}$$

are the well known Lax-Oleinik and Hopf formula which require convexity for H and u_0 respectively. In Appendix the reader can find an extensive discussion about these formulae, their relationship and possible extensions.

When H is convex, the Lax-Oleinik formula is

$$u(x, t) = \sup_{y \in \mathbb{R}^d} \left[u_0(y) - t H^*(\frac{y - x}{t}) \right], \tag{3.58}$$

where, given a convex function $w : \mathbb{R}^d \to \mathbb{R}$, $w^*(q) = \sup[\langle q, p \rangle - w(p)]$ is its Legendre transform.

The Hopf formula, which is the "dual" of the Lax-Oleinik one, says that, if u_0 is convex, then

$$u(x, t) = \sup_{p \in \mathbb{R}^d} \left[\langle p, x \rangle + t H(p) - u_0^*(p) \right]. \tag{3.59}$$

In general, neither formula extends to the solutions of

$$u_t = H(Du)\dot{\xi} \text{ in } Q_T \quad u(\cdot, 0) = u_0, \tag{3.60}$$

except in time intervals where the path is either increasing or decreasing in which case it is possible to change time.

Indeed, if $H \in C(\mathbb{R}^d)$, $\xi \in C^1$ and u_0 convex, the natural extension of (3.59) should be

$$\sup_{p \in \mathbb{R}^d} \left[\langle p, x \rangle + \xi(t) H(p) - u_0^*(p) \right].$$

The formula above is a subsolution, as the "sup" of solutions $\langle p, x \rangle + B(t) H(p) - u_0^*(p)$, but, in general, is not a solution of (3.60). The heuristic reason is that shocks are not reversible.

For example, if $H(p) = |p|$ and $u_0(x) = |x|$, then

$$\sup_{p \in \mathbb{R}} (px + \xi(t)|p| - |\cdot|^*(p)) = (|x| + \xi(t))_+.$$

On the other hand, the following is true.

Proposition 7.1 *The unique viscosity solution of* (3.60) *with* $\xi \in C^1$, $\xi(0) = 0$, $H(p) = |p|$ *and* $u_0(x) = |x|$ *is*

$$u(x, t) = \max \left[(|x| + \xi(t))_+, (\max_{0 \leq s \leq t} \xi(s))_+ \right]. \tag{3.61}$$

Although the regularity of ξ is used in the proof of (3.61), the actual formula extends by density to arbitrary continuous ξ's.

It is possible to give two different proofs for (3.61). One is based on dividing $[0, T]$ into intervals where $\dot{\xi}$ is positive or negative and iterating the Hopf formula. The second is a direct justification that (3.61) is the viscosity solution to the problem. The details can be found in [71].

From the analysis point of view, the difficulty is related to the fact that when the signal changes sign, the convexity properties of Hamiltonian also change. This leads to the possibility of using the formulae provided by the interpretation of the solution as the value function of a two-player, zero-sum differential games, which I briefly recall next.

Assume that

$$H(p) = \sup_{\alpha \in A} \inf_{\beta \in B} (\langle f(\alpha, \beta), p \rangle + h(\alpha, \beta)),$$

where, for simplicity, the sets A and B are assumed to be compact subsets of $\mathbb{R}^p$ and $\mathbb{R}^q$ and $f : A \times B \to \mathbb{R}^d$ and $h : A \times B \to \mathbb{R}$ are bounded; note that any Lipschitz continuous Hamiltonian H can be written as a max/min of linear maps.

It was shown in Evans and Souganidis [26] that the unique viscosity solution of the initial value problem

$$u_t = H(Du) \text{ in } Q_T \quad u(\cdot, 0) = u_0,$$

admits the representation

$$u(x, t) = \sup_{\alpha \in \Gamma(T-t)} \inf_{z \in N(T-t)} \left\{ \int_{T-t}^{T} h(\alpha[z](s), z(s))ds + u(x(T)) \right\},$$

where $N(T - t)$ is the set of controls $z : [T - t, T] \to B$, $\Gamma(T - t)$ is the set of nonanticipating strategies which map B-valued controls to A-valued ones, and $(x(s))_{s \in [T-t,T]}$ is the solution of the ode

$$\dot{x} = f(\alpha[z](s), z(s)) \quad x(T - t) = x.$$

An attempt to extend this formula to (3.59) meets immediately difficulties. Assume, for example, that $\xi \in C^1$. Then

$$H(p)\dot{\xi}(t) = H(p)\dot{\xi}(t)_+ - H(p)\dot{\xi}(t)_-,$$

and it easy to check that, in general, it is not possible to find compact sets C and D, vectors $f_\pm : C \times D \to \mathbb{R}^d$ and scalars $h_\pm : C \times D \to \mathbb{R}$ such that

$$H(p)\dot{\xi}(t) = \sup_{c \in C} \inf_{d \in D} \langle ((f_+(c, d)\dot{\xi}(t)_+ + f_-(c, d)\dot{\xi}(t)_-), p \rangle$$

$$+ (h_+(c, d)\dot{\xi}(t)_+ + h_-(c, d)\dot{\xi}(t)_-)).$$

Of course, if the above holds, then the solution of (3.59) is given by the formula

$$u(x, t) = \sup_{\alpha \in \Gamma(T-t)} \inf_{z \in N(T-t)} \int_{T-t}^{T} h_+(\alpha[z](s), z(s))\dot{\xi}(T-s)_+))$$

$$+ h_-(\alpha[z](s), z(s)\dot{\xi}(T-s)_-)ds + u(x(T)),$$

where for the control $z \in N(T-t)$ and strategy $a \in \Gamma(T-t)$, $x(s)_{s \in [T-t, T]}$ solves

$$\dot{x}(s) = f_+(\alpha[z](s), z(s))\dot{\xi}(T-s)_+ + f_-(\alpha[z](s), z(s))\dot{\xi}(T-s)_- \quad x(T-t) = x.$$

3.7.2 Pathwise Solutions for Nonsmooth Hamiltonians

When H is less regular than in Theorem 6.2, it is also possible to prove the unique extension property for the solution operator for smooth paths, but the argument is different and does not rely on inverting the characteristics. It is, however, possible to use the general strategy summarized in Theorem 6.3 to identify the conditions on H that will allow for the extension to exist for the initial value problem

$$du = \sum_{i=1}^{m} H^i(Du) \cdot dB_i \text{ in } Q_T \quad u(\cdot, 0) = u_0. \tag{3.62}$$

The main result is stated next.

Theorem 7.1 *The solution operator of (3.62) with smooth paths has a unique extension to continuous paths if and only if H^i is the difference of two convex functions for $i = 1, \ldots, m$.*

Identifying the class of Hamiltonians H which can be as written as the difference of two convex functions is a difficult question.

When $d = 1$, a necessary and sufficient condition for H to be the difference of two convex functions is that $H' \in$ BV. Indeed in this case, in the sense of distributions, $H'' = H_1'' - H_2''$ with H_1'' and H_2'' nonnegative distributions and, hence, locally bounded measures. Conversely, if $H' \in$ BV, then $H'' = (H'')_+ - (H'')_-$.

When $d \geq 2$, if $H = H_1 - H_2$ with H_1, H_2 convex, then, as above, $DH \in$ BV. The converse is, however, false. Functions with gradients in BV may not have directional derivatives at every point, while differences of convex functions do.

Finally, if $H \in C^{1,1}$, then H is clearly the difference of convex functions. Indeed since, for some $c > 0$, $D^2 H \geq -2cI$, then $H = H_1 - H_2$ with $H_1(p) = H(p) + c|p|^2$ and $H_2(p) = c|p|^2$.

The proof of Theorem 7.1 is divided in several parts and requires a number of ingredients which are developed next.

Proposition 7.2 *Assume that the extension operator exists for all continuous paths. Then H must be the difference of two convex functions.*

Proof In what follows, I assume for simplicity that $m = 1$ and the problem is set in Q_1.

The necessity follows from the criterion summarized in Theorem 6.3. Since the extension must hold for any continuous path, it is possible to construct a sequence of paths satisfying the assumptions of Theorem 6.3 such that (3.56) implies that H must be the difference of two convex functions.

Consider a partition of $[0, 1]$ of $2n$ intervals of length $1/2n$ and define the piecewise linear paths $B_n : [0, 1] \to \mathbb{R}$ with slope $\dot{B}_n = \pm\mu$ and, for definiteness, assume that $\dot{B}_n = \mu$ in the first interval. It follows that

$$\sup_{t \in [0,1]} |B_n(t)| \leq \frac{\mu}{2n} \quad \text{and, if} \quad \mu/2n \to 0, \quad B_n \to 0 \text{ in } C([0, \infty)).$$

Fix $\lambda > 0$ and let $v_n^\lambda : \overline{Q}_1 \to \mathbb{R}$ be the solution of

$$v_{n,t}^\lambda = H(Dv_n^\lambda)\dot{B}_n \text{ in } Q_1 \quad v_n^\lambda(z, 0) = \lambda|z|. \tag{3.63}$$

Assume that, for some $\delta > 0$, $\mu = 2n\delta$. The claim follows if it is shown that the v_n^λ's blow up, as $n \to \infty$, if H is not the difference of two convex functions in a ball of radius λ.

Recall that, in each time interval of length $1/2n$, the equation in (3.63) are

$$\text{either} \quad v_{n,t} = 2n\delta H(Dv_n) \quad \text{or} \quad v_{n,t} = -2n\delta H(Dv_n),$$

or, after rescaling,

$$\text{either} \quad U_{n,t} = \delta H(DU_n) \text{ in } \overline{Q}_1 \quad \text{or} \quad V_{n,t} = -\delta H(DV^n) \text{ in } \overline{Q}_1;$$

here, for notational simplicity, I omit the explicit dependence on λ.

The V_n's are constructed by a repeated iteration of Hopf's formula. This procedure yields sequences $(V_{2k+1}^*)_{k=0}^\infty$ and $(V_{2k}^*)_{k=0}^\infty$ which, as $k \to \infty$, either blow up or converge, uniformly in $\overline{B}_\lambda$, to $\overline{V}_1^*$ and $\overline{V}_2^*$ respectively.

In the latter case, it follows that

$$\overline{V}_2^* = (\overline{V}_1^* - \delta H)^{**} \quad \text{and} \quad \overline{V}_1^* = (\overline{V}_2^* + \delta H)^{**},$$

and, therefore,

$$\delta H = \overline{V}_1^* - \overline{V}_2^*,$$

which yields that H is the difference of two convex functions.

If the sequences $(V_{2k+1}^*)_{k=0}^{\infty}$ and $(V_{2k}^*)_{k=0}^*$ blow up, then a diagonal argument, in the limit $\delta \to 0$, shows that (3.56) cannot hold.

Indeed, since, for each $\delta > 0$ and as $k \to \infty$, $V_{2k+1}^* \to -\infty$ and $V_{2k}^* \to -\infty$ in $\overline{B}_\lambda$, choosing $\delta = 1/m$ along a sequence $k_m \to \infty$ yields $V_{2k_m}^* \leq -1$.

Going back to the original scaled problem, it follows that $v_{k_m} \leq -1$, while $B_{k_m} \to 0$ in $C([0, \infty))$ and $v_{k_m}(0, 0) = 0$.

The next step is to show that, if H is the difference of two convex functions, then there exists a unique extension of the solution operator with B smooth to the class of merely continuous B.

The main difficulties are the lack of differentiability of B and how to control the oscillation of the solutions with respect to time. This was actually already exploited in the proof of Proposition 7.2. For the sufficiency, it is important to obtain a more explicit estimate.

Controlling the cancellations due to the oscillations of the paths is very much related to the irreversibility of the equations due to the formation of shocks. "Some memory", however, remains resulting in cancellations taking place as it can be seen in the next result.

Consider the initial value problems

$$u_t = \sum_{i=1}^{m} H^i(Du)\dot{B}^i \text{ in } Q_T \quad u(\cdot, 0) = u_0, \tag{3.64}$$

and

$$v_t^i = H^i(Dv^i)\dot{B}^i \text{ in } Q_T \quad v^i(\cdot, 0) = v_0^i, \tag{3.65}$$

where, for each $i = 1, \ldots, m$,

$$H^i \in C(\mathbb{R}^d), \quad B^i \in C^1([0, \infty)) \quad \text{and} \quad u_0, v_0^i \in BUC(\mathbb{R}^d). \tag{3.66}$$

It is known that both initial value problems in (3.64) and (3.65) have unique viscosity solutions. In the statement below, $S_{H^i}(t)v$ is the solution of (3.65) with $\dot{B}^i \equiv 1$ at time $t > 0$.

Theorem 7.2 *Assume, in addition to (3.66), that, for each $i = 1, \ldots, m$, H^i is convex and $DH^i(p_i)$ exists for some $p_i \in \mathbb{R}^d$, and let $u \in BUC(\overline{Q}_T)$ be the* *viscosity solution of (3.64). Then, for all* $(x, t) \in \overline{Q}_T$,

$$
\prod_{i=1}^{m} S_{H^i}(-\min_{0 \leq s \leq t} B^i(s)) u_0 \left(x + \sum_{i=1}^{m} DH^i(p_i)(\min_{0 \leq s \leq t}(B^i(s) - B^i(t))) \right)
$$

$$
+ \sum_{i=1}^{m} H^i(p_i)(\min_{0 \leq s \leq t}(B^i(s) - B^i(t))) \leq u(x, t)
$$

$$
\leq \prod_{i=1}^{m} S_{H^i}(\max_{0 \leq s \leq t} B^i(s)) u_0 \left(x + \sum_{i=1}^{m} DH^i(p_i)(\min_{0 \leq s \leq t}(B^i(s) - B^i(t))) \right)
$$

$$
- \sum_{i=1}^{m} H^i(p_i)(\max_{0 \leq s \leq t}(B^i((s) - B^i(t))).
$$

(3.67)

The proof of the estimated above, which is complicated, is based on repeated use of the Lax-Oleinik and Hopf formulae. The details can be found in [71].

The following remark is useful for what follows.

If, in addition,

$$
\min H^i = 0 \quad \text{for} \quad i = 1, \dots, m,
$$

then the claim can be simplified considerably to read

$$
\prod_{i=1}^{m} S_{H^i}(\max_{0 \leq s \leq t} B_s^{i,-}) u_0(x) \leq u(x, t) \leq \prod_{i=1}^{m} S_{H^i}(\max_{0 \leq s \leq t} B_s^{i,+}) u_0(x). \quad (3.68)
$$

The bounds in (3.67) are sharp. Indeed, recall that in the particular case

$$
H(p) = |p|, \quad u_0(x) = |x| \quad \text{and} \quad m = 1,
$$

it was already claimed that the solution of (3.64) is given by

$$
u(x, t) = \max \left[(|x| + B(t))_+, \max_{0 \leq s \leq t} B(s))_+ \right].
$$

Evaluating the formula at $x = 0$ yields that the upper bound in Proposition 7.2 is sharp, since, in this case,

$$
S_H(\max_{0 \leq s \leq t} B(s)) u_0(0) = \max_{0 \leq s \leq t} B(s) \text{ and } \max(B_+(t), \max_{0 \leq s \leq t} B(s)) = \max_{0 \leq s \leq t} B(s).
$$

Using Theorem 7.2 it is now possible to prove the sufficient part of Theorem 7.2.

Proposition 7.3 *Assume that, for each other* $i = 1, \ldots, m$, $H^i \in C(\mathbb{R}^d)$ *is the difference of two convex functions. Then the solution operator of* (3.63) *on the class of smooth paths has a unique extension to the space of continuous paths.*

Proof The proof is based again on Theorem 6.3. Fix $\lambda > 0$, let $B_n = (B_1^n, \ldots B_m^n) \in C_0([0, \infty), \mathbb{R}^m) \cap C^1((0, \infty); \mathbb{R}^m)$ be sequence of signals such that, as $n \to \infty$, $B^n \to 0$ in $C([0, \infty), \mathbb{R}^m)$, and consider the solution $\phi^n \in \mathrm{BUC}(\overline{Q}_\infty)$ of

$$\phi_t^n = \sum_{i=1}^m H^i(D\phi^n)\dot{B}_i^n \text{ in } Q_\infty \quad \phi^n(z, 0) = \phi_0(z) = \lambda|z|. \tag{3.69}$$

It shown here that the assumption on H yields, for each $\lambda > 0$ and $T > 0$,

$$\lim_n \max_{(z,t)\in\overline{Q}_T} (\phi^n(z, t) - \lambda|z|) = 0.$$

For each $i = 1, \ldots, m$, $H^i = H_1^i - H_2^i$ with H_1^i, H_2^i convex. To simplify the presentation, it is assumed that each H_1^i and H_2^i has minimum 0 which is attained at $p = 0$. Then, it is possible to use (3.67).

Rewriting (3.69) as

$$\phi_t^n = \sum_{i=1}^m H^{i,1}(D\phi^n)B_i^n + \sum_{i=1}^m H^{i,2}(D\phi^n)(-\dot{B}_i^n),$$

and using (3.68) yields, for all $x \in \mathbb{R}^d$,

$$\prod_{i=1}^m S_{H^{i,1}}(-\min_{0\leq s\leq t} B_i^n(s)) \prod_{i=1}^m S_{H^{i,2}}(\max_{0\leq s\leq t} B_i^n(s))\phi_0(x)$$

$$\leq \phi^{\lambda,\varepsilon,\eta}(x, t) \leq \prod_{i=1}^M S_{H^{i,1}}(\max_{0\leq s\leq t} B_i^n(s)) \prod_{i=1}^m S_{H^{i,2}}(-\min_{0\leq s\leq t} B_i^n(s))\phi_0(x),$$

and the claim now follows since, $\lim_{\varepsilon,\eta\to 0}(\max_{i=1,:m} \max_{s\in[0,T]} |B_i^n|) = 0$.

Another consequence of the "cancellation" estimates of Theorem 7.2 is an explicit error estimate between two solutions with different signals.

In what follows, for $k = 1, 2$, $u^k \in \mathrm{BUC}(\overline{Q}_\infty)$, $B^k \in C_0([0, \infty); \mathbb{R}^m)$ and $u_0^k \in C^{0,1}(\mathbb{R}^d)$ is the solution of the initial value problem

$$u_t^k = \sum_{i=1}^m H^i(Du^k)\dot{B}_i^k \text{ in } Q_\infty \quad u^k(\cdot, 0) = u_0^k. \tag{3.70}$$

In (3.70), the solution is either a classical viscosity solution if the signal is smooth, or the function obtained by the extension operator if B^k is continuous.

Theorem 7.3 *Assume that, for each $1, \ldots, m$, $H^i \in C(\mathbb{R}^d)$ is the difference of two nonnegative convex functions $H^{i,1}$, $H^{i,2}$. For $k = 1, 2$, $B^k \in C_0([0, \infty); \mathbb{R}^m)$ and $u_0^k \in C^{0,1}(\mathbb{R}^d)$. Let $u^k \in BUC(\overline{Q}_\infty)$ be the solution of (3.70). There exists $C > 0$ depending on $\|u_0^k\|$ and $\|Du_0^k\|$ and the growth of H^i's such that, for all $t > 0$,*

$$\sup_{x \in \mathbb{R}^d} |u^1(x, t) - v^1(x, t)| \leq C \max_{i=1,\ldots,m} \max_{0 \leq s \leq t} |B_i^1(s) - B_i^2(s)| + \sup_{x \in \mathbb{R}^d} |u_0^1(x) - u_0^2(x)|.$$

Proof Only the estimate for $\sup(u^1 - u^2)$ is shown here. The one for $\sup(u_2 - u_1)$ follows similarly. Moreover, the claim is proven under the additional assumption that the signals are smooth. The general case follows by density.

Let $L = \max_{k=1,2} \|Du_0^k\|$. Since the Hamiltonians are x-independent, it is immediate from the contraction property that, for all $t \geq 0$, $u^1(\cdot, t), u^2(\cdot, t) \in C^{0,1}(\mathbb{R}^d)$ and $\max_{k=1,2} \|Du^k(\cdot, t)\| \leq L$. The standard comparison estimate for viscosity solutions implies that, for all $(x, t) \in \overline{Q}_T$,

$$u^1(x, t) - u^2(x, t) - \phi^L(x, x, t) \leq \sup_{x,y \in \mathbb{R}^d} \left[u_0^1(x) - u_0^2(y) - L|x - y| \right] \leq 0,$$

where ϕ^L is the solution of the usual doubled equation with $\phi^L(x, y, 0) = L|x - y|$.

Basic estimates from the theory of viscosity solutions yields that, for any $\tau > 0$ and $w \in C^{0,1}(\mathbb{R}^d)$ with $\|Dw\| \leq L$,

$$\max_i \|S_{H^i}(\tau)w - w\| \leq \left(\max_i \max_{|p| \leq L} |H^i(p)| \right) \tau .$$

It follows that

$$\phi^L(x, x, t) \leq L|x - x| + m \max_{1 \leq i \leq m} [\max_{|p| \leq L} |H^i(p)| \max_{0 \leq s \leq t} |B_i^1(s) - B_i^2(s)] $$

$$= m \max_{1 \leq i \leq m} [\max_{|p| \leq L} |H^i(p)| \max_{0 \leq s \leq t} |\xi^{i,\varepsilon}(s) - \zeta^{i,\eta}(s)|].$$

Combining the upper bounds for $u^1(x, t) - u^2(x, t)$ and $\phi^L(x, x, t)$ gives the claim.

3.7.3   Control of Cancellations for Spatially Dependent Hamiltonians

It is both interesting and important for the study of qualitative properties of the pathwise solutions, see, for example, Sect. 3.8 of the notes, to extend the results about the cancellations to spatially dependent Hamiltonians $H = H(p, x)$ and the initial value problem

$$du = H(Du, x) \cdot d\xi \text{ in } Q_\infty. \tag{3.71}$$

The basic cancellation estimate reduces to whether if, for any $u \in \mathrm{BUC}(\mathbb{R}^d)$ and any $a > 0$,

$$S_H(a)S_{-H}(a)u \leq u \leq S_{-H}(a)S_H(a)u, \tag{3.72}$$

where $S_{\pm H}$ is the solution operator of (3.71) with Hamiltonians $\pm H$.

A consequence of a counterexample of Gassiat [33] presented in the next section is that such a result cannot be expected for nonconvex Hamiltonians, since it would imply a domain of dependence property which is shown in [33] not to hold for a very simple nonconvex problem.

A first step towards an affirmative result was shown some time ago by Lions and the author. This was extended lately by Gassiat et al. [35] who established the following.

Theorem 7.4 *Fix* $\xi \in C_0([0, \infty); \mathbb{R})$ *and assume that*

$$\begin{cases} H = H(p, x) : \mathbb{R}^d \times \mathbb{R}^d \to \mathbb{R} \text{ is convex and} \\ \\ \text{Lipschitz continuous in } p \text{ uniformly in } x. \end{cases} \tag{3.73}$$

Then (3.72) *holds.*

Proof The result is shown for $\xi \in C^1([0, \infty))$. The general conclusion follows by density. Moreover, since the arguments are identical, I only work with the inequality on the left.

For notational simplicity, I assume that $\|D_p H\| = 1$. If L is the Legendre transform of H, it follows that

$$H(p, x) = \sup_{B_1(0)} \{\langle p, v \rangle - L(v, x)\}.$$

Let $\mathcal{A} = L^\infty(\mathbb{R}_+; \overline{B}_1(0))$. The control representation of the solution u of (3.71) (see, for example, Lions [58]) with $\xi_t \equiv t$ and $u_0 \in \mathrm{BUC}(\mathbb{R}^d)$ gives

$$S_H(t)u_0(x) = \sup_{q \in \mathcal{A}} \left\{ u_0(X(t)) \right.$$

$$\left. - \int_0^t L(q(s), X(s))ds : X(0) = x, \ \dot{X}(s) = q(s) \text{ for } s \in [0, t] \right\},$$

and

$$S_{-H}(t)u_0(y) = \inf_{r \in \mathcal{A}} \left\{ u_0(Y(t)) \right.$$

$$\left. + \int_0^t L(r(s), Y(s))ds : Y(0) = y, \ \dot{Y}(s) = -r(s) \text{ for } s \in [0, t] \right\}.$$

It follows that

$$S_H(t) \circ S_H(-t)u_0(x) = \sup_{q \in \mathcal{A}} \inf_{r \in \mathcal{A}} \left\{ u_0(Y(t)) + \int_0^t L(r(s), Y(s))ds \right.$$

$$- \int_0^t L(q(s), X(s))ds :$$

$$\left. Y(0) = X(t), \ \dot{Y}(s) = -r(s), \ X(0) = x, \ \dot{X}(s) = q(s) \text{ for } s \in [0, t] \right\}.$$

Given $q \in \mathcal{A}$ choose $r(s) = q(t - s)$ in the infimum above. Since $Y(s) = X(t - s)$, it follows that

$$S_H(t) \circ S_H(-t)u_0(x) \leq \sup_{q \in \mathcal{A}} \left\{ u_0(X(0)) + \int_0^t L(q(t - s), X(t - s))ds \right.$$

$$- \int_0^t L(q(s), X(s))ds : \quad X(0) = x, \ \dot{X}(s) = q(s) \text{ for } s \in [0, t] \right\}$$

$$= u_0(x).$$

As a matter of fact, Lions and Souganidis came up recently with a more refined form of (3.72), which is stated below without proof.

Theorem 7.5 *Fix $\xi \in C_0([0, \infty); \mathbb{R})$ and assume (3.73). For every $a, b, c > 0$ such that $b \leq \min(a, c)$,*

$$S_H(c)S_H(b)S_H(a) = S_H(a + c - b).$$

3.7.4 The Interplay Between the Regularity of the Hamiltonians and the Paths

The classical theory of viscosity solutions applies when $H \in C$ and $B \in C^1$; actually it is possible to consider $B \in C^{1,1}$ or even discontinuous B as long as $\dot{B} \in L^1$. It was also shown here that, when $H \in C^{1,1}$ or, more generally, if H is the difference of two convex (or half-convex) functions, there exists a unique extension for any $B \in C([0, \infty))$.

Arguments similar to the ones presented next yield a unique extension for $B \in C^{0,\alpha}([0, \infty))$ with $\alpha \in (0, 1)$ and $H \in C^{2(1-\alpha)+\varepsilon}(\mathbb{R}^d)$ for $\varepsilon > 0$; recall that, for any $\beta \in [0, \infty)$, $C^\beta(\mathbb{R}^d)$ is the space $C^{[\beta], \beta-[\beta]}(\mathbb{R}^d)$. It is not clear, however, if the additional ε-regularity is necessary.

The conclusion resembles nonlinear interpolation. Indeed, consider the solution mapping $T(B, H) = u$, which is a bounded map from $C^1 \times C^0$ into C and $C^0 \times C^2$ into C. Typically, if T is bilinear, abstract interpolation results would imply that T must be a bounded map from $C^{0,\alpha} \times C^{[2(-1-\alpha)], 2(1-\alpha)-[2(1-\alpha)]}$ into C. But T is far from being bilinear.

Next it is stated without proof (see [71] for the details) that, in the particular case $\alpha = 1/2$, it is possible to have a unique extension if $H \in C^{1,\delta}$ for $\delta > 0$. Of course, the goal is to show that is enough to have $H \in C^1$ or even $H \in C^{0,1}$. Questions related to the issues described above are studied in an ongoing work by Lions et al. [68].

A sequence $(B_n)_{n \in \mathbb{N}}$ in $C^1([0, \infty))$ is said to approximate $B \in C([0, \infty))$ in $C^{0, 1/2}$ if, as $n \to \infty$,

$$B_n \to B \text{ in } C([0, \infty)) \quad and \quad \sup_n \|\dot{B}_n\|_\infty \|B_n - B\|_\infty < \infty.$$

Given $B \in C^{0,1/2}([0, \infty))$, it is possible to find at least two classes of such approximations. The first uses convolution with a suitable smooth kernel, while the second relies on finite differences.

Let $\rho_n(t) = n\rho(nt)$ with ρ a smooth nonnegative kernel with compact support in $[-1, 1]$ such that $\int z\rho(z)dz = 0$ and $\int \rho(z)dz = 1$, and consider the smooth function $B_n = B * \rho_n$. If $C = (\|\rho'\| + \|\rho\| + 1)[B]_{0, 1/2}$, then

$$\|\dot{B}_n\| \leq C\sqrt{n} \quad and \quad \|B_n - B\| \leq C/\sqrt{n}.$$

For the second approximation, subdivide $[0, T]$ into intervals of length $\Delta = T/n$ and construct B_n by a linear interpolation of $(B_{k\Delta})_{k=1,\dots,n}$. Then

$$|\dot{B}_n| = \frac{|B_{(k+1)\Delta} - B_{k\Delta}|}{\Delta} \leq \frac{[B]_{0, 1/2}}{\sqrt{\Delta}} = C\sqrt{n} \quad and$$

$$\|B - B_n\| \leq [B]_{0, 1/2}\sqrt{\Delta} = \frac{C}{\sqrt{n}}.$$

The next result says that $C^{0,\frac{1}{2}}$-approximations of $C^{0,\frac{1}{2}}$ paths yield a unique extension for $H \in C^{1,\delta}(\mathbb{R}^d)$ with $\delta > 0$. As a matter of fact the result not only gives an extension but also an estimate. For the proof I refer to [71].

Theorem 7.6 *Assume that $B \in C^{0,1/2}([0,\infty))$ and $H \in C^{1,\delta}(\mathbb{R}^d)$ for some $\delta > 0$, and fix $T > 0$ and $u_0 \in BUC(\mathbb{R}^d)$. For any $(\xi_n)_{n\in\mathbb{N}}, (\zeta^m)_{m\in\mathbb{N}} \in C_0([0,\infty))$ and $u_{0,n}, v_{0,m} \in BUC(\mathbb{R}^d)$, which are respectively $C^{0,1/2}$-approximations of B and u_0 in $BUC(\mathbb{R}^d)$, let $u_n, v_m \in BUC(\overline{Q}_T)$ be the solutions of the corresponding initial value problems. Then there exists $u \in BUC(\overline{Q}_T)$ such that, as $n, m \to \infty$, $u_n, v_m \to u$ in $BUC(\overline{Q}_T)$. Moreover, if $\|u_{0,n} - u_0\| \leqq Cn^{-\beta}$ for some $\alpha, \beta > 0$, then there exist $\gamma > 0$ and $C >$ such that $|u_n - u| \leqq Cn^{-\gamma}$ in $\overline{Q}_T$.*

A discussion follows about the need to have conditions on H. The key step in the proof of Theorem 7.6 can be reformulated as follows. Let $(B_n)_{n\in\mathbb{N}}$ be a sequence of C^1-functions such that, as $n \to \infty$, $B_n \to 0$ and $\sup_n \|B_n\|_\infty \|\dot{B}_n\|_\infty < \infty$, and consider the solution v_n of $v_{n,t} = H(Dv_n)\dot{B}_n$ in Q_T and $v_n(x,0) = \lambda|x|$. As before, it suffices to show that, for each fixed $T > 0$ and for all $(x,t) \in \overline{Q}_T$,

$$\lim_{n\to\infty} \sup_{(x,t)\in\overline{Q}_T} [v_n(x,t) - \lambda|x|] \to 0 .$$

Next let $\dot{B}$ be piecewise constant such that, for $t_i = \frac{T}{k}i$,

$$\dot{B} = \Delta_1 \text{ in } [t_{2k}, t_{2k+1}] \quad \text{and} \quad \dot{B} = -\Delta_2 \text{ in } [t_{2k+1}, t_{2k}],$$

and, for simplicity, take $\lambda = 1$. Arguments similar to the ones earlier in this section and the fact that v_k is convex, since $v_k(\cdot, 0)$ is, yield a sequence $w_k = v_k^*$ such that

$$w_0 = 0\mathbb{1}_{\{|p|\leq 1\}} + \infty\mathbb{1}_{\{|p|>1\}}$$

and

$$w_{2k+1} = (w_{2k} + \Delta_{2k}H)^{**} \text{ and } w_{2k} = (w_{2k-1} - \Delta_{2k-1}H)^{**}$$

$$\text{where} \quad \Delta_i = k\Big[B(\frac{(i+1)T}{k}) - B(\frac{iT}{k})\Big].$$

The convexity of the v_k's and Hopf's formula implies that the sequence w_k is decreasing. Then convergence will follow if there is a lower bound for the w_k's.

Consider next the particular case $H(p) = |p|^\theta$ and assume that, for all i, $\Delta^i = \sqrt{k}$.

If $\tilde{w}_k$ is constructed similarly to w_k but with $\Delta_i \equiv 1$, it is immediate that $w_k = k^{-1/2}\tilde{w}_k$, and, since $\tilde{w}_{k+1} = ((\tilde{w}_k \pm |p|^\theta)^{**} \mp |p|^\theta)^{**}$, it follows that $\tilde{w}_{k+1} \leqq \tilde{w}_k$ and $\tilde{w}_k = +\infty$ if $|p| > 1$.

Let $m_k = -\inf_{|p|<1} w_k(p)$. Since H is not the difference of two convex functions if $\theta \in (0, 1/2)$, it must be that $m_k \to \infty$ as $k \to \infty$.

It turns out, and this is tedious computation, that there exists $c > 0$ such that $\tilde{w}_k \leqq -ck^{1-\theta}$.

It follows that, if $\theta < 1/2$,

$$w_k = k^{-1/2}\tilde{w}_k \leqq -ck^{1/2-\theta} \to -\infty \quad \text{as} \quad k \to \infty .$$

The above calculations show that, if $H \in C^{0,\alpha}(\mathbb{R}^d)$ with $\alpha \in (0, \frac{1}{2})$ and $\sup_n \|B_n\|_{C^{0,\frac{1}{2}}} < \infty$, then there is blow up, and, hence, not a good solution. On the other hand, if $H \in C^{0,\frac{1}{2}}(\mathbb{R}^d)$, there is no blow up.

3.8 Qualitative Properties

Recently there has been great interest in the study and understanding of various qualitative properties of the solutions. In this section, I focus mainly on the initial problem

$$du = H(Du, x) \cdot dB \text{ in } Q_\infty \quad u(\cdot, 0) = u_0, \tag{3.74}$$

and I discuss the following three qualitative behaviors: domain of dependence and finite speed of propagation , intermittent regularizing effect and regularity, and long time behavior of the pathwise solutions.

3.8.1 Domain of Dependence and Finite Speed of Propagation

Given that the pathwise solutions are obtained as uniform limits of solutions of hyperbolic equations with domain of dependence and finite speed of propagation property, it is natural to ask if this property remains true in the limit.

In the context of the "non-rough" viscosity solutions, it is known that, if H is Lipschitz continuous with constant L, and $u^1, u^2 \in BUC(\overline{Q_T})$ solve the initial value problems

$$u_t^1 = H(Du^1) \text{ in } Q_T \quad u^1(\cdot, 0) = u_0^1 \quad \text{and} \quad u_t^2 = H(Du^2) \text{ in } Q_T \quad u^2(\cdot, 0) = u_0^2,$$

then

$$\text{if } u_0^1 = u_0^2 \text{ in } B(0, R), \quad \text{then } u^1(\cdot, t) = u^2(\cdot, t) \text{ in } B(0, R - Lt).$$

The first positive but partial result in this direction for pathwise solutions was proved [71]. The claim is the following.

Proposition 8.1 *Assume that $H = H_1 - H_2$ with H_1 and H_2 convex and bounded from below, and $u_0 \in C^{0,1}(\mathbb{R}^d)$. Let L be the Lipschitz constant of H_1 and H_2 in $B(0, \|Du\|)$ and consider the solution $u \in BUC(\overline{Q}_T)$ of (3.74). If, for some $A \in \mathbb{R}$ and $R > 0$,*

$$u(\cdot, 0) \equiv A \ \text{ in } B(0, R),$$

then

$$u(\cdot, t) \equiv A \ \text{ in } \ B(0, R - L(\max_{0 \leq s \leq t} B(s) - \min_{0 \leq s \leq t} B(s))).$$

Proof Without loss of generality, the problem may be reduced to Hamiltonians with the additional property

$$H_1, H_2 \ \text{nonnegative and } \ H_1(0) = H_2(0) = 0. \tag{3.75}$$

As long as $R > L(\max_{0 \leq s \leq t} B(s) - \min_{0 \leq s \leq t} B(s))$, and, since $H_1(0) = H_2(0) = 0$, the finite speed of propagation of the initial value problem with $B(t) = t$ yields

$$S_{H_1}(\max_{0 \leq s \leq t} B^{\pm}(s))u_0 = S_{H_2}(\max_{0 \leq s \leq t} B^{\pm}(s))u_0 = A,$$

and the claim then follows using the estimate in Theorem 7.2.

The following example in [33] shows that, when the Hamiltonian is neither convex nor concave, the initial value problem does not have the finite speed of propagation property.

Fix $T > 0$ and $\xi \in C_0([0, \infty); \mathbb{R})$. The total variation $V_{0,T}(\xi)$ of ξ in $[0, T]$ is

$$V_{0,T}(\xi) := \sup_{(t_0, \ldots, t_n) \in \mathcal{P}} \sum_{i=0}^{n-1} |\xi(t_{i+1}) - \xi(t_i)|,$$

where $\mathcal{P} = \{0 = t_0 < t_1 < \cdots, t_n = T\}$ is a partition of $[0, T]$.

The result is stated next.

Proposition 8.2 *Given $\xi \in C_0([0, T]; \mathbb{R})$, let $u \in BUC(\mathbb{R}^2 \times [0, T])$ be the solution of*

$$du = (|u_x| - |u_y|) \cdot d\xi \ \text{in } \mathbb{R}^2 \times [0, T] \quad u(x, y, 0) = |x - y| + \Theta(x, y), \tag{3.76}$$

with $\Theta \in BUC(\mathbb{R}^2)$ nonnegative and such that, for some $R > 0$, $\Theta(x, y) \geq R$ if $\min(x, y) \geq R$. *Then*

$$u(0, 0, T) \geq \left(\sup_{(t_0, \ldots, t_n) \in \mathcal{P}} \frac{\sum_{i=0}^{n-1} |\xi(t_{i+1}) - \xi(t_i)| - R}{n} \right)_+ \wedge 1. \qquad (3.77)$$

In particular, $u(0, 0, T) > 0$ as soon as $V_{0,T}(\xi) > R$.

If ξ is a Brownian motion, then $V_{0,T}(\xi) = +\infty$ for all $T > 0$. Then (3.76) implies there is no finite speed of propagation property for any $R > 0$.

Proof (The Proof of Proposition 8.2) The argument is based on the differential games representation formula discussed earlier in the notes, which is possible to have for the very special Hamiltonian considered here.

Arguing by density, I assume that $\xi \in C^1$. A simple calculation shows that, for all $p, q \in \mathbb{R}$,

$$(|p| - |q|)\dot{\xi}(t) = \max_{|a| \leq 1} \min_{|b| \leq 1} \left\{ (a\dot{\xi}(t)_+ + b\dot{\xi}(t)_-)p + (b\dot{\xi}(t)_+ + a\dot{\xi}(t)_-)q \right\}.$$

It follows that, for any $T > 0$,

$$u(0, 0, T) = \sup_{\alpha \in \Gamma(T)} \inf_{z \in N(T)} J(\alpha[z], z),$$

where, for each pair (w, z) of controls in $[0, T]$,

$$J(w, z) = |x^{w,z}(T) - y^{w,z}(T)| + \Theta(x^{w,z}(T), y^{w,z}(T)),$$

and

$$\dot{x}^{w,z}(s) = w(s)\dot{\xi}(T - s)_+ + z(s)\dot{\xi}(T - s)_- \quad x^{w,z}(0) = 0,$$

$$\dot{y}^{w,z}(s) = z(s)\dot{\xi}(T - s)_+ + w(s)\dot{\xi}(T - s)_- \quad y^{w,z}(0) = 0.$$

I refer to [33] for the rest of the argument, which is based on the choice, for each partition of $[0, T]$, of a suitable pair of strategy and control, and the assumption on Θ.

Motivated by the general question and the partial result and counterexample discussed above, Gassiat et al. [35] considered the case of convex, spatially dependent Hamiltonians. Using the cancellation property discussed in the previous section, it is proven in [35] that, in this setting, there is a finite speed of propagation. This required the use of what is known as "skeleton" of the path. The details are presented next.

Given $\xi \in C([0, T])$, if arg $\min_{[a,b]}$ (resp. arg $\max_{[a,b]}$) denotes the set of minima (resp. maxima) points of ξ on the interval $[a, b] \subseteq [0, T]$, the sequence $(\tau_i)_{i \in \mathbb{Z}}$ of successive extrema of ξ is defined by

$$\tau_0 = \sup\left\{ t \in [0, T] : \xi(t) = \max_{0 \leq s \leq T} \xi(s) \text{ or } \xi(t) = \min_{0 \leq s \leq T} \xi(s) \right\}, \qquad (3.78)$$

where, for all $i \geq 0$,

$$\tau_{i+1} = \begin{cases} \sup \arg \max_{[\tau_i, T]} \xi & \text{if } \xi(\tau_i) < 0, \\ \sup \arg \min_{[\tau_i, T]} \xi & \text{if } \xi(\tau_i) > 0, \end{cases} \qquad (3.79)$$

and, for all $i \leq 0$,

$$\tau_{i-1} = \begin{cases} \inf \arg \max_{[0, \tau_i]} \xi & \text{if } \xi(\tau_i) < 0, \\ \inf \arg \min_{[0, \tau_i]} \xi & \text{if } \xi(\tau_i) > 0. \end{cases} \qquad (3.80)$$

The skeleton (resp. full skeleton) or reduced (resp. fully reduced) path $R_{0,T}(\xi)$ (resp. $\tilde{R}_{0,T}(\xi)$) of $\xi \in C_0([0, T])$ are defined as follows.

Definition 8.1 Let $\xi \in C([0, T])$.

(i) The reduced path $R_{0,T}(\xi)$ is a piecewise linear function which agrees with ξ on $(\tau_i)_{i \in \mathbb{Z}}$.

(ii) The fully reduced path $\tilde{R}_{0,T}(\xi)$ is a piecewise linear function agreeing with ξ on $(\tau_{-i})_{i \in \mathbb{N}} \cup \{T\}$.

(iii) A path $\xi \in C_0([0, T])$ is reduced (resp. fully reduced) if $\xi = R_{0,T}(\xi)$ (resp. $\xi = \tilde{R}_{0,T}(\xi)$).

Note that the reduced and the fully reduced paths coincide prior to the global extremum τ_0. While the reduced path captures the max-min fluctuations also after τ_0, the fully reduced path is affine linear on $[\tau_0, T]$ and, in this sense, is more "reduced".

Throughout the discussion, it is assumed that

$$\begin{cases} H : \mathbb{R}^d \times \mathbb{R}^d \to \mathbb{R} \text{ is convex and} \\ \text{Lipschitz continuous with constant } L \text{ in the first argument.} \end{cases} \qquad (3.81)$$

The speed of propagation of (3.74) at time T is defined by

$$\rho_H(\xi, T) = \sup\Big\{ R \geq 0 : \text{ there exist solutions } u^1, u^2 \text{ of (3.74) and } x \in \mathbb{R}^d, \qquad (3.82)$$

$$\text{such that } u^1(\cdot, 0) = u^2(\cdot, 0) \text{ in } B_R(x) \text{ and } u^1(x, T) \neq u^2(x, T) \Big\}.$$

To keep track of the dependence of the solution on the path, in what follows I use the notation u^ξ for the solution of (3.74) with path ξ. The main observation is that

$$u^\xi(\cdot, T) = u^{R_{0,T}(\xi)}(\cdot, T), \qquad (3.83)$$

which immediately implies the following result about the speed of propagation.

Theorem 8.1 *Assume* (3.81). *Then, for all* $\xi \in C([0, T])$,

$$\rho_H(\xi, T) \le L \, \|R_{0,T}(\xi)\|_{TV([0,T])}. \qquad (3.84)$$

The second main result of [35] concerns the total variation of the reduced path of a Brownian motion. To state it, it is necessary to introduce the random variable $\theta : [0, \infty) \to [0, \infty)$ given by

$$\theta(a) = \inf\{t \ge 0 : \ \max_{[0,t]} B - \min_{[0,t]} B = a\}, \qquad (3.85)$$

which is the first time that the range, that is $\max - \min$ of a Brownian motion equals a.

It is proved in [35], where I refer for the details, that the length of the reduced path is a random variable with almost Gaussian tails. It is also shown that if the range, that is, the maximum minus the minimum of B, is fixed instead of the time horizon T, then the length has Poissonian tails.

Theorem 8.2 *Let* B *be a Brownian motion and fix* $T > 0$. *Then, for each* $\gamma \in (0, 2)$, *there exists* $C = C(\gamma, T) > 0$ *such that, for any* $x \ge 2$,

$$\mathbb{P}\left(\|R_{0,T}(B)\|_{TV([0,T])} \ge x\right) \le C \exp\left(-Cx^\gamma\right), \qquad (3.86)$$

and

$$\lim_{x \to \infty} \frac{\ln \mathbb{P}\left(\|R_{0,\theta(1)}(B)\|_{TV([0,\theta(1)])} \ge x\right)}{x \ln(x)} = -1. \qquad (3.87)$$

A related result, proving that the expectation of the total variation of the so-called piecewise linear oscillating running max/min function of Brownian motion is finite, has been obtained independently by Hoel et al. in [42].

The following remark shows that upper bound in Theorem 8.1 is actually sharp.

Proposition 8.3 *Let* $H(p) = |p|$ *on* $\mathbb{R}^d$ *with* $d \ge 1$. *Then, for all* $T > 0$ *and* $\xi \in C_0([0, T]; \mathbb{R})$,

$$\rho_H(\xi, T) \ge \|\tilde{R}_{0,T}(\xi)\|_{TV([0,T])}. \qquad (3.88)$$

When $d = 1$, then

$$\rho_H(\xi, T) = \|\tilde{R}_{0,T}(\xi)\|_{TV([0,T])}.$$

Here I only sketch the proof of the first result.

Proof (A Sketch of the Proof of Theorem 8.1) The first step is (3.72). The second is a monotonicity property for piecewise linear paths. Let $\xi_t = 1_{t\in[0,t_1]}(a_0 t) + 1_{t\in[t_1,T]}(a_1(t - t_1) + a_0 t_1)$ and, for $s < t$, set $\xi_{s,t} = \xi_t - \xi_s$.

If $a_0 \geq 0$ and $a_1 \leq 0$ (resp. $a_0 \leq 0$ and $a_1 \geq 0$), then

$$S_H^\xi(0, T) \geq S_H(\xi_{0,T}) \qquad (resp. \ S_H^\xi(0, T) \leq S_H(\xi_{0,T}).) \tag{3.89}$$

Since the claim is immediate if $a_0 = 0$ or $a_1 = 0$, next it is assumed that $a_0 > 0$ and $a_1 < 0$.

If $\xi_{0,T} \leq 0$, then

$$S_H(a_1(T - t_1)) = S_{-H}(-a_1(T - t_1)) = S_{-H}(-a_1(T - t_1) - a_0 t_1) \circ S_{-H}(a_0 t_1)$$
$$= S_{-H}(-\xi_{0,T}) \circ S_{-H}(a_0 t_1) = S_H(\xi_{0,T}) \circ S_H(-a_0 t_1),$$

and, hence, in view of (3.73),

$$S_H^\xi(0, T) = S_H(\xi_{0,T}) \circ S_H(-a_0 t_1) \circ S_H(a_0 t_1) \geq S_H(\xi_{0,T}).$$

If $\xi_{0,T} \geq 0$ then, again, (3.73) yields

$$S_H^\xi(0, T) = S_H(a_1(T - t_1)) \circ S_H(-a_1(T - t_1) + a_0 t_1 + a_1(T - t_1))$$
$$= S_H(a_1(T - t_1)) \circ S_H(-a_1(T - t_1)) \circ S_H(a_0 t_1 + a_1(T - t_1))$$
$$\leq S_H(\xi_{0,T}).$$

For the second inequality, note that $S_{-H}^{-\xi}(0, T) = S_H^\xi(0, T)$, $S_{-H}(-t) = S_H(t)$. It then follows from the first part that

$$S_H^\xi(0, T) = S_{-H}^{-\xi}(0, T) \geq S_{-H}(-\xi_{0,T}) = S_H(\xi_{0,T}).$$

The next observation provides the first indication of the possible reduction encountered when using the max or min of a given path. For the statement, given a piecewise linear path ξ, set

$$\tau_{max} = \sup \left\{ t \in [0, T] : \xi_t = \max_{s \in [0,T]} \xi_s \right\}$$

and

$$\tau_{min} = \inf \left\{ t \in [0, T] : \xi_t = \min_{s \in [0,T]} \xi_s \right\}.$$

Lemma 8.1 *Fix a piecewise linear path ξ. Then*

$$S_H^\xi(\tau_{max}, T) \circ S_H(\xi_{0,\tau_{max}}) \leq S_H^\xi(0, T) \leq S_H(\xi_{\tau_{min},T}) \circ S_H^\xi(0, \tau_{min}).$$

Proof Since the proofs of both inequalities are similar, I only show the details for the first.

Without loss of generality, it is assumed that $\text{sign}(\xi_{t_{i-1},t_i}) = -\text{sign}(\xi_{t_i,t_{i+1}})$ for all $[t_{i-1}, t_{i+1}] \subseteq [0, \tau_{max}]$. It follows that, if $\xi_{|[0,\tau_{max}]}$ is linear, then $S_H^\xi(0, \tau_{max}) = S_H(\xi_{0,\tau_{max}})$.

If not, since $\xi_{0,\tau_{max}} \geq 0$, there is an index j such that $\xi_{t_{j-1},t_{j+1}} \geq 0$ and $\xi_{t_{j-1},t_j} \leq 0$. It then follows from (3.89) that

$$S_H^\xi(0, \tau_{max}) \leq S_H^{\tilde{\xi}}(0, \tau_{max}),$$

where $\tilde{\xi}$ is piecewise linear and coincides with ξ for all $t \in \{t_i : i \neq j\}$.

A simple iteration yields $S_H^\xi(0, \tau_{max}) \leq S_H(\xi_{0,\tau_{max}})$, and, since $S_H^\xi(0, T) = S_H^\xi(\tau_{max}, T) \circ S_H^\xi(0, \tau_{max})$, this concludes the proof.

The previous conclusions and lemmata are combined to establish the following monotonicity result.

Corollary 8.1 *Let ξ, ζ be piecewise linear, $\xi(0) = \zeta(0)$, $\xi(T) = \zeta(T)$ and $\xi \leq \zeta$ on $[0, T]$. Then*

$$S_H^\xi(0, T) \leq S_H^\zeta(0, T). \tag{3.90}$$

Proof Assume that ξ and ζ are piecewise linear on each interval $[t_i, t_{i+1}]$ on a common partition $0 = t_0 \leq \ldots \leq t_N = T$ of $[0, T]$.

If $N = 2$, then, for all $\gamma \geq 0$ and all $a, b \in \mathbb{R}$,

$$S_H(a + \gamma) \circ S_H(b - \gamma) \leq S_H(a) \circ S_H(b). \tag{3.91}$$

If $a \geq 0$, this follows from the fact that, in view of (3.89),

$$S_H(\gamma) \circ S_H(b - \gamma) \leq S_H(b).$$

If $a + \gamma \leq 0$, then again (8.1) yields

$$S_H(a) \circ S_H(b) = S_H(a + \gamma) \circ S_H(-\gamma) \circ S_H(b) \geq S_H(a + \gamma) \circ S_H(b - \gamma).$$

Finally, if $a \leq 0 \leq a + \gamma$ we have

$$S_H(a) \circ S_H(b) \geq S_H(a + b) \geq S_H(a + \gamma) \circ S_H(b - \gamma).$$

The proof for $N > 2$ follows by induction on N. Let ρ be piecewise linear on the same partition and coincide with ζ on t_0, t_1, and with ξ on $t_2, \ldots, t_N$. The induction hypothesis then yields

$$S_H^\xi(0, t_2) \leq S_H^\rho(0, t_2) \text{ and } S_H^\rho(t_1, T) \leq S_H^\zeta(t_1, T)$$

from which we deduce

$$S_H^\xi(0, T) \leq S_H^\rho(0, T) \leq S_H^\zeta(0, T).$$

To complete the study of the cancellations, it is necessary to use a density argument, which, itself, requires a result about the uniform continuity of the solutions with respect to the paths. Such a result was shown earlier in the notes for spatially-independent Hamiltonians which are the difference of two convex functions and for spatially dependent under some additional conditions on the joint dependence but not convexity. The most general result available without additional assumptions other than convexity was obtained in [73]. Here it is stated without a proof.

Theorem 8.3 *Assume* (3.73). *Then, for each* $u_0 \in BUC(\mathbb{R}^d)$ *and* $T \geq 0$, *the family*

$$\left\{ S_H^\xi(0, T)u_0 : \quad \xi \text{ piecewise linear} \right\}$$

has a uniform modulus of continuity.

An immediate consequence is the following extension result which is stated as a corollary without proof; see [35] for the details.

Corollary 8.2 *The map* $\xi \mapsto S_H(\xi)$ *is uniformly continuous in the sup-norm in the sense that, if* $(\xi^n)_{n \in \mathbb{N}}$ *is a sequence of piecewise-linear functions on* $[0, T]$ *with* $\lim_{n,m \to \infty} \|\xi^n - \xi^m\|_{\infty, [0,T]} = 0$, *then, for all* $u \in BUC(\mathbb{R}^d) \times (0, \infty)$,

$$\lim_{n,m \to \infty} \|S_H^{\xi^n}(0, T)u - S_H^{\xi^m}(0, T)u\|_\infty = 0. \tag{3.92}$$

Combining all the results above completes the proof.

3.8.2 Stochastic Intermittent Regularization

A very interesting question is whether there is some kind of stochastic regularization-type property for the pathwise solutions of

$$du = H(Du) \cdot d\zeta \text{ in } Q_\infty. \tag{3.93}$$

It is assumed that

$$H \in C^2(\mathbb{R}^d) \text{ is uniformly convex}, \tag{3.94}$$

which implies that there exist $\Theta \geq \theta > 0$ such that, for all $p \in \mathbb{R}^d$ and in the sense of symmetric matrices,

$$\theta I \leq D^2 H(p) \leq \Theta I. \tag{3.95}$$

The upper bound in (3.95) can be relaxed when dealing with Lipschitz solutions of (3.93).

Motivated by a recent observation of Gassiat and Gess [34] for the very special case that $H(p) = (1/2)|p|^2$, recently Lions and the author [80] investigated this question. A summary of these results is presented next without proofs. The details can be found in [80].

The possible intermittent regularizing results follow from iterating regularizing and propagation of regularity-type results for the "non rough" problem

$$u_t = \pm H(Du) \text{ in } Q_\infty. \tag{3.96}$$

It turns out that the quantity to measure the regularizing effects is the symmetric matrix

$$F(p) = \sqrt{D^2 H(p)},$$

the reason being that, if, for example, u is a smooth solution of (3.96), then a simple calculation yields that the matrix $W(x, t) = F(Du(x, t))$ satisfies the matrix valued problem

$$W_t = DH(Du)DW \pm |DW|^2.$$

The first claim is about the regularizing effect of (3.96). In what follows all the inequalities and solutions below should be understood in the viscosity sense.

Theorem 8.4 *Assume* (3.94). *If* $u \in BUC(\mathbb{R}^d \times (0, \infty))$ *is a solution of* $u_t = H(Du)$
(*resp.*

$$u_t = -H(Du)) \text{ in } \mathbb{R}^d \times (0, \infty) and, for some C \in (0, \infty],$$

$$- F(Du(\cdot, 0))D^2u(\cdot, 0)F(Du(\cdot, 0)) \leq CI \text{ in } \mathbb{R}^d, \tag{3.97}$$

(*resp.*

$$- F(Du(\cdot, 0))D^2u(\cdot, 0)F(Du(\cdot, 0)) \geq -CI \text{ in } \mathbb{R}^d), \tag{3.98}$$

then, for all $t > 0$,

$$- F(Du(\cdot, t))D^2u(\cdot, t)F(Du(\cdot, t)) \leq \frac{C}{1 + Ct}I \text{ in } \mathbb{R}^d, \tag{3.99}$$

(*resp.*

$$- F(Du(\cdot, t))D^2u(\cdot, t)F(Du(\cdot, t)) \geq -\frac{C}{1 + Ct}I \text{ in } \mathbb{R}^d). \tag{3.100}$$

Estimates (3.99) and (3.100) are sharper versions of the classical regularizing effect-type results for viscosity solutions (see Lions [58], Lasry and Lions [52]), which say that, if $u_t = H(Du)$ (resp. $u_t = -H(Du)$) in Q_∞, and, for some $C \in (0, \infty]$, $-D^2u(\cdot, 0) \leq CI$ (resp. $-D^2u(\cdot, 0) \geq -CI$) in $\mathbb{R}^d$, then, for all $t > 0$,

$$- D^2u(\cdot, t) \leq \frac{C}{1 + \theta Ct}I \text{ in } \mathbb{R}^d \tag{3.101}$$

(*resp.*

$$- D^2u(\cdot, t) \geq -\frac{C}{1 + \theta Ct}I \text{ in } \mathbb{R}^d.) \tag{3.102}$$

Note that, when $C = \infty$, that is, no assumption is made on $u(\cdot, 0)$, then (3.97) and (3.98) reduce to

$$- F(Du(\cdot, t))D^2u(\cdot, t)F(Du(\cdot, t))$$
$$\leq \frac{1}{t} \left(\text{resp.} - F(Du(\cdot, t))D^2u(\cdot, t)F(Du(\cdot, t)) \geq -\frac{1}{t} \right), \tag{3.103}$$

which are sharper versions of (3.101) and (3.102), in the sense that they do not depend on θ, of the classical estimates

$$-D^2u(\cdot,t) \le \frac{1}{\theta t} \quad (\text{resp. } D^2u(\cdot,t) \ge -\frac{1}{\theta t}).$$

To continue with the propagation of regularity result, I first recall that it was shown in [52] that, if u solves $u_t = H(Du)$ (resp. $u_t = -H(Du)$) in $\mathbb{R}^d \times [0,\infty)$, with H satisfying (3.95), then,

$$\text{if } -D^2u(\cdot,0) \ge -CI, \text{ then } -D^2u(\cdot,t) \ge -\frac{C}{(1-\Theta Ct)_+}, \tag{3.104}$$

(resp.

$$\text{if } -D^2u(\cdot,0) \le CI, \text{ then } -D^2u(\cdot,t) \le \frac{C}{(1-\Theta Ct)_+}.) \tag{3.105}$$

The new propagation of regularity result depends on the dimension. In what follows, it is said that $H : \mathbb{R}^d \to \mathbb{R}$ is quadratic, if there exists a symmetric matrix A which satisfies (3.95) such that

$$H(p) = (Ap, p).$$

Theorem 8.5 *Assume* (3.94) *and let* $u \in BUC(\overline{Q}_\infty)$ *solve* $u_t = H(Du)$ *(resp.* $u_t = -H(Du))$ *in* Q_∞. *Suppose that either* $d = 1$ *or* H *is quadratic. If, for some* $C > 0$,

$$-F(Du(\cdot,0))D^2u(\cdot,0)F(Du(\cdot,0)) \ge -CI \text{ in } \mathbb{R}^d, \tag{3.106}$$

(resp.

$$-F(Du(\cdot,0))D^2u(\cdot,0)F(Du(\cdot,0)) \le CI \text{ in } \mathbb{R}^d), \tag{3.107}$$

then, for all $t > 0$,

$$-F(Du(\cdot,t))D^2u(\cdot,t)F(Du(\cdot,t)) \ge -\frac{C}{(1-Ct)_+}I \text{ in } \mathbb{R}^d, \tag{3.108}$$

(resp.

$$-F(Du(\cdot,t))D^2u(\cdot,t)F(Du(\cdot,t)) \le \frac{C}{(1-Ct)_+}I \text{ in } \mathbb{R}^d.) \tag{3.109}$$

The result for $d \ge 2$ and general H required more regularity for the initial condition.

Theorem 8.6 *Assume that $d > 1$ and that H satisfies (3.94) but is not quadratic. Let $u \in BUC(\mathbb{R}^d \times [0, \infty))$ solve $u_t = H(Du)$ (resp. $u_t = -H(Du)$) in $\mathbb{R}^d \times (0, \infty)$ and assume that $u(\cdot, 0) \in C^{1,1}(\mathbb{R}^d)$. If, for some $C > 0$,*

$$- F(Du(\cdot, 0))D^2 u(\cdot, 0)F(Du(\cdot, 0)) \geq -CI \text{ in } \mathbb{R}^d, \tag{3.110}$$

(resp.

$$- F(Du(\cdot, 0))D^2 u(\cdot, 0)F(Du(\cdot, 0)) \leq CI \text{ in } \mathbb{R}^d), \tag{3.111}$$

then, for all $t > 0$,

$$- F(Du(\cdot, t))D^2 u(\cdot, t)F(Du(\cdot, t)) \geq -\frac{C}{(1 - Ct)_+}I \text{ in } \mathbb{R}^d, \tag{3.112}$$

(resp.

$$- F(Du(\cdot, t))D^2 u(\cdot, t)F(Du(\cdot, t)) \leq \frac{C}{(1 - Ct)_+}I \text{ in } \mathbb{R}^d.) \tag{3.113}$$

It turns out that the assumption that $u(\cdot, 0) \in C^{1,1}(\mathbb{R}^d)$ if $d > 1$ and H is not quadratic is necessary to have estimates like (3.112) and (3.113). This is the claim of the next result.

Theorem 8.7 *Assume (3.94) and $d > 1$. If (3.112) holds for all solutions $u \in BUC(\overline{Q_\infty})$ of $u_t = H(Du)$ (resp. $u_t = -H(Du)$) in Q_∞ with $u \in C^{0,1}(\mathbb{R}^d)$ satisfying (3.110) (resp. (3.111)), then the map $\lambda \to (D^2 H(p + \lambda\xi)\xi^\perp, \xi^\perp)$ must be concave (resp. convex). In particular, both estimates hold without any restrictions on the data if and only if H is quadratic.*

The motivation behind Theorems 8.4–8.6 is twofold. The first is to obtain as sharp as possible regularity results for solutions of (3.96). The second is to obtain intermittent regularity results for (3.93), like the ones obtained in [34] in the specific case that $H(p) = \frac{1}{2}|p|^2$, where, of course, $\theta = \Theta = 1$, $F(Du)D^2 uF(Du) = D^2 u$ and the "new" estimates are the same as the old ones, that is, (3.104) and (3.105), which hold without any regularity conditions.

The regularity results of [34] follow from an iteration of (3.99), (3.100), (3.104) and (3.105). As shown next, the iteration scheme cannot work when H is not quadratic unless $d = 1$.

To explain the problem, I consider the first two steps of the possible iteration for $u \in BUC(Q_\infty)$ solving

$$u_t = H(Du) \text{ in } \mathbb{R}^d \times (0, a], \quad u_t = -H(Du) \text{ in } \mathbb{R}^d \times (a, a + b]$$

$$\text{and} \quad u_t = H(Du) \text{ in } \mathbb{R}^d \times (a + b, a + b + c].$$

If the only estimates available were (3.99), (3.100), (3.104) and (3.105), we find, after some simple algebra, that

$$D^2u(\cdot, a) \geq -\frac{1}{\theta a}I, \quad D^2u(\cdot, a+b) \geq -\frac{1}{(\theta a - \Theta b)_+}I \quad \text{and}$$

$$D^2u(\cdot, a+b+c) \geq -\frac{1}{(\theta a - \Theta b)_+ + \theta c}I.$$

It is immediate that the above estimates cannot be iterated unless there is a special relationship between the time intervals and the convexity constants, something which will not be possible for arbitrary continuous paths ξ.

If it were possible, as is the case when $d = 1$, to use the estimates of Theorem 8.6 without any regularity restrictions, then Theorems 8.4–8.6 would imply

$$\mathcal{W}(a) \geq -\frac{1}{a}I, \quad \mathcal{W}(a+b) \geq -\frac{1}{(a-b)_+}I \text{ and } \mathcal{W}(a+b+c) \geq -\frac{1}{(a-b)_+ + c}I,$$

which can be further iterated, since the estimates are expressed only in terms of increments ζ.

Before turning to the intermittent regularity results, it is necessary to make some additional remarks. For the sake of definiteness, I continue the discussion in the context of the example above. Although $u(\cdot, a)$ may not be in $C^{1,1}$, it follows from (3.104) and (3.111) that, for some $h \in (0, b]$ and $t \in (a, a+h)$, $u(\cdot, t) \in C^{1,1}$. There is no way, however, to guarantee that $h = b$. Moreover, as was shown in [80], in general, it is possible to have u and $h > 0$ such that $u_t = -H(Du)$ in $\mathbb{R}^d \times (-h, 0]$, $u_t = H(Du)$ in $\mathbb{R}^d \times (0, h]$, $u(\cdot, t) \in C^{1,1}$ for $t \in (-h, 0) \cup (0, h)$ and $u(\cdot, 0) \notin C^{1,1}$. The implication is that when $d > 1$ and H is not quadratic, there is no hope to obtain after iteration smooth solutions.

To state the results about intermittent regularity, it is convenient to introduce the running maximum and minimum functions $M : [0, \infty) \to \mathbb{R}$ and $m : [0, \infty) \to \mathbb{R}$ of a path $\zeta \in C_0([0, \infty); \mathbb{R})$ defined respectively by

$$M(t) = \max_{0 \leq s \leq t} \zeta(t) \quad \text{and} \quad m(t) = \min_{0 \leq s \leq t} \zeta(t). \tag{3.114}$$

Theorem 8.8 *Assume* (3.94) *and either* $d = 1$ *or* H *is quadratic when* $d > 1$, *fix* $\zeta \in C_0([0, T); \mathbb{R})$ *and let* $u \in BUC(\overline{Q}_\infty$ *be a solution of* (3.93). *Then, for all* $t > 0$,

$$-\frac{1}{M(t) - \zeta(t)} \leq -F(Du(\cdot, t))D^2u(\cdot, t)F(Du(\cdot, t)) \leq \frac{1}{\zeta(t) - m(t)}. \tag{3.115}$$

Note that when (3.115) holds, then, at times t such that $m(t) < \zeta(t) < M(t)$, $u(\cdot, t) \in C^{1,1}(\mathbb{R}^d)$ and (3.115) implies that, for all $t > 0$,

$$|F(Du(\cdot, t))D^2 u(\cdot, t))F(Du(\cdot, t))| \leq \max\left[\frac{1}{\zeta(t) - m(t)}, \frac{1}{M(t) - \zeta(t)}\right].$$

(3.116)

When, however, (3.115) is not available, the best regularity estimate available, which is also new, is a decay on the Lipschitz constant $\|Du\|$.

Theorem 8.9 *Assume* (3.94), *fix* $\zeta \in C_0([0, T); \mathbb{R})$, *and let* $u \in BUC(\overline{Q}_\infty$ *be a solution of* (3.93). *Then, for all* $t > 0$,

$$\|Du(\cdot, t)\| \leq \sqrt{\frac{2\|u(\cdot, t)\|}{\theta(M(t) - m(t))}}.$$

(3.117)

It follows from (3.117) that, for any $t > 0$ such that $m(t) < M(t)$, any solution of (3.93) is actually Lipschitz continuous.

An immediate consequence of the estimates in Theorems 8.9 and 8.8, which is based on well known properties of the Brownian motion (see, for, example, Peres [88]) is the following observation.

Theorem 8.10 *Assume that* ζ *is a Brownian motion and H satisfies* (3.94). *There exists a random uncountable subset of* $(0, \infty)$ *with no isolated points and of Hausdorff measure* $1/2$, *which depends on* ζ, *off of which, any stochastic viscosity solution of* (3.93) *is in* $C^{0,1}(\mathbb{R}^d)$ *with a bound satisfying* (3.117). *If $d = 1$ or H is quadratic, for the same set of times, the solution is in* $C^{1,1}(\mathbb{R}^d)$ *and satisfies* (3.115).

3.8.3 Long Time Behavior of the "Rough" Viscosity Solutions

I begin with a short introduction about the long time behavior of solutions of Hamilton-Jacobi equations. In order to avoid technicalities due to the behavior of the solutions at infinity, throughout this subsection, it is assumed that solutions are periodic functions in $\mathbb{T}^d$.

To explain the difficulties, I first look at two very simple cases. In the first case, fix some $p \in \mathbb{R}^d$ and consider the linear initial value problem

$$du = (p, Du) \cdot d\zeta \text{ in } Q_\infty \quad u(\cdot, 0) = u_0.$$

Its solution is $u(x, t) = u_0(x + p\zeta(t))$, and clearly it is not true that $u(\cdot, t)$ has, as $t \to \infty$, a uniform limit.

The second example is about (3.93) with H satisfying (3.94), and $\dot{\xi} > 0$ and $\lim_{t \to \infty} \xi(t) = \infty$. Since

$$u(x, t) = \sup_{y \in \mathbb{R}^d} \left[u_0(y) - tH^*(\frac{x - y}{\xi(t)}) \right],$$

it is immediate that, as $t \to \infty$ and uniformly in x, $u(x, t) \to \sup u$.

The intermittent regularizing results yield information about the long time behavior of the solutions of (3.93) under the rather weak assumption that

$$H \in C(\mathbb{R}^d) \text{ is convex and } H(p) > H(0) = 0 \text{ for all } p \in \mathbb{R}^d \setminus \{0\}. \quad (3.118)$$

Theorem 8.11 *Assume* (3.118), *fix* $\zeta \in C_0([0, T); \mathbb{R})$, *and let* $u \in BUC(\overline{Q}_\infty$ *be a space periodic solution of* (3.93). *If there exists* $t_n \to \infty$ *such that* $M(t_n) - m(t_n) \to \infty$, *then there exists* $u_\infty \in \mathbb{R}$ *such that, as* $t \to \infty$ *and uniformly in space,* $u(\cdot, t) \to u_\infty$.

In the particular case that ξ is a standard Brownian motion the long time result is stated next.

Theorem 8.12 *Assume* (3.118). *For almost every Brownian path* B, *if* $u \in BUC(\overline{Q}_\infty)$ *is a periodic solution of* $du = H(Du) \cdot dB$ *in* Q_∞, *there exists a constant* $u_\infty = u_\infty(B, u(\cdot, 0))$ *such that, as* $t \to \infty$ *and uniformly in* $\mathbb{R}^d$, $u(\cdot, t) \to u_\infty$. *Moreover, the random variable is, in general, not constant.*

Proof The contraction property and the fact that $H(0) = 0$ yield that the family $(u(\cdot, t))_{t \geq 0}$ is uniformly bounded.

It is assumed next that the Hamiltonian satisfies (3.94). It follows from the intermittent regularizing property, the a.s. properties of the running max and min of the Brownian motion, and the fact that the Lipschitz constant of the solutions decreases in time that, as $t \to \infty$, $\|Du(\cdot, t)\| \to 0$.

In view of the periodicity, it follows that, along subsequences $s_n \to \infty$, the $u(\cdot, s_n)$'s converge uniformly to constants.

It remains to show that the whole family converges to the same constant. This is again a consequence of the intermittent regularizing result and the fact that the periodicity, the contraction property of the solutions of (3.93) and $H(0) = 0$ yield that

$$t \to \max_{x \in \mathbb{R}^d} u(x, t) \quad \text{is nonincreasing,}$$

and   (3.119)

$$t \to \min_{x \in \mathbb{R}^d} u(x, t) \quad \text{is nondecreasing.}$$

It remains to remove the assumption that the Hamiltonians satisfy (3.94). Indeed, if (3.118) holds, H can be approximated uniformly by a sequence $(H_m)_{m \in \mathbb{N}}$

of Hamiltonians satisfying (3.94). Let u_m be the solution of the (3.49) with Hamiltonian H_m and same initial datum. Since, as $m \to \infty$, $u_m \to u$ uniformly in Q_T for all $T > 0$, it follows that, for all $t > 0$,

$$\int_{\mathbb{T}} H(Du(x,t))dx \leq \liminf_{m\to\infty} \int_{\mathbb{T}} H(Du_m(x,t))dx.$$

Choose the sequence t_n and s_n as before to conclude.

I conclude with an example that shows that, in the stochastic setting, the limit constant u_∞ must be random.

Consider the initial value problem

$$du = |u_x| \cdot dB \text{ in } Q_\infty \quad u(\cdot,0) = u_0, \tag{3.120}$$

with u_0 a 2-periodic extension on $\mathbb{R}$ of $u_0(x) = 1 - |x - 1|$ on $[0,2]$. Let c be the limit as $t \to \infty$ of u. Since $1 - u_0(x) = u_0(x + 1)$ and $-B$ is also a Brownian motion with the same law as B, if $\mathcal{L}(f)$ denotes the law of the random variable f, it follows that

$$\mathcal{L}(c) = \mathcal{L}(1 - c). \tag{3.121}$$

If the limit c of the solution of (3.120) is deterministic, then (3.121) implies that $c = 1/2$. It is shown next that this is not the case.

Recall that the pathwise solutions are Lipschitz with respect to paths. Indeed, if u, v are two pathwise solutions of (3.120) with paths respectively B, ξ and $u(\cdot,0) \equiv v(\cdot,0)$, then there exists $L > 0$, which depends on $\|u_x(\cdot,0)\|$ such that, for any $T > 0$,

$$\max_{x\in\mathbb{R},t\in[0,T]} |u(x,t) - v(x,t)| \leq L \max_{t\in[0,T]} |\zeta(t) - \xi(t)|. \tag{3.122}$$

Next fix $T = 2$ and use (3.122) to compare the solutions of (3.120) with $\zeta \equiv B$ and $\xi(t) \equiv t$ and $\zeta \equiv B$ and $\xi(t) \equiv -t$.

When $\xi(t) \equiv t$ (resp. $\xi \equiv -t$) the solution v of (3.120) is given by

$$v(x,t) = \max_{|y|\leq t} u_0(x + y) \quad (\text{resp. } v(x,t) = \min_{|y|\leq t} u_0(x + y)).$$

It is then simple to check that, if $\xi(t) \equiv t$, then $v(\cdot,2) \equiv 1$, while, when $\xi(t) \equiv -t$, $v(x,2) = 0$.

Fix $\varepsilon = 1/4L$ and consider the events

$$A_+ := \left\{ \max_{t\in[0,2]} |B(t) - t| < \varepsilon \right\} \quad \text{and} \quad A_- := \left\{ \max_{t\in[0,2]} |B(t) + t| > \varepsilon \right\}.$$

Of course,

$$\mathbb{P}(A_+) > 0 \quad \text{and} \quad \mathbb{P}(A_-) > 0.$$

Then (3.122) implies

$$u(x, 2) \geq 1 - L\varepsilon = 3/4 \text{ on } A_+ \quad \text{and} \quad u(x, 2) \leq L\varepsilon = 1/4 \text{ on } A_-.$$

It follows that the random variable c cannot be constant since in A_+ it must be bigger than $3/4$ and in A_- smaller than $1/4$.

In an upcoming publication [36] we are visiting this problem and obtain in a special case more information about u_∞.

3.9 Pathwise Solutions for Fully Nonlinear, Second-Order PDE with Rough Signals and Smooth, Spatially Homogeneous Hamiltonians

Consider the initial value problem

$$\begin{cases} du = F(D^2u, Du, u, x, t) \, dt + \sum_{i=1}^{m} H^i(Du) \cdot dB_i \text{ in } Q_\infty \\ u(\cdot, 0) = u_0, \end{cases} \tag{3.123}$$

with

$$H = (H^1, \ldots, H^m) \in C^2(\mathbb{R}^d; \mathbb{R}^m), \tag{3.124}$$

$$B = (B_1, \ldots, B_m) \in C_0([0, \infty); \mathbb{R}^m) \tag{3.125}$$

and

$$F \text{ is degenerate elliptic.} \tag{3.126}$$

The case of "irregular" Hamiltonians requires different arguments. Spatially dependent regular Hamiltonians are discussed later.

An important question is if the Hamiltonian's can depend on u and Du at the same time. The theory for Hamiltonians depending only on u was developed in Sect. 3.4. The case where H depends both on u and Du is an open problem with the exception of a few special cases, like, for example, linear dependence on u and p, which are basically an exercise.

The theory of viscosity solutions for equations like (3.123) with $H \equiv 0$ is based on using smooth test functions to test the equation at appropriate points. As already discussed earlier this cannot be applied directly to (3.123).

Recall that, when H is sufficiently regular, it is possible to construct, using the characteristics, local in time smooth solutions to (3.46). These solutions, for special initial data, play the role of the smooth test functions for (3.123).

Definition 9.1 Fix $B \in C([0, \infty); \mathbb{R}^m)$ and $T > 0$. A function $u \in BUC(\overline{Q}_T)$ is a pathwise subsolution (resp. supersolution) of (3.123) if, for any maximum (resp. minimum) $(x_0, t_0) \in Q_\infty$ of $u - \Phi - \psi$, where $\psi \in C^1((0, \infty))$ and Φ is a smooth solution of $d\Phi = \sum_{i=1}^m H^i(D\Phi) \circ dB^i$ in $\mathbb{R}^d \times (t_0 - h, t_0 + h)$ for some $h > 0$, then

$$\psi'(t_0) \leqq F(D^2\Phi(x_0, t_0), D\Phi(x_0, t_0), u(x_0, t_0), x_0, t_0) \qquad (3.127)$$

$$(\text{resp.} \quad \psi'(t_0) \geqq F(D^2\Phi(x_0, t_0), D\Phi(x_0, t_0), u(x_0, t_0), x_0, t_0).) \qquad (3.128)$$

Finally, $u \in BUC(\overline{Q}_T)$ is a solution of (3.123) if it is both a subsolution and supersolution.

As in the. classical "non rough" theory, it is possible to have upper-semicontinuous subsolutions, lower-semicontinuous supersolutions and discontinuous solutions. For simplicity, this is avoided here. Such weaker "solutions" are used to carry out the Perron construction in Sect. 3.11.

Although somewhat natural, the definition introduces several difficulties at the technical level. One of the advantages of the theory of viscosity solutions is the flexibility associated with the choice of the test functions. This is not, however, the case here. As a result, it is necessary to work very hard to obtain facts which were almost trivial in the deterministic setting. For example, in the definition, it is often useful to assume that the max/min is strict. Even this fact, which is trivial for classical viscosity solutions, in the current setting requires a more work.

It is also useful to point out the relationship between the approach used for equations with linear dependence on Du and the above definition. Heuristically, in Definition 9.1, one inverts locally the characteristics in an attempt to "eliminate" the bad term involving dB. Since the problem is nonlinear and u is not regular, it is, of course, not possible to do this globally. In a way consistent with the spirit of the theory of viscosity solutions, this difficulty is overcome by working at the level of the test functions, where, of course, it is possible to invert locally the characteristics. The price to pay for this is that the test functions used here are very robust and not as flexible as the ones used in the classical deterministic theory. This leads to several technical difficulties, since all the theory has to be revisited.

The fact that Definition 9.1 is good in the sense that it agrees with the classical (deterministic) one if $B \in C^1$, is left as an exercise. There are also several other preliminary facts about short time behavior, etc., which are omitted.

The emphasis here is on establishing a comparison principle and some stability properties. The existence follows either by a density argument or by Perron's method. The latter was established lately in a very general setting by Seeger [100] for $m \geq 1$.

The next result is about the stability properties of the pathwise viscosity solutions. Although it can be stated in a much more general form using "half relaxed limits" and lower- and upper-semicontinuous envelopes, here it is presented in a simplified form.

Proposition 9.1 *Let F_n, F be degenerate elliptic, H_n, $H \in C^2(\mathbb{R}^d; \mathbb{R}^m)$, B_n, $B \in C([0, \infty); \mathbb{R}^m)$ be such that $\sup_{i,n} \|D^2 H_{i,n}\| < \infty$ and, as $n \to \infty$ and locally uniformly, $F_n \to F$, $H_n \to H$ in $C^2(\mathbb{R}^d; \mathbb{R}^m)$, and $B_n \to B$ in $C([0, \infty); \mathbb{R}^m)$. If u_n is a pathwise solution of (3.123) with nonlinearity F_n, Hamiltonian H_n and path B_n and $u_n \to u$ in $C(\overline{Q_T})$, then u is a pathwise solution of (3.123).*

The assumptions that $H_n \to H$ in $C^2(\mathbb{R}^d; \mathbb{R}^m)$ instead of just in $C(\mathbb{R}^d)$ and $\sup_n \|D^2 H_n\| < \infty$ are not needed for the "deterministic" theory. Here they are dictated by the nature of the test functions.

Proof (Proof of Proposition 9.1) Let $(x_0, t_0) \in \mathbb{R}^d \times (0, T]$ be a strict maximum of $u - \Phi - \psi$ where $\psi \in C^1((0, \infty))$ and, for some $h > 0$, Φ is a smooth solution of (3.46) in $(t_0 - h, t_0 + h)$.

Let Φ_n be the smooth solution of

$$\Phi_{nt} = H_n(D\Phi_n)\dot{B}_n \text{ in } \mathbb{R}^d \times (t_0 - h_n, t_0 + h_n) \quad \Phi_n(\cdot, t_0) = \Phi(\cdot, t_0).$$

The assumptions on the H_n and B_n imply that, as $n \to \infty$, $\Phi_n \to \Phi$, $D\Phi_n \to D\Phi$ and $D^2\Phi_n \to D^2\Phi$ in $C(\mathbb{R}^d \times (t_0 - h', t_0 + h'))$, for some, uniform in n, $h' \in (0, h)$; note that this is the place where $H_n \to H$ in $C^2(\mathbb{R}^d)$ and $\sup_n \|D^2 H_n\| < \infty$ are used.

Let (x_n, t_n) be a maximum point of $u_n - \Phi_n - \psi$ in $\mathbb{R}^d \times [t_0 - h', t_0 + h']$. Since (x_0, t_0) is a strict maximum of $u - \Phi - \psi$, there exists a subsequence such that $(x_n, t_n) \to (x_0, t_0)$. The definition of viscosity solution then gives

$$\psi'(t_n) \leq F(D^2\Phi_n(x_n, t_n), D\Phi_n(x_n, t_n), u_n(x_n, t_n), x_n, t_n).$$

Letting $n \to \infty$ yields the claim.

The next result is the comparison principle for pathwise viscosity solutions of the first-order initial value problem, that is,

$$du = \sum_{i=1}^{m} H^i(Du) \cdot dB_i \text{ in } Q_\infty \quad u(\cdot, 0) = u_0. \tag{3.129}$$

Theorem 9.1 *Assume that (3.124), (3.125) and $u_0 \in BUC(\mathbb{R}^d)$. Then (3.129) has a unique pathwise solution $u \in BUC(\overline{Q_\infty})$ which agrees with the "solution" obtained from the extension operator.*

The proof follows from the arguments used to prove the next result about the extension operator for (3.123) which is stated next, hence it is omitted.

The next result is about the extension operator for (3.123). As before, it is shown that the solutions to initial value problems (3.123) with smooth time signal approximating the given rough one form a Cauchy family in $BUC(\overline{Q}_T)$ and, hence, all converge to the same function which is a pathwise viscosity solution to (3.123).

The next result provides an extension from smooth to arbitrary continuous paths B. For simplicity the dependence of F on u, x and t is omitted.

Theorem 9.2 *Assume* (3.124)–(3.126) *and fix* $u_0 \in BUC(\mathbb{R}^d)$ *and* $B \in C_0([0, \infty); \mathbb{R}^m)$. *Consider two families* $(\zeta_\varepsilon)_{\varepsilon > 0}$, $(\xi_\eta)_{\eta > 0}$ *in* $C_0([0, \infty); \mathbb{R}^m) \cap C^1([0, \infty); \mathbb{R}^m)$ *and* $(u_{0,\varepsilon})_{\varepsilon > 0}$, $(v_{0,\eta})_{\eta > 0} \in BUC(\mathbb{R}^d)$ *such that, as* $\varepsilon, \eta \to 0$, ζ_ε *and* ξ_η *converge to* B *in* $C([0, \infty); \mathbb{R}^m)$ *and* $u_{0,\varepsilon}$ *and* $v_{0,\eta}$ *converge to* u_0 *uniformly in* $\mathbb{R}^d$. *Let* $(u_\varepsilon)_{\varepsilon > 0}, (v_\eta)_{\eta > 0} \in BUC(\overline{Q}_\infty)$ *be the unique viscosity solutions of* (3.123) *with signal and initial datum* ζ_ε, $u_{0,\varepsilon}$ *and* ξ_η, $v_{0,\eta}$ *respectively. Then, for all* $T > 0$, *as* $\varepsilon, \eta \to 0$, $u_\varepsilon - v^\eta \to 0$ *uniformly in* $\overline{Q}_T$. *In particular, the family* $(u_\varepsilon)_{\varepsilon > 0}$ *is Cauchy in* $BUC(\overline{Q}_T)$ *and all approximations converge to the same limit.*

Proof Fix $T > 0$ and consider the doubled initial value problem

$$
\begin{cases}
dZ^{\lambda, \varepsilon, \eta} = \sum_{i=1}^m H(D_x Z^{\lambda, \varepsilon, \eta}) \dot{\xi}_{i,\varepsilon} - \sum_{i=1}^m H(-D_y Z^{\lambda, \varepsilon, \eta}) \dot{\zeta}_{i,\eta}, & \text{in } \mathbb{R}^d \times \mathbb{R}^d \times (0, T) \\[2mm]
Z^{\lambda, \varepsilon, \eta}(x, y, 0) = \lambda |x - y|^2.
\end{cases}
$$

$$(3.130)$$

It is immediate that $Z^{\lambda, \varepsilon, \eta}(x, y, t) = \Phi^{\lambda, \varepsilon, \eta}(x - y, t)$, where

$$
\Phi_t^{\lambda, \varepsilon, \eta} = \sum_{i=1}^m H(D_z \Phi^{\lambda, \varepsilon, \eta})(\dot{\xi}_{i,\varepsilon} - \dot{\zeta}_{i,\eta}) \text{ in } Q_T \quad \Phi^{\lambda, \varepsilon, \eta}(z, 0) = \lambda |z|^2. \quad (3.131)
$$

As discussed earlier, there exists $T^{\lambda, \varepsilon, \eta} > 0$ such that $\Phi^{\lambda, \varepsilon, \eta}$ is given by the method of characteristics in $\mathbb{R}^d \times [0, T^{\lambda, \varepsilon, \eta})$ and

$$
\lim_{\varepsilon, \eta \to 0} T^{\lambda, \varepsilon, \eta} = \infty \text{ and } \lim_{\varepsilon, \eta \to 0} \sup_{(z,t) \in \mathbb{R}^d \times [0,T]} \left(\Phi^{\lambda, \varepsilon, \eta}(z) - \lambda |z|^2 \right) = 0. \quad (3.132)
$$

The conclusion will follow as soon as it established that

$$
\lim_{\lambda \to \infty} \overline{\lim_{\varepsilon, \eta \to 0}} \sup_{(x,y) \in \mathbb{R}^{2N}, t \in [0,T]} (u^\varepsilon(x, t) - v^\eta(y, t) - \lambda |x - y|^2) = 0. \quad (3.133)
$$

Consider next the function

$$
\Psi^{\lambda, \varepsilon, \eta}(x, y, t) = u^\varepsilon(x, t) - v^\eta(y, t) - \Phi^{\lambda, \varepsilon, \eta}(x - y, t).
$$

The classical theory of viscosity solutions (see [17]) yields that the map

$$t \longmapsto M^{\lambda,\varepsilon,\eta}(t) = \sup_{x,y\in\mathbb{R}^d} [u^\varepsilon(x,t) - v^\eta(y,t) - \Phi^{\lambda,\varepsilon,\eta}(x-y,t)]$$

is nonincreasing in $[0, T^{\lambda,\varepsilon,\eta})$.

Hence, for $x, y \in \mathbb{R}^d$ and $t \in [0, T^{\lambda,\varepsilon,\eta})$,

$$u^\varepsilon(x,t) - v^\eta(y,t) - \Phi^{\lambda,\varepsilon,\eta}(x-y,t) \leqq \sup_{x,y\in\mathbb{R}^d} (u_0^\varepsilon(x) - v_0^\eta(y) - \lambda|x-y|^2).$$

The claim now follows from the assumptions on u_0^ε and v_0^η.

The uniqueness of the pathwise viscosity solutions of (3.123) is considerably more complicated than the one for (3.129). This is consistent with the deterministic theory, where the uniqueness theory of viscosity solutions for second-order degenerate, elliptic equations is by far more complex than the one for Hamilton-Jacobi equations. For the same reasons as for the existence, I will present the argument omitting the dependence on u, x and t.

The proof follows the general strategy outlined in the "User's Guide". The actual arguments are, however, different and more complicated.

Recall that in the background of the "deterministic" proof are the so called sup- and inf-convolutions. These are particular regularizations that yield approximations which have parabolic expansions almost everywhere and are also subsolutions and supersolutions of the nonlinear pde.

This is exactly where the pathwise case becomes different. The "classical" sup-convolutions and inf-convolutions of pathwise viscosity solutions do not have parabolic expansions. To deal with this serious difficulty, it is necessary to change the sup-convolutions and inf-convolutions by replacing the quadratic weights by short time smooth solutions of the first-order part of (3.123). The new regularizations have now parabolic expansions—the reader should think that the new weights remove the "singularities" due to the roughness of B.

Theorem 9.3 *Assume* (3.124)–(3.126). *Let* $u, v \in BUC(\overline{Q}_\infty)$ *be respectively a viscosity subsolution and supersolution of* (3.123). *Then, for all* $t \geq 0$,

$$\sup_{x\in\mathbb{R}^d} (u - v)(x,t) \leqq \sup_{x\in\mathbb{R}^d} (u - v)(x,0). \tag{3.134}$$

Proof To simplify the presentation below it is assumed that $m = 1$. Recall that, for any $\phi \in C^3(\mathbb{R}^d \times \mathbb{R}^d) \cap C^{0,1}(\mathbb{R}^d \times \mathbb{R}^d)$, there exists some $a > 0$ such that the doubled initial value problem

$$dU = [H(D_xU)-H(-D_yU)]\circ dB \text{ in } \mathbb{R}^d\times(t_0-a, t_0+a) \quad U(x,y,t_0) = \phi(x,y),$$

has a smooth solution which, for future use, is denoted by $S_H^d(t - t_0, t_0)\phi$.

If ϕ is of separated form, that is, $\phi(x, y) = \phi_1(x) + \phi_2(y)$, making if necessary, the interval of existence smaller, it is immediate that

$$S^d(t - t_0, t_0)\phi(x, y) = S_H^+(t - t_0, t_0)\phi_1(x) + S_H^-(t - t_0, t_0)\phi_2(y),$$

where, as before, $S_{H\pm}^{\pm}$ denote the smooth short time solution operators to $du = \pm H(Du) \cdot dB$.

Moreover, for any $\lambda > 0$ and $t, t_0 \in \mathbb{R}$, it is obvious that

$$S^d(t - t_0, t_0)(\lambda| \cdot - \cdot |^2)(x, y) = \lambda|x - y|^2.$$

Finally, again for smooth solutions,

$$S_H^-(t - t_0, t_0)\phi_2(y) = -S_H^+(t - t_0, t_0)(-\phi_2)(y).$$

Fix $\mu > 0$. The claim is that, for large enough λ,

$$\Phi(x, y, t) = u(x, t) - u(y, t) - \lambda|x - y|^2 - \mu t$$

cannot have a maximum in $\mathbb{R}^d \times \mathbb{R}^d \times (0, T]$. This leads to the desired conclusion as in the classical proof of the maximum principle.

Arguing by contradiction, it is assumed that there exists $(x_\lambda, y_\lambda, t_\lambda) \in \mathbb{R}^d \times \mathbb{R}^d \times (0, T]$ such that, for all $(x, y, t) \in \mathbb{R}^d \times \mathbb{R}^d \times [0, T]$,

$$\Phi(x, y, t) = u(x, t) - v(y, t) - S^d(t - t_\lambda, t_\lambda)(\lambda| \cdot - \cdot |^2)(x, y) - \mu t$$
$$\leqq \Phi(x_\lambda, y_\lambda, t_\lambda). \tag{3.135}$$

To handle the behavior at infinity and assert the existence of a maximum, it is necessary to consider $S^d(t - t_\lambda, t_\lambda)[\lambda| \cdot - \cdot |^2 + \beta v(\cdot)](x, y)$ instead of $S^d(t - t_\lambda)[\lambda| \cdot - \cdot |^2]$ in (3.135), for $t - t_\lambda$ small, $\beta \to 0$, and a smooth approximation $v(x)$ of $|x|$. Since this adds some tedious details which may obscure the main ideas of the proof, below it is assumed that a maximum exists.

Elementary computations and a straightforward application of the Cauchy-Schwarz inequality yield, for all $\varepsilon > 0$ and $\xi, \eta \in \mathbb{R}^d$,

$$|x - y|^2 - |x_\lambda - y_\lambda|^2$$
$$\leqq 2\langle x_\lambda - y_\lambda, x - x_\lambda - \xi \rangle$$
$$- 2\langle x_\lambda - y_\lambda, y - y_\lambda - \eta \rangle + 2\langle x_\lambda - y_\lambda, \xi - \eta \rangle \tag{3.136}$$
$$+ (2 + \varepsilon^{-1})(|x - x_\lambda - \xi|^2 + |y - y_\lambda - \eta|^2)$$
$$+ (1 + 2\varepsilon)|\xi - \eta|^2.$$

Let

$$p_\lambda = \lambda(x_\lambda - y_\lambda), \quad \lambda_\varepsilon = \lambda(2 + \varepsilon^{-1}) \quad \text{and} \quad \beta_\varepsilon = \lambda(1 + 2\varepsilon).$$

The comparison of local in time smooth solutions of stochastic Hamilton-Jacobi equations, which are easily obtained by the method of characteristics, and the facts explained before the beginning of the proof yield that the function

$$\begin{aligned}
\Psi(x, y, \xi, \eta, t) = {}& u(x, t) - v(y, t) \\
& - S_H^+(t - t_\lambda, t_\lambda)(2\langle p_\lambda, \cdot - x_\lambda - \xi\rangle \\
& + \lambda_\varepsilon| \cdot -x_\lambda - \xi|^2)(x) \\
& - S_H^-(t - t_\lambda, t_\lambda)(-2\langle p_\lambda, \cdot - y_\lambda - \eta\rangle \\
& + \lambda_\varepsilon| \cdot -y_\lambda - \eta|^2)(y) \\
& - 2\langle p_\lambda, \xi - \eta\rangle - \beta_\varepsilon|\xi - \eta|^2 - \mu t
\end{aligned}$$

achieves, for $h \leq h_0 = h_0(\lambda, \varepsilon^{-1})$, its maximum in $\mathbb{R}^d \times \mathbb{R}^d \times \mathbb{R}^d \times \mathbb{R}^d \times (t_\lambda - h, t_\lambda + h)$ at $(x_\lambda, y_\lambda, 0, 0, t_\lambda)$.

Note that here it is necessary to take $t - t_\lambda$ sufficiently small to have local in time smooth solutions for the doubled as well as the H and $-H(-)$ equations given by the characteristics.

For $t \in (t_\lambda - h, t_\lambda + h)$ define the modified sup- and inf-convolutions

$$\bar{u}(\xi, t) = \sup_{x \in \mathbb{R}^d} [u(x, t) - S_H^+(t - t_\lambda, t_\lambda)(2\langle p_\lambda, \cdot - x_\lambda - \xi\rangle + \lambda_\varepsilon| \cdot -x_\lambda - \xi|^2)(x)]$$

and

$$\underline{v}(\eta, t) = \inf_{y \in \mathbb{R}^d} [v(y, t) + S_H^-(t - t_\lambda, t_\lambda)(-2\langle p_\lambda, \cdot - y_\lambda - \eta\rangle + \lambda_\varepsilon| \cdot -y_\lambda - \eta|^2)(y)].$$

It follows that, for $\delta > 0$,

$$G(\xi, \eta, t) = \bar{u}(\xi, t) - \underline{v}(\eta, t) - (\beta_\varepsilon + \delta)|\xi - \eta|^2 - 2\langle p_\lambda, \xi - \eta\rangle - \mu t$$

attains its maximum in $\mathbb{R}^d \times \mathbb{R}^d \times (t_\lambda - h, t_\lambda + h)$ at $(0, 0, t_\lambda)$.

Observe next that there exists a constant $K_{\varepsilon,\lambda} > 0$ such that, in $\mathbb{R}^d \times (t_\lambda - h, t_\lambda + h)$,

$$D_\xi^2 \bar{u} \geq -K_{\varepsilon,\lambda}, \quad D_\eta^2 \underline{v} \leq K_{\varepsilon,\lambda}, \quad \bar{u}_t \leq K_{\varepsilon,\lambda}, \quad \text{and} \quad \underline{v}_t \geq -K_{\varepsilon,\lambda}. \tag{3.137}$$

with the inequalities understood both in the viscosity and distributional sense.

The one sided bounds of $D_\xi^2 \bar{u}$ and $D_\eta^2 \underline{v}$ are an immediate consequence of the definition of $\bar{u}$ and $\underline{v}$ and the regularity of the kernels, which imply that, for some $K_{\varepsilon,\lambda} > 0$ and in $\mathbb{R}^d \times (t_\lambda - h, t_n + h)$,

$$|D_\xi^2 S_H^+ (\cdot - t_\lambda, t_\lambda)(2\langle p_\lambda, \cdot - x_\lambda - \xi \rangle + \lambda_\varepsilon | \cdot -x_\lambda - \xi|^2)|$$

$$+ |D_\eta^2 S_H^- (\cdot - t_\lambda, t_\lambda)(-2\langle p_\lambda, \cdot - y_\lambda - \eta \rangle + \lambda_\varepsilon | \cdot -y_\lambda - \eta|^2)| \leq K_{\varepsilon,\lambda} \,.$$

The bound for $\bar{u}_t$ is shown next; the argument for $\underline{v}_t$ is similar. Note that, in view of the behavior of B, such a bound cannot be expected to hold for u_t. Indeed take $F \equiv 0$ and $H \equiv 1$, in which case $u(x,t) = B_t$.

Assume that, for some smooth function g and for ξ fixed, the map $(\xi, t) \rightarrow \bar{u}(\xi, t) - g(t)$ has a max at $\hat{t}$. It follows that

$$(x, t) \mapsto u(x, t) - S_H^+(t - t_\lambda, t_\lambda) \left(2\langle p_\lambda, \cdot - x_\lambda - \xi \rangle + \lambda_\varepsilon | \cdot -\xi|^2 \right)(x) - g(t)$$

has a max at $(\hat{x}, \hat{t})$, where $\hat{x}$ is a point where that supremum in the definition of $\bar{u}(\xi, t)$ is achieved, that is,

$$\bar{u}(\xi, \hat{t}) = u(\hat{x}, \hat{t}) - S_H^+(\hat{t} - t_\lambda, t_\lambda) \left(2\langle p_\lambda, \cdot - x_\lambda - \xi \rangle + \lambda_\varepsilon | \cdot -\xi|^2 \right)(\hat{x}) \,.$$

In view of the definition of the pathwise viscosity sub-solution, it follows that there exists some $\hat{K}_{\varepsilon,\lambda}$ depending on $K_{\varepsilon,\lambda}$ and H, such that $g(\hat{t}) \leq \hat{K}_{\varepsilon,\lambda}$, and, hence, the claim follows.

The one-sided bounds (3.137) yield the existence of $p_n, q_n, \xi_n, \eta_n \in \mathbb{R}^d$ and $t_n > 0$ such that, as $n \rightarrow \infty$,

1. $(\xi_n, \eta_n, t_n) \rightarrow (0, 0, t_\lambda)$, $p_n, q_n \rightarrow 0$,
2. the map $(\xi, \eta, t) \rightarrow \bar{u}(\xi, t) - \underline{v}(\eta, t) - \beta_\varepsilon |\xi - \eta|^2 - \langle p_n, \xi \rangle - \langle q_n, \eta \rangle$

   $- 2\langle p_\lambda, \xi - \eta \rangle - \mu t$ has a maximum at (ξ_n, η_n, t_n),
3. $\bar{u}$ and $\underline{v}$ have parabolic second-order expansions from above and below at (ξ_n, t_n) and (η_n, t_n) respectively, that is, there exist $a_n, b_n \in \mathbb{R}$ such that

$$\bar{u}(\xi, t) \leq \bar{u}(\xi_n, t_n) + a_n(t - t_n) + (D_\xi \bar{u}(\xi_n, t_n), \xi - \xi_n)$$

$$+ \frac{1}{2}(D_\xi^2 \bar{u}(\xi_n, t_n)(\xi - \xi_n), \xi - \xi_n) + o(|\xi - \xi_n|^2 + |t - t_n|) \,,$$

and

$$\underline{v}(\eta, t) \geq \underline{v}(\eta_n, t_n) + b_n(t - t_n) + (D_\eta \underline{v}(\eta_n, t_n), \eta - \eta_n)$$

$$+ \frac{1}{2}(D_\eta^2 \underline{v}(\eta_n, t_n)(\eta - \eta_n), \eta - \eta_n) + o(|\eta - \eta_n|^2 + |t - t_n|) \,,$$

and, finally,

4. $a_n = b_n + \mu$, $D_\xi \bar{u}(\xi_n, t_n) = p_n + 2p_\lambda + 2\beta_\varepsilon(\xi_n - \mu_n)$, $D_\eta \underline{v}(\xi_n, t_n) = -q_n + 2p_\lambda 2\beta_\varepsilon(\xi_n - \eta_n)$ and $D_\xi^2 \bar{u}(\xi_n, t_n) \le D_\eta^2 \underline{v}(\eta_n, t_n)$.

It follows that, for some $\theta > 0$ fixed, $t < t_n$, (ξ, t) near (ξ_n, t_n) and (η, t) near (η_n, t_n), the maps

$$(x, \xi, t) \to u(x, t) - S_H^+(t - t_\lambda, t_\lambda)(2\langle p_\lambda, \cdot - x_\lambda - \xi \rangle + \lambda_\varepsilon | \cdot - x_\lambda - \xi |^2)(x) - \Phi(\xi, t),$$

and

$$(y, \eta, t) \to v(y, t) + S_H^-(t - t_\lambda, t_\lambda)(-2\langle p_\lambda, \cdot - y_\lambda - \eta \rangle + \lambda_\varepsilon | \cdot - \eta |^2)(y) - \Psi(\eta, t),$$

attain respectively a maximum at (x_n, ξ_n, t_n) and a minimum at (y_n, η_n, t_n), where

$$\Phi(\xi, t) = \bar{u}(\xi_n, t_n) + (\bar{u}_t(\xi_n, t_n) - \theta)(t - t_n) + (D_\xi \bar{u}(\xi_n, t_n), \xi - \xi_n)$$

$$+ \frac{1}{2}((D_\xi^2 \bar{u}(\xi_n, t_n) + \theta I)(\xi - \xi_n), \xi - \xi_n)$$

and

$$\Psi(\eta, t) = \underline{v}(\eta_n, t_n) + (\underline{v}_t(\eta_n, t_n) + \theta)(t - t_n) + (D_\eta \underline{v}(\eta_n, t_n), \eta - \eta_n)$$

$$+ \frac{1}{2}((D_\eta^2 \underline{v}(\eta_n, t_n) + \theta I)(\eta - \eta_n), \eta - \eta_n).$$

Next, for sufficiently small $r > 0$, let $\mathcal{B}(\xi_n, t_n, r_n) = B(\xi_n, r) \times (t_n - r, t_n]$ and define

$$\overline{\Phi}(x, t) = \inf \big[\Phi(\xi, t) + S_H^+(t - t_\lambda, t_\lambda)(2\langle p_\lambda, \cdot - x_\lambda - \xi \rangle$$

$$+ \lambda_\varepsilon | \cdot - x_\lambda - \xi |^2)(x) : (\xi, t) \in \mathcal{B}(\xi_n, t_n, r_n) \big],$$

and

$$\underline{\Psi}(y, t) = \sup \big[\Psi(\eta, t) - S_H^-(t - t_\lambda, t_\lambda)(-2\langle p_\lambda, \cdot - y_\lambda - \eta \rangle$$

$$+ \lambda_\varepsilon | \cdot - y_\lambda - \eta |^2)(x) : (\xi, t) \in \mathcal{B}(\xi_n, t_n, r_n) \big].$$

It follows that $u - \overline{\Phi}$ and $v - \underline{\Psi}$ attain a local max at (x_n, t_n) and a local min at (y_n, t_n). Moreover, $\overline{\Phi}$ and $\underline{\Psi}$ are smooth solutions of $du = H(Du) \cdot dB$ for (x, t) near (x_n, t_n) and $dv = -H(-D_y v) \cdot dB$ for (y, t) near (y_n, t_n). This last assertion for $\overline{\Phi}$ and $\underline{\Psi}$ follows, using the inverse function theorem, from the fact that, at (x_n, t_n) and (y_n, t_n), there exists a unique minimum in the definition of $\overline{\Phi}$ and $\underline{\Psi}$. This in turn comes from the observation that for $\lambda > \lambda_0$, at (ξ_n, x_n, t_n) and (η_n, y_n, t_n),

$$D^2 \overline{\Phi}(\xi_n, t_n) + (\lambda_\varepsilon + \theta)I > 0 \quad \text{and} \quad D^2 \underline{\Psi}(\eta_n, t_n) - (\lambda_\varepsilon + \theta)I < 0.$$

Finally, elementary calculations also yield that

$$D_\xi^2 \Phi(\xi_n, t_n) \geqq D_x^2 \bar{\Phi}(x_n, t_n) \quad \text{and} \quad D_\eta^2 \Psi(\eta_n, t_n) \leqq D_y^2 \underline{\Psi}(y_n, t_n) .$$

Applying now the definitions of the pathwise subsolution and supersolution to u and v respectively, yields

$$\bar{u}_t(\xi_n, t_n) - \theta \leq F(D_x^2 \bar{\Phi}(x_n, t_n), D_x \bar{\Phi}(x_n, t_n))$$

$$\leq F(D_\xi^2 \Phi(\xi_n, t_n), D_\xi \Phi(\xi_n, t_n))$$

$$= F(D_\xi^2 \bar{u}(\xi_n, t_n) + \theta I, D_\xi \bar{u}(\xi_n, t_n))$$

and

$$\underline{v}_t(\eta_n, t_n) + \theta \geq F(-D_\xi^2 \underline{v}(\xi_n, t_n) - \theta I, D_\eta \underline{v}(\eta_n, t_n)).$$

Hence

$$\mu - 2\theta \leq a_n - b_n - 2\theta$$

$$\leq \sup[F(A + \theta I, p + p_n)$$

$$- F(A - \theta I, p + q_n) : |p_n|, |q_n| \leq n^{-1}, |A| \leq K_{\varepsilon, \lambda}].$$

The conclusion now follows choosing $\varepsilon = (2\lambda)^{-1}$ and letting $\lambda \to \infty$ and $\theta \to 0$.

It is worth remarking that, in the course of the previous proof, it was shown that, for $0 < h \leq \hat{h}_0$, with $\hat{h}_0 = \hat{h}_0(\lambda, \varepsilon) \leq h_0$, $\bar{u}$ (resp. $\underline{v}$) is a viscosity subsolution (resp. supersolution) of

$$\bar{u}_t \leq F(D_\xi^2 \bar{u}, D_\xi \bar{u}) \quad (\text{resp. } \underline{v}_t \geq F(D_\eta^2 \underline{v}, D_\eta \underline{v})) \text{ in } \mathbb{R}^d \times (t_\lambda - h, t_\lambda + h).$$

3.10 Pathwise Solutions to Fully Nonlinear First and Second Order pde with Spatially Dependent Smooth Hamiltonians

3.10.1 The General Problem, Strategy and Difficulties

The next step in the development of the theory is to consider spatially dependent Hamiltonians and, possibly, multiple paths.

Most of this section is about pathwise solutions of initial value problems of the form

$$du = F(D^2u, Du, u, x) + H(Du, x) \cdot dB \quad \text{in } Q_\infty \quad u(\cdot, 0) = u_0, \qquad (3.138)$$

with only one path and, as always, F degenerate elliptic.

Extending the theory to equations with multiple rough time dependence had been an open problem until very recently, when Lions and Souganidis [72] came up with a way to resolve the difficulty. A brief discussion about this appears at the end of this section.

Finally, to study equations for nonsmooth Hamiltonians, it is necessary to modify the definition of the solution using now as test functions solutions of the doubled equations constructed for non smooth Hamiltonians. The details appear in [71].

The strategy of the proof of the comparison is similar to the one followed for spatially homogeneous Hamiltonians. The pathwise solutions are defined using as test functions smooth solutions of

$$du = H(Du, x) \cdot dB \quad \text{in } \mathbb{R}^d \times (t_0 - h, t_0 + h), \quad u(\cdot, t_0) = \phi, \qquad (3.139)$$

which under the appropriate assumptions on H exist for each $t_0 > 0$ and smooth ϕ in $(t_0 - h, t_0 + h)$ for some small h.

The aim in this section is to prove that pathwise solutions are well posed. To avoid many technicalities, the discussion here is restricted to Hamilton-Jacobi initial value problems

$$du = H(Du, x) \cdot dB \quad \text{in } Q_T \quad u(\cdot, 0) = u_0. \qquad (3.140)$$

The general problem (3.138) is studied using he arguments of this and the previous sections; some details can be found in [100].

Similarly to the spatially homogeneous case, the main technical issue is to control the length of the interval of existence of smooth solutions of the doubled equation with quadratic initial datum and smooth approximations to ζ^ε and ξ^η of the path B, that is,

$$\begin{cases} dz = H(D_xz, x) \cdot d\zeta^\varepsilon - H(-D_yz, y) \cdot d\xi^\eta \\ \qquad \qquad \text{in } \mathbb{R}^d \times \mathbb{R}^d \times (t_0 - h, t_0 + h), \\ z(x, y, 0) = \lambda|x - y|^2. \end{cases} \qquad (3.141)$$

As already discussed earlier, the most basic estimate is that $h = O(\lambda^{-1})$, which, as is explained below, is too small to carry out the comparison proof. The challenge, therefore, is to take advantage of the cancellations, due to the special form of the initial datum as well as of the doubled Hamiltonian, to obtain smooth solutions in a longer time interval.

Since the smooth solutions to (3.141) are constructed by the method of characteristics, the technical issue is to control the length of the interval of invertibility of
the characteristics. This can be done by estimating the interval of time in which
the Jacobian does not vanish. It is here that using a single path helps, because,
after a change of time, the problem reduces to studying the analogous question for
homogeneous in time odes.

To further simplify the presentation, the "rough" problem discussed in the sequel
is not (3.141) but rather the doubled equation with the rough path, that is

$$
\begin{cases}
dw = H(D_x w, x) \cdot dB - H(-D_y w, y) \cdot dB \\
\qquad\qquad \text{in } \mathbb{R}^d \times \mathbb{R}^d \times (t_0 - h, t_0 + h) \\
w(x, y, 0) = \lambda'|x - y|^2.
\end{cases}
\tag{3.142}
$$

In what follows, to avoid cumbersome expressions, $\lambda = 2\lambda'$

The short time smooth solutions of (3.142) are given by $w(x, y, t) =
U(x, y, B(t) - B(t_0))$, where U is the short time smooth solutions to the "non-
rough" doubled initial value problem

$$
\begin{cases}
U_t = H(D_x U, x) - H(-D_y U, y) \text{ in } \mathbb{R}^d \times \mathbb{R}^d \times (-T^*, T^*) \\
U(x, y, 0) = \lambda'|x - y|^2,
\end{cases}
\tag{3.143}
$$

and $T^* > 0$ and h are such that $\sup_{s \in (t_0 - h, t_0 + h)} |B(s) - B(t_0)| \le T^*$.

The smooth solutions of (3.143) are constructed by inverting the map $(x, y) \to
(X(x, y, t), Y(x, y, t))$ of the corresponding system of characteristics, that is

$$
\begin{cases}
\dot{X} = -D_p H(P, X) \quad \dot{Y} = -D_q H(Q, Y), \\
\dot{P} = D_x H(P, X) \quad \dot{Q} = D_y H(Q, Y), \\
\dot{U} = H(P, X) - \langle D_p H(P, X), P \rangle \\
\qquad - H(Q, Y) + \langle D_q H(Q, Y), Q \rangle \\
X(x, y, 0) = x \quad Y(x, y, 0) = y, \\
P(x, y, 0) = Q(x, y, 0) = \lambda(x - y) \\
U(x, y, 0) = \lambda'|x - y|^2.
\end{cases}
\tag{3.144}
$$

A crude estimate, which does not take into account the special form of the system
and the initial data, gives that the map $(x, y) \mapsto (X(x, y, t), Y(x, y, t))$ is invertible
at least in a time interval of length $O(\lambda^{-1})$ with the constant depending on $\|H\|_{C^2}$.
This implies that the characteristics of (3.142) are invertible as long as

$$
\sup_{s \in (t_0 - h, t_0 + h)} |B(s) - B(t_0)| \le O(\lambda^{-1}).
$$

It turns out, as it is shown below, that this interval is not long enough to yield a comparison for the pathwise solutions. Taking, however, advantage of the special structure of (3.143) and (3.144) and under suitable assumptions on H and its derivatives, it is possible to improve the estimate of the time interval.

The discussion next aims to explain the need of intervals of invertibility that are longer than $O(\lambda^{-1})$, and serves as a blueprint for the strategy of the actual proof.

Assume that u and v are respectively a subsolution and a supersolution of (3.139). As in the x-independent case, it is assumed that, for some $\alpha > 0$ and $\lambda > 0$, (x_0, y_0, t_0) with $t_0 > 0$ is a maximum point of

$$u(x, t) - v(y, t) - \lambda'|x - y|^2 - \alpha t.$$

Then, for $h > 0$ and all x, y,

$$u(x, t_0 - h) - v(y, t_0 - h) \leq \lambda'|x - y|^2 - \alpha h + u(x_0, t_0) - v(y_0, t_0) - \lambda'|x_0 - y_0|^2 .$$

Since $w(x, y, t) = u(x, t) - v(y, t)$ solves the doubled equation (3.139), to obtain the comparison it is enough to compare w with the small time smooth solution z to (3.142) starting at $t_0 - h$.

It follows that

$$u(x_0, t_0) - v(y_0, t_0) \leq w(x_0, y_0, t_0) + u(x_0, t_0) - v(y_0, t_0) - w(x_0, y_0, t_0 - h) - \alpha h,$$

and, hence,

$$\alpha \leq \frac{w(x_0, y_0, t_0) - w(x_0, y_0, t_0 - h)}{h}.$$

Recall that h depends on λ and, to conclude, this dependence must be such that

$$\limsup_{\lambda \to \infty} \frac{w(x_0, y_0, t_0) - w(x_0, y_0, t_0 - h)}{h} \leq 0.$$

On the other hand, it will be shown that, if z is a smooth solution to (3.142), then

$$w(x_0, y_0, t_0) - w(x_0, y_0, t_0 - h) \lesssim \sup_{s \in (t_0 - h, t_0 + h)} |B(s) - B(t_0)| h^{-1} \lambda^{-\frac{1}{2}}.$$

Combining the last two statements implies that, to get a contradiction, $h = h(\lambda)$ must be such that

$$\limsup_{\lambda \to \infty} \sup_{s \in (t_0 - h, t_0 + h)} |B(s) - B(t_0)| h^{-1} \lambda^{-\frac{1}{2}} = 0. \tag{3.145}$$

The next argument indicates that there is indeed a problem if the smooth solutions of the "deterministic" doubled problem exist only for times of order $O(\lambda^{-1})$.

Indeed in this case, the proof of the comparison argument outlined above, yields

$$\alpha h \lesssim o(1)|B(t_0) - B(t_0 - h)|\lambda^{-1/2},$$

and, if $B \in C^{0,\beta}([0, \infty))$, it follows that $h^\beta \approx \lambda^{-1}$, and the above inequality yields $\alpha \lesssim o(1)h^{\frac{3\beta}{2}-1}$ in which case it is not possible to obtain a contradiction, if $\beta < 2/3$, which, of course, is the case for Brownian paths.

It appears, at least for the moment formally, that for this case the Brownian case "optimal" interval of existence is $O(\lambda^{-1/2})$. Indeed if this is the case then we must have $|B(t_0) - B(t_0 - h)| \approx \lambda^{-\frac{1}{2}}$, and, hence, $h \approx \lambda^{-1}$. This leads to $\alpha \lesssim o(1)$ and, hence, a contradiction.

3.10.2   Improvement of the Interval of Existence of Smooth Solutions

The problem is to find longer than $O(\lambda^{-1})$ intervals of existence of smooth solution of the doubled deterministic Hamilton-Jacobi equation (3.143).

Two general sets of conditions will be modeled by two particular classes of Hamiltonians, namely separated and linear H's.

To give the reader a flavor of the type of arguments that will be involved, it is convenient to begin with "separated" Hamiltonians of the form

$$H(p, x) = H(p) + F(x), \tag{3.146}$$

in which case the doubled equation and its characteristics are

$$\begin{cases} U_t = H(D_x U) - H(-D_y U) + F(x) - F(y) \\ \qquad\qquad \text{in } \mathbb{R}^d \times \mathbb{R}^d \times (-T, T) \\ U(x, y, 0) = \lambda'|x - y|^2, \end{cases} \tag{3.147}$$

and

$$\begin{cases} \dot{X} = -DH(P) \quad \dot{Y} = -DH(Q), \\ \dot{P} = DF(X) \quad \dot{Q} = DF(Y), \\ \dot{U} = H(P) - \langle DH(P), P \rangle - H(Q) + \langle DH(Q), Q \rangle \\ \qquad\qquad +F(X) - F(Y), \\ X(0) = x \quad Y(0) = y \quad P(0) = Q(0) = \lambda(x - y) \\ U(0) = \lambda'|x - y|^2. \end{cases} \tag{3.148}$$

Let $J(t)$ denote the Jacobian of the map $(x, y) \mapsto (X(x, y, t), Y(x, y, t))$ at time
t. In what follows, to avoid the rather cumbersome notation involving determinants,
all the calculations below are presented for $d = 1$, that is $x, y \in \mathbb{R}$.

It follows that

$$J = \frac{\partial X}{\partial x} \frac{\partial Y}{\partial y} - \frac{\partial X}{\partial y} \frac{\partial Y}{\partial x} \quad \text{and} \quad J(0) = 1.$$

The most direct way to find an estimate for the time of existence of smooth solutions
is, for example, to obtain a bound for the first time t_λ such that $J(t_\lambda) = \frac{1}{2}$, and, for
this, it is convenient to calculate and estimate the derivatives of J with respect to
time at $t = 0$.

Hence, it is necessary to derive the odes satisfied by $\frac{\partial X}{\partial x}, \frac{\partial X}{\partial y}, \frac{\partial Y}{\partial x}$ and $\frac{\partial Y}{\partial y}$.

Writing $\frac{\partial X}{\partial \alpha}, \frac{\partial Y}{\partial \alpha}, \frac{\partial P}{\partial \alpha}$, and $\frac{\partial Q}{\partial \alpha}$ with $\alpha = x$ or y, differentiating (3.148) and
omitting the subscripts for the derivatives of H and F yields the systems

$$\begin{cases} \dfrac{\partial \dot{X}}{\partial \alpha} = -D^2 H(P) \dfrac{\partial P}{\partial \alpha}, \\[2em] \dfrac{\partial \dot{P}}{\partial \alpha} = D^2 F(X) \dfrac{\partial X}{\partial \alpha}, \end{cases} \qquad \begin{cases} \dfrac{\partial \hat{X}}{\partial x}(x, y, 0) = 1, \\[1em] \dfrac{\partial \hat{X}}{\partial y}(x, y, 0) = 0, \\[1.5em] \dfrac{\partial \hat{P}}{\partial x}(x, y, 0) = \lambda, \\[1em] \dfrac{\partial \hat{P}}{\partial y}(x, y, 0) = -\lambda, \end{cases}$$

and

$$\begin{cases} \dfrac{\partial \dot{Y}}{\partial \alpha} = -D^2 H(Q) \dfrac{\partial Q}{\partial \alpha}, \\[2em] \dfrac{\partial \dot{Q}}{\partial \alpha} = D^2 F(Y) \dfrac{\partial Y}{\partial \alpha}, \end{cases} \qquad \begin{cases} \dfrac{\partial \hat{Y}}{\partial x}(x, y, 0) = 0, \\[1em] \dfrac{\partial \hat{Y}}{\partial Y}(x, y, 0) = 1, \\[1.5em] \dfrac{\partial \hat{Q}}{\partial x}(x, y, 0) = \lambda, \\[1em] \dfrac{\partial \hat{Q}}{\partial y}(x, y, 0) = -\lambda. \end{cases}$$

Proposition 10.1 *Assume that $DH, DF, D^2 F, D^2 H, |D^3 H|(1 + |p|)$ and $D^4 H$
are bounded. If t_λ is the first time that $J(t_\lambda) = 1/2$, then, for some uniform constant
$c > 0$ which depends on the bounds on H, F and their derivatives, and for all*

$x, y \in \mathbb{R}^d$,

$$t_\lambda \geq c \min(1, \lambda^{-1/3}).$$

Proof Straightforward calculations that take advantage of the separated form of the Hamiltonian yield

$$\dot{J} = -D^2 H(P) \left(\frac{\partial P}{\partial x} \frac{\partial Y}{\partial y} - \frac{\partial P}{\partial y} \frac{\partial Y}{\partial x} \right) - D^2 H(Q) \left(\frac{\partial X}{\partial x} \frac{\partial Q}{\partial y} - \frac{\partial X}{\partial y} \frac{\partial Q}{\partial x} \right)$$

and

$$\ddot{J} has = - \left(D^2 H(P) D^2 F(X) + D^2 H(Q) D^2 F(Y) \right) J$$

$$+ 2D^2 H(P) D^2 H(Q) \left(\frac{\partial P}{\partial x} \frac{\partial Q}{\partial y} - \frac{\partial P}{\partial y} \frac{\partial Q}{\partial x} \right)$$

$$- D^3 H(P) DF(X) \left(\frac{\partial P}{\partial x} \frac{\partial Y}{\partial y} - \frac{\partial P}{\partial y} \frac{\partial Y}{\partial x} \right)$$

$$- D^3 H(Q) DF(Y) \left(\frac{\partial X}{\partial x} \frac{\partial Q}{\partial y} - \frac{\partial X}{\partial y} \frac{\partial Q}{\partial x} \right).$$

To simplify the expressions for $\dot{J}$ and $\ddot{J}$, it is convenient to write $\dfrac{\partial X}{\partial \alpha}, \dfrac{\partial Y}{\partial \alpha}, \dfrac{\partial P}{\partial \alpha}$ and $\dfrac{\partial Q}{\partial \alpha}$ in terms of the solutions $(\eta_1, \psi_1, \eta_2, \psi_2)$ of the linearized system

$$\begin{cases} \dot{\xi}_1 = -D^2 H(P)\phi_1, & \xi_1(0) = 1, \\ \dot{\phi}_1 = D^2 F(X)\xi_1, & \phi_1(0) = 0, \\ \dot{\xi}_2 = -D^2 H(P)\phi_2, & \xi_2(0) = 0, \\ \dot{\phi}_2 = D^2 F(X)\xi_2, & \phi_2(0) = 1, \end{cases} \quad \text{and} \quad \begin{cases} \dot{\eta}_1 = -D^2 H(Q)\psi_1, & \eta_1(0) = 1, \\ \dot{\psi}_1 = D^2 F(Y)\eta_1, & \psi_1(0) = 0, \\ \dot{\eta}_2 = -D^2 H(Q)\psi_2, & \eta_2(0) = 0, \\ \dot{\psi}_2 = D^2 F(Y)\eta_2, & \psi_2(0) = 1, \end{cases}$$

which are bounded in $[0, 1]$ and satisfy

$$\begin{cases} \xi_1(t) = 1 + O(1)t, & \xi_2(t) = O(1)t, \\ \phi_1(t) = O(1)t, & \phi_2(t) = 1 + O(1)t, \end{cases}$$

and

$$\begin{cases} \eta_1(t) = 1 + O(1)t, & \eta_2 = O(1)t, \\ \psi_1(t) = O(1)t, & \psi_2 = 1 + O(1)t, \end{cases}$$

where $O(1)$ denotes different quantities for each functions which are uniformly bounded in $[0, 1]$; note that the assumption that $D^2 H$ and $D^2 F$ are bounded is used here.

A direct substitution yields

$$\begin{cases} \frac{\partial X}{\partial x} = \xi_1 + \lambda \xi_2 & \frac{\partial X}{\partial y} = -\lambda \xi_2, \\[2mm] \frac{\partial P}{\partial x} = \phi_1 + \lambda \phi_2 & \frac{\partial P}{\partial y} = -\lambda \phi_2, \end{cases} \quad \text{and} \quad \begin{cases} \frac{\partial Y}{\partial x} = \lambda \eta_2 & \frac{\partial Y}{\partial y} = \eta_1 - \lambda \eta_2, \\[2mm] \frac{\partial Q}{\partial x} = \lambda \psi_2 & \frac{\partial Q}{\partial y} = \psi_1 - \lambda \psi_2. \end{cases}$$

Using the observations above gives

$$\frac{\partial P}{\partial x} \frac{\partial P}{\partial y} - \frac{\partial P}{\partial y} \frac{\partial Q}{\partial x} = (\phi_1 + \lambda \phi_2)(\psi_1 - \lambda \psi_2) - (-\lambda \phi_2) \lambda \psi_2$$

$$= \phi_1 \psi_1 + 2\lambda(\phi_2 \psi_1 - \phi_1 \psi_2) = O(1)(1 + 2\lambda t),$$

since

$$\phi_1 \psi_1 = O(1) \quad \text{and} \quad \phi_2 \psi_1 - \phi_1 \psi_2 = (1 + O(1)t)O(1)t - O(1)t(1 + O(1)t)$$

$$= O(1)t \, .$$

Similarly, since

$$\phi_2 \eta_1 - \phi_1 \eta_2 = (1 + O(1)t)(1 + O(1)t) - O(1)t O(1)t = 1 + O(1)t \quad \text{and}$$

$$\xi_2 \psi_1 - \xi_1 \psi_2 = O(1)t O(1)t - (1 + O(1)t)(1 + O(1)t) = O(1)t - 1,$$

it follows that

$$\frac{\partial P}{\partial x} \frac{\partial Y}{\partial y} - \frac{\partial P}{\partial y} \frac{\partial Y}{\partial x} = (\phi_1 + \lambda \phi_2)(\eta_1 - \lambda \eta_2) - (-\lambda \phi_2) \lambda \eta_2$$

$$= \phi_1 \xi_1 + \lambda(\phi_2 \eta_1 - \phi_1 \eta_2) = O(1)(1 + \lambda t) + \lambda,$$

and

$$\frac{\partial X}{\partial x} \frac{\partial Q}{\partial y} - \frac{\partial X}{\partial y} \frac{\partial Q}{\partial x} = (\xi_1 + \lambda \xi_2)(\psi_1 - \psi_2) - (-\lambda \xi_2) \lambda \psi_2$$

$$= \xi_1 \psi_1 + \lambda(\xi_2 \psi_1 - \xi_1 \psi_2) = O(1)(1 + \lambda t) - \lambda \, .$$

Inserting all the above in the expression for $\ddot{J}$ yields

$$\ddot{J} = O(1)J + O(1)(1 + \lambda t) + \lambda(D^3 H(Q)DF(Y) - D^3 H(P)DF(X)).$$

Set

$$A := (D^3 H(Q) - D^3 H(P))DF(Y) \quad \text{and} \quad D := D^3 H(P)(DF(Y) - DF(X)).$$

It is immediate that

$$\lambda |A| \leqq \lambda O(\|DF\|_\infty |Q - P|) = \lambda O(1)t,$$

with the last estimate following from the observation that

$$(P - Q)(t) = \lambda(x - y) + \int_0^t DF(X(s))ds - (\lambda(x - y)t \int_0^t DF(Y(s))ds$$

$$= \int_0^t (DF(X(s)) - DF(Y(s)))ds = O(1)t.$$

As far as D is concerned, observe that

$$|D| \leqq \|D^3 H\|_\infty |X - Y| \lesssim \frac{|X - Y|}{1 + |P|},$$

and recall that

$$|X - Y| \leqq |x - y| + O(1)t \quad \text{and} \quad |P| = |\lambda(x - y) + O(1)t|.$$

Hence,

$$\lambda |D| \lesssim \left[\frac{\lambda |x - y|}{1 + |\lambda(x - y) + O(1)t|} + \frac{\lambda O(1)t}{1 + |P|} \right]$$

$$\lesssim \left[\frac{|P(0)|}{1 + |P(0) + O(1)t|} + \frac{\lambda O(1)t}{1 + |P|} \right];$$

the second term in the bound above comes from $\lambda O(1)t$, while an additional argument is needed for the first.

Choose $t \leq t_1$ so that the $O(1)t$ term in P is such that $|O(1)t| \leq \frac{1}{2}$. If $|P(0)| \leqq 1$, then

$$\frac{|P(0)|}{1 + |P(0) + O(1)t|} \leqq 1$$

while, if $|P(0)| > 1$,

$$1 + |P(0) + O(1)t| \geqq 1 + |P(0)| - |O(1)t| \geqq |P(0)| + \frac{1}{2}$$

and

$$\frac{|P(0)|}{1+|P(t)|} \le \frac{|P(0)|}{\frac{1}{2}+|P(0)|} \le 1.$$

Combining the estimates on λA and λD gives

$$\ddot{J} = O(1)J + O(1)\lambda t + O(1).$$

It is also immediate that

$$J(0) = 1 \quad \text{and} \quad \dot{J}(0) = 0;$$

this is another place where the separated form of the Hamiltonian and the symmetric form of the test function play a role.

It follows there exists $s_\lambda \in (0, t_\lambda)$ such that

$$\frac{1}{2} = 1 + \frac{1}{2}t_\lambda^2 \ddot{J}(s_\lambda)$$

and, hence,

$$|t_\lambda^2 \ddot{J}(s_\lambda)| = 1,$$

which implies

$$1 \lesssim t_\lambda^2(1 + \lambda t_\lambda).$$

It follows that

$$1 \lesssim \lambda t_\lambda^3 + t_\lambda^2,$$

and the claim is proved.

Having established a longer than $O(\lambda^{-1})$ interval of existence for the solution U_λ of (3.147), it is now possible to obtain the following comparison result for pathwise solutions to Hamilton-Jacobi equations with separated Hamiltonians.

Theorem 10.1 Let $u \in BUC(\overline{Q}_T)$ and $v \in BUC(\overline{Q}_T)$ be respectively a subsolution and a supersolution of (3.139) in Q_T with H as in (3.146), that is, of separated form, satisfying the assumptions of Proposition 10.1. Moreover, assume that $B \in C^{0,\beta}([0, \infty])$ with $\beta \ge 2/5$. Then, for all $t \in [0, T]$,

$$\sup_{x \in \mathbb{R}^d} (u(x, t) - v(x, t)) \le \sup_{x \in \mathbb{R}^d} (u(\cdot, 0) - v(\cdot, 0)).$$

The following lemma, which is stated without a proof since it is rather classical, will be used in the proof of Theorem 10.1.

Lemma 10.1 *Assume that $H \in C(\mathbb{R}^d)$ and $F \in C^{0,1}(\mathbb{R}^d)$ and let U_λ be the viscosity solution of the doubled equation $w_t = H(D_x w) + F(x) - H(-D_y w, y) - F(y)$ in Q_T with initial datum $\lambda|x - y|^2$. Then, for all $x, y \in \mathbb{R}^d$ and $t \in [0, T]$,*

$$|U_\lambda(x, y, t) - \lambda|x - y|^2| \leqq t\|DF\| \, |x - y| \, .$$

Proof (The Proof of Theorem 10.1) Assume that (x_0, y_0, t_0) with $t_0 > 0$ is a maximum point of $u(x, t) - v(y, t) - \lambda|x - y|^2 - \alpha t$. Repeating the arguments at the end of the previous subsection and using Lemma 10.1 yields

$$\alpha h \leqq \|DF\|_\infty |x_0 - y_0| \, |B(t_0) - B(t_0 - h)| \, . \tag{3.149}$$

Recall that, in view of Proposition 10.1, the above inequality holds as long as

$$|B(t_0) - B(t_0 - h)| \lesssim \lambda^{-1/3} \, .$$

Since $B \in C^{0,\beta}([0, \infty])$, $h = h(\lambda)$ can be chosen so that

$$\lambda^{-1} \approx h^{-3\beta}.$$

Moreover, $(x_0, y_0) \in \mathbb{R}^d \times \mathbb{R}^d$ being a maximum of $u(x, t_0) - v(y, t_0) - \lambda|x - y|^2$ yields $\lambda|x_0 - y_0|^2 \leqq \max(\|u\|, \|v\|)$ and, if ω is the modulus of continuity of u, $\lambda|x_0 - y_0|^2 \leqq \omega(\lambda^{-1/2} \max(\|u\|, \|v\|)^{1/2}) = O(1)$, and, hence, $|x_0 - y_0| \lesssim \lambda^{-1/2}$.

Inserting all the observations above in (3.149), gives $\alpha h \lesssim o(1)h^{\frac{5\beta}{2}}$, and, thus, $\alpha \lesssim o(1)h^{\frac{5\beta}{2}-1}$, which leads to a contradiction as $\lambda \to \infty$.

Note that it is possible to assume less on B in Theorem 10.1, if more information is available about the modulus of continuity of either u or v.

For Hamiltonians that are not of separated form, the situation is more complicated. Indeed the "canonical" assumption on H for the deterministic theory is that, for some modulus ω_H and all $x, y, p \in \mathbb{R}^d$,

$$|H(p, x) - H(p, y)| \leqq \omega_H(|x - y|(1 + |p|)). \tag{3.150}$$

On the other hand, the proof of the comparison yields

$$\lambda|x_0 - y_0|^2 \leqq 2 \max(\|u\|, \|v\|),$$

and

$$\lambda|x_0 - y_0|^2 \leqq \max(\omega_u(|x_0 - y_0|), \omega_v(|x_0 - y_0|)) \, ,$$

and, hence,

$$|x_0 - y_0|^2 \leqq \lambda^{-1} \max(\omega_u, \omega_v)(2(\lambda^{-1} \max(\|u\|, \|v\|))^{1/2}) \,.$$

If either u or v is Lipschitz continuous, then the above estimate can be improved to

$$|x_0 - y_0| \leqq \min(\|Du\|, \|Dv\|)\lambda^{-1} \,.$$

The next technical result replaces Lemma 10.1. Its proof is again classical and it is omitted.

Lemma 10.2 *Assume that H satisfies (3.150) with $\omega_H(r) = Lr$. Let U_λ be the viscosity solution of the doubled initial value problem (3.143). Then there exists $C > 0$ depending on L such that, for all $x, y \in \mathbb{R}^d$ and $t \in [0, T]$,*

$$|U_\lambda(x, y, t) - \lambda' e^{Ct}|x - y|^2| \leqq (e^{Ct} - 1)|x - y| \,. \tag{3.151}$$

If, in addition $|x - y| \lesssim \lambda^{-1}$, then

$$|U_\lambda(x, y, t) - \lambda' e^{Ct}|x - y|^2| \lesssim t\lambda^{-1} \,.$$

A discussion follows about how to "increase" the length of the interval of existence of solutions given by the method of characteristics for H's which are not separated. To keep the notation simple, it is again convenient to argue for $d = 1$.

The characteristic odes for the deterministic doubled pde (3.147) are

$$\begin{cases} \dot{X} = -D_p H(P, X) \quad \dot{Y} = -D_q H(Q, Y), \\ \dot{P} = D_x H(P, X) \quad \dot{Q} = D_y H(Q, Y), \\ \dot{U} = H(P, X) - \langle D_p H(P, X, P) \rangle - H(Q, Y) + \langle D_Q H(Q, Y)Q \rangle, \\ X(0) = x \quad Y(0) = y \quad P(0) = Q(0) = \lambda(x - y) \quad U(0) = \lambda'|x - y|^2. \end{cases}$$

Recall that the Jacobian is given by

$$J = \frac{\partial X}{\partial x} \frac{\partial Y}{\partial y} - \frac{\partial X}{\partial y} \frac{\partial Y}{\partial x},$$

and, for $\alpha = x$ or y,

$$\begin{cases} \dfrac{\partial X}{\partial \alpha} = -D_p^2 H(P, X)\dfrac{\partial P}{\partial \alpha} - D_{px}^2 H(P, X)\dfrac{\partial X}{\partial \alpha} \,, \quad \dfrac{\partial X}{\partial \alpha}(0) = \begin{cases} 1 & \text{if } \alpha = x \\ 0 & \text{if } \alpha \neq x \end{cases}, \\[4mm] \dfrac{\partial P}{\partial \alpha} = D_{px}^2 H(P, X)\dfrac{\partial P}{\partial \alpha} + D_x^2 H(P, X)\dfrac{\partial X}{\partial \alpha} \,, \quad \dfrac{\partial P}{\partial \alpha}(0) = \begin{cases} \lambda & \text{if } \alpha = x \\ -\lambda & \text{if } \alpha = y \end{cases}, \end{cases}$$

and

$$
\begin{cases}
\dfrac{\partial Y}{\partial \alpha} = -D_q^2 H(Q,Y)\dfrac{\partial Q}{\partial \alpha} - D_{qy}^2 H(Q,Y)\dfrac{\partial Y}{\partial \alpha}, & \dfrac{\partial Y}{\partial \alpha}(0) = \begin{cases} 0 & \text{if } \alpha = x \\ 1 & \text{if } \alpha = y \end{cases}, \\[4mm]
\dfrac{\partial Q}{\partial \alpha} = D_{yq}^2 H(Q,Y)\dfrac{\partial Q}{\partial \alpha} + D_y^2 H(Q,Y)\dfrac{\partial Y}{\partial \alpha}, & \dfrac{\partial Q}{\partial \alpha}(0) = \begin{cases} \lambda & \text{if } \alpha = x \\ -\lambda & \text{if } \alpha = y \end{cases}.
\end{cases}
$$

It is also convenient to consider, for $i = 1, 2$ and $z = x - y$, the linearized auxiliary systems

$$
\begin{cases}
\dot{\xi}_i = -D_{pp}^2 H(P,X)(1+\lambda|z|)\phi_i - D_{xp}^2 H(P,X)\xi_i, \\[2mm]
\xi_1(0) = 1, \ \xi_2(0) = 0 \\[2mm]
\dot{\phi}_i = D_{xp}^2 H(P,X)\phi_i + \dfrac{D_{xx}^2 H(P,X)\xi_i}{1+\lambda|z|} \\[2mm]
\phi_1(0) = 0, \ \phi_1(0) = \dfrac{1}{1+\lambda|z|},
\end{cases}
\tag{3.152}
$$

and

$$
\begin{cases}
\dot{\eta}^i = -D_q^2 H(Q,Y)(1+\lambda|z|)\psi_i - D_{qy}^2 H(Q,Y)\eta^i, \\[2mm]
\eta_1(0) = 1, \ \eta_2(0) = 0, \\[2mm]
\dot{\psi}_i = D_{qy}^2 H(Q,Y)\psi_i + D_{yy}^2 H(Q,Y)\dfrac{\eta^i}{1+\lambda|z|}, \\[2mm]
\psi_1(0) = 0, \ \psi_2(0) = \dfrac{1}{1+\lambda|z|}.
\end{cases}
\tag{3.153}
$$

It is immediate that

$$
\begin{cases}
\dfrac{\partial X}{\partial x} = \xi_1 + \lambda\xi_2, & \dfrac{\partial X}{\partial y} = -\lambda\xi_2, & \dfrac{\partial Y}{\partial x} = -\lambda\eta_2, & \dfrac{\partial Y}{\partial x} = -\lambda\eta_2, \\[3mm]
\dfrac{\partial P}{\partial x} = (\phi_1 + \lambda\phi_2)(1+\lambda|z|) & \dfrac{\partial Q}{\partial x} = -\lambda\psi_2(1+\lambda|z|), \\[3mm]
\dfrac{\partial P}{\partial y} = -\lambda\phi_2(1+\lambda|z|) & \dfrac{\partial Q}{\partial y} = (\psi_1 + \lambda\psi_2)(1+\lambda|z|).
\end{cases}
$$

Assume next that, for all $p, x \in \mathbb{R}^d$,

$$|D_{xp}^2 H(p, x)| \lesssim 1, \quad |D_p^2 H(p, x)| \lesssim 1, \quad (1 + |p|)|D_x^2 H(p, x)| \lesssim 1$$

$$|D_{xxp}^3 H(p, x)| \lesssim 1 \quad (1 + |p|)|D_{xpp}^3 H(p, x)| \lesssim 1$$

$$(1 + |p|)^2 |D_p^3 H(p, x)| \lesssim 1.$$

$$(3.154)$$

It follows that there exists $C = C(T) > 0$ such that, for all $t \in [-T, T]$,

$$|\xi_1(t)| \leq C, \quad |\eta_1(t)| \leq C, \quad |\xi_2(t)| \leq \frac{Ct}{1 + \lambda|z|}$$

$$\text{and} \quad |\eta_2(t)| \leq \frac{Ct}{1 + \lambda|z|}.$$

$$(3.155)$$

Consider the matrices

$$A^x = \begin{pmatrix} -D_{xp}^2 H(P, X) & -D_{pp}^2 H(P, X)(1 + \lambda|z|) \\ \dfrac{D_x^2 H(P, X)}{1 + \lambda|z|} & D_{xp}^2 H(P, X) \end{pmatrix}$$

and

$$A^y = \begin{pmatrix} -D_{yq}^2 H(Q, Y) & -D_{qq}^2 H(Q, Y)(1 + \lambda|z|) \\ \dfrac{D_{yy}^2 H(Q, Y)}{1 + \lambda|z|} & D_{yq}^2 H(Q, Y). \end{pmatrix}$$

The next lemma, which is stated without proof, is important for the development of the rest of the theory here as well as for the theory of pathwise conservation laws.

Lemma 10.3 *Assume that, in addition to (3.154), for all $p, x \in \mathbb{R}^d$, $|D_p H(p, x)|$ and $(1 + |p|)^{-1}|D_x H(p, x)|$ are bounded. Then there exist $\varepsilon_0 > 0$ and $C > 0$ such that, for all $t \in (0, \varepsilon_0)$,*

$$\|A^x - A^y\| \leq C|z|.$$

Lemma 10.3 implies that, for all $t \in (0, \varepsilon_0)$,

$$|\xi_1 - \eta_1| \leq C|z|t \quad \text{and} \quad |\xi_2 - \eta_2| \leq \frac{C|z|t}{1 + \lambda|z|},$$

$$(3.156)$$

and, since

$$\lambda(\xi_2 \eta_1 - \xi_1 \eta_2) = \lambda(\xi_2 - \eta_2)\eta_1 + \lambda\eta_2(\eta_1 - \xi_1),$$

it follows from (3.155) and (3.156) that

$$|\lambda(\xi_2\eta_1 - \xi_1\eta_2)| \lesssim Ct. \tag{3.157}$$

Similar arguments allow to obtain an interval of invertibility of the characteristics that is uniform in λ, and, hence, a $O(1)$-interval of existence of smooth solutions of the doubled equation if either one of the following three groups of possible assumptions hold for all $(x, p) \in \mathbb{R}^d \times \mathbb{R}^d$:

$$\begin{cases} |D_x^2 H| \lesssim 1, \quad |D_{xxx}^3 H| \lesssim 1, \quad |D_{xp}^2 H| \lesssim 1, \\ |D_{xxp}^3 H|(1+|p|) \lesssim 1, |D_p^2 H| \lesssim 1, \\ |D_{xpp}^3 H|(1+|p|) \lesssim 1, \quad |D_p^3 H|(1+|p|) \lesssim 1. \end{cases} \tag{3.158}$$

$$\begin{cases} |D_x^2 H| \lesssim 1, \quad |D_x^3 H| \lesssim (1+|p|), \quad |D_{xp}^2 H| \lesssim 1, \\ |D_{xxp}^3 H| \lesssim 1, \quad |D_p^2 H|(1+|p|) \lesssim 1, \\ |D_{ppx}^3 H|(1+|p|) \lesssim 1, \quad |D_p^3 H|(1+|p|^2) \lesssim 1. \end{cases} \tag{3.159}$$

$$\begin{cases} |D_p^2 H| \lesssim 1, \quad |D_{xp}^2 H| \lesssim 1, \quad |D_x^2 H|(1+|p|) \lesssim 1, \\ |D_{xxp}^3 H|(1+|p|^2) \lesssim 1, |D_{xpp}^3 H| \lesssim 1, \\ |D_x^3 H|(1+|p|) \lesssim 1. \end{cases} \tag{3.160}$$

Note that (3.158) contains the split variable case, and linear-type Hamiltonians are a special case of (3.159).

Calculations similar to the ones used in the split variable case yield

$$|\xi_2\eta_1 - \eta_2\xi_1| \lesssim t^2,$$

and, as was already seen, $t_\lambda = \lambda^{-1/3}$. Note that, if $|DH|$, $|D^2 H|$ and $|D^3 H|$ are all bounded, then $t_\lambda = \lambda^{-1/2}$.

3.10.3   The Necessity of the Assumptions

An important question is whether conditions like the ones stated above are actually necessary to have well posed problems for Hamiltonians that depend on p, x. That some conditions are needed is natural since the argument is based on inverting characteristics and, hence, staying away from shocks. In view of this, assumptions that control the behavior of H and its derivatives for large $|p|$ are to be expected.

On the other hand, some of the restrictions imposed are due to the specific choice of the initial datum of the doubled equation, which, in principle, does not "interact well" with the cancellation properties of the given H.

Consider, for example, the Hamiltonian

$$H(p, x) = F(a(x)p),$$ (3.161)

with

$$a, F \in C^2(\mathbb{R}) \cap C^{0,1}(\mathbb{R}) \quad \text{and} \quad a > 0.$$ (3.162)

The characteristics are

$$\dot{X} = -F'(a(X)P)a(X) \quad \text{and} \quad \dot{P} = F'(a(X)P)a'(X)P.$$

Let $\phi \in C^2(\mathbb{R})$ be such that $\phi' = \dfrac{1}{a}$, $\hat{X} = \phi(X)$ and $\hat{P} = a(X)P$. Then

$$\dot{\hat{X}} = \phi'(X)\dot{X} = -a^{-1}(X)F'(a(X)P)a(X) = -F'(\hat{P})$$

and

$$\dot{\hat{P}} = a'(X)\dot{X}P + a(X)\dot{P} = -a'(X)F'(\hat{P})a(X)P + a(X)F'(a(X)P)a'(X)P = 0.$$

The observations above yield that it is better to use $\lambda|\phi^{-1}(x) - \phi^{-1}(y)|^2$ instead of $\lambda|x - y|^2$ in the comparison proof.

At the level of the pde

$$du = F(a(x)u_x) \cdot dB,$$

the above transformation yields that, if $u(x, t) = U(\phi(x), t)$, then

$$dU = F(U_x) \cdot dB,$$

a problem which is, of course, homogeneous in space, and, hence, as already seen, there is a $O(1)$-interval of existence for the doubled pde.

This leads to the question if it is possible to find, instead of $\lambda|x - y|^2$, an initial datum for the doubled pde, which is still coercive, and, in the mean time, better adjusted to the structure of the doubled equation. This is the topic of the next subsection.

3.10.4   Convex Hamiltonians and a Single Path

The example discussed was the motivation behind several works which eventually
led to a new class of well-posedness results in the case of a single path and convex
Hamiltonians.

The first result in this direction which applied to quadratic Hamiltonians
corresponding to Riemannian metrics is due to Friz et al. [30]. A more general
version of the problem (positively homogeneous and convex in p Hamiltonians)
was studied in [98]. The final and definitive results, which apply to general convex
in p Hamiltonians with minimal regularity conditions, were obtained by Lions and
Souganidis [73]. These results are sketched next.

To keep the ideas simple, here I only discuss the first-order problem

$$du = H(Du, x) \cdot dB \text{ in } Q_T \quad u(\cdot, 0) = u_0. \tag{3.163}$$

To motivate the question, I recall that the basic step of any comparison proof for
viscosity solutions is to maximize functions like $u(x, t) - v(y, t) - \lambda|x - y|^2$. The
properties of $\lambda|x - y|^2$ used in the proofs are that

$$\begin{cases} D_x L_\lambda = -D_y L_\lambda, \ L_\lambda \geq 0, \ L_\lambda(x, x) = 0 \text{ and} \\ L_\lambda(x, y) \to \infty \text{ if } |x - y| > 0. \end{cases} \tag{3.164}$$

The difficulty is that in the spatially dependent problems this choice of L_λ leads
to expressions like $H(\lambda(x - y), x) - H(\lambda(x - y), y)$ and, hence, error terms that
are difficult to estimate when dealing with rough signals.

To circumvent this problem it seems to be natural to ask if it is possible to replace
$\lambda|x - y|^2$ by some $L_\lambda(x, y)$ that has similar continuity and coercivity properties and
is better suited to measure the "distance" between $H(\cdot, x)$ and $H(\cdot, y)$.

In particular, it is necessary to find $L_\lambda : \mathbb{R}^d \times \mathbb{R}^d \to \mathbb{R}$ such that

$$H(D_x L_\lambda, x) = H(-D_y L_\lambda, y),$$

$$L_\lambda \gtrsim -\lambda^{-1}, L_\lambda(x, y) \xrightarrow[\lambda \to \infty]{} \infty \text{ if } x \neq y,$$

$$L_\lambda(x, x) \xrightarrow[\lambda \to \infty]{} 0, \text{ and } L_\lambda \in C^1_{x,y} \text{ in a neighborhood of } \{x = y\}. \tag{3.165}$$

It turns out (see [73]) that this is possible if H is convex or, more generally, if
there exists H_0 convex such that the pair H, H_0 is an involution, that is, $\{H, H_0\} = 0$. Here I concentrate on the convex case.

Given H convex with Legendre transform L, define

$$L_\lambda(x, y) = \inf\left\{ \int_0^{\lambda^{-1}} L(-\dot{x}(s), x(s))ds : \right.$$

$$\left. x(0) = x, \ x(\lambda^{-1}) = y, \ x(\cdot) \in C^{0,1}([0, \lambda^{-1}]) \right\}.$$

It follows, see, for example, Crandall et al. [16] and Lions [58], that $L_\lambda(x, y) = \overline{L}(x, y, \lambda^{-1})$, where $\overline{L}$ is the unique solution of

$$\overline{L}_t + H(D_x\overline{L}, x) = 0 \ \text{ in } \ \mathbb{R}^d \times (0, \infty) \quad \overline{L}(x, y, 0) = \delta_{\{y\}}(x)$$

$$\overline{L}_t + H(-D_y\overline{L}, y) = 0 \ \text{ in } \ \mathbb{R}^d \times (0, \infty) \quad \overline{L}(x, y, 0) = \delta_{\{x\}}(y),$$

where $\delta_A(x) = 0$ if $x \in A$ and $\delta_A(x) = \infty$ otherwise.

Note that, at least formally, the above imply that

$$H(D_x L_\lambda, x) = H(-D_y L_\lambda, y).$$

From the remaining properties in (3.165) the most challenging one is the regularity.

I summarize next without proofs the main result of [73]. In what follows v and μ denote respectively constants for lower and upper bounds.

The assumption on $L : \mathbb{R}^d \times \mathbb{R}^d \to \mathbb{R}$ is that there exist positive constants q, v, μ and $C \geq 0$ such that, for all $\xi \in \mathbb{R}^d$,

$$\begin{cases} v|p|^q - C \leq L \leq \mu|p|^q + C, \\ |D_x L| \leq \mu|p|^q + C, \ \ |D_{px}L| \leq \mu|p|^{q-1} + C, \end{cases} \tag{3.166}$$

and

$$\begin{cases} v|p|^{q-2}|\xi|^2 \leq \langle D^2 L_p\xi, \xi\rangle \leq (\mu|p|^{q-2} + C)|\xi|^2, \\ |D^2 L_x| \leq \mu|p|^q + C; \end{cases} \tag{3.167}$$

notice that it is important that $D_p^2 L$ is positive definite.

The result is stated next.

Theorem 10.2 *Assume* (3.166) *and* (3.167). *Then:*

(i) *If $q \leq 2$, then there exists λ_0 such that, if $\lambda > \lambda_0$, $L_\lambda \in C_{x,y}^1(\{|x - y| < \lambda^{-1}\})$.*

(ii) *If $q > 2$ and $C > 0$, then, in general, (i) above is false, and, in fact, $\overline{L}(x, x, \lambda^{-1})$ may not be differentiable for any λ.*

(iii) *If $q > 2$ and $C = 0$, then there exists λ_0 such that, if $\lambda > \lambda_0$, $L_\lambda \in C_{x,y}^1(\{|x - y| < \lambda^{-1}\})$.*

(iv) *In all cases, $\overline{L}$ is semiconcave in both x and y.*

It follows that, when $q \leq 2$ or $q > 2$ and $C = 0$, the pathwise solutions of the stochastic Hamilton-Jacobi initial value problem are well posed. The result extends to the full second order problem, because the semiconcavity is enough to carry out the details.

3.10.5 Multiple Paths

I sketch here briefly the strategy that Lions and the author developed in [72] to establish the well-posedness of the pathwise solutions in the multi-path spatially dependent setting with Brownian signals. The argument is rather technical and to keep the ideas as simple as possible I only discuss the first-order problem

$$du = \sum_{i=1}^{m} H^i(Du, x) \cdot dB_i \quad \text{in } Q_T \quad u(\cdot, 0) = u_0, \tag{3.168}$$

and provide some hints about the difficulties and the methodology.

As in the single-path case, the main step is to obtain a sufficiently long interval of existence of smooth solutions of the doubled initial value problem

$$\begin{cases} dU = \sum_{i=1}^{m} \left[H^i(D_x U, x) - H^i(-D_y U, y) \right] \cdot dB_i \\ \qquad \text{in } \mathbb{R}^d \times \mathbb{R}^d \times (0, T], \\ U(x, y, 0) = \lambda |x - y|^2. \end{cases} \tag{3.169}$$

The semiformal argument presented earlier suggests that it is necessary to have an interval of existence of order $\lambda^{-\alpha}$ for an appropriately chosen small $\alpha > 0$ which depends on the properties of the path. This was accomplished by reverting to the "non rough" time homogeneous doubled equation, something that is not possible for (3.169).

The new methodology developed in [72] consists of several steps. The first is to provide a large deviations-type estimate about the error, in terms of powers of λ^{-1}, between the stochastic characteristics and their linearizations and the Jacobian, and their second-order expansion in terms of B and its Levy areas. This would be straightforward, if it were not for the fact that the error must be uniform in (x, y) such that $|x - y| = O(\lambda^{-1/2})$.

Next I describe this problem for the solution S of a stochastic differential equation $dS = \sigma(S)dB$ with $S(0) = s$. The aim is to obtain an exponentially small estimate for the probability of the event that $\sup_{s \in K} |S(t) - (s + \sigma(s)B(s) + (1/2)\sigma\sigma'(s)B^2(t))| > \lambda^{-\beta}$, where is A is a subset of $\mathbb{R}$ which may depend on λ. In other words we need an estimate for the probability of the sup instead of the sup of the probability. Obtaining such a result requires a new approach based on estimating L^p-norms of events for large p.

Having such estimates allows for a local in time comparison result off a set of exponentially small probability in terms of λ^{-1}. An "algebraic" iteration of this local comparison provides the required result at the limit $\lambda \to \infty$.

3.11 Perron's Method

Perron's method is a general way to obtain solutions of equations which satisfy a comparison principle. The general argument is that the maximal subsolution is actually a solution. The idea is that, at places where it fails to be a solution, a subsolution can be strictly increased and maintain the subsolution property. This is a local argument which has been carried out successfully for "deterministic" viscosity solutions. This locality creates, however, serious technical difficulties in the rough path setting due to the rigidity of the test functions.

In this section I discuss this method in the context of the simplified initial value problem

$$du = F(D^2u, Du)\, dt + \sum_{i=1}^{m} H^i(Du, x) \cdot dB_i \quad \text{in} \quad Q_T \quad u(\cdot, 0) = u_0, \quad (3.170)$$

where $u_0 \in BUC(\mathbb{R}^d)$, $T > 0$, and $B = (B^1, \ldots, B^m)$ is a Brownian path. The method can be a extended to problems with F depending also on (x, t) and B a geometric rough path that is α-Hölder continuous for some $\alpha \in (1/3, 1/2]$. For details I refer to [100].

Throughout the discussion it is assumed that

$$\begin{cases} F : \mathcal{S}^d \times \mathbb{R}^d \to \mathbb{R} \text{ is continuous, bounded for bounded} \\ (X, p) \in \mathcal{S}^d \times \mathbb{R}^d \quad \text{and degenerate elliptic,} \end{cases} \quad (3.171)$$

and the Hamiltonians are sufficiently regular, for example,

$$H \in C_b^4(\mathbb{R}^d \times \mathbb{R}^d; \mathbb{R}^m), \quad (3.172)$$

to allow for the construction of local-in-time, C^2 in space solutions of $dw = \sum_{i=1}^{m} H^i(Du, x) \cdot dB_i$.

As mentioned in Sect. 3.4, if the Poisson brackets of the $\{H^i\}$ vanish, for example, if $m = 1$ or there is no spatial dependence, then it suffices to have $H \in C_b^2(B_R \times \mathbb{R}^d; \mathbb{R}^m)$ for all $R > 0$.

The result is stated next.

Theorem 11.1 *Assume* (3.171) *and* (3.172). *Then* (3.170) *has a unique solution* $u \in BUC(Q_T)$, *which is given by*

$$u(x,t) = \sup\{v(x,t) : v(\cdot, 0) \leq u_0$$

$$\text{and } v \text{ is a subsolution of } (3.170)\}. \qquad (3.173)$$

As has been discussed earlier, more assumptions are generally required for F, H, and B in order for the comparison principle to hold. This is especially the case when H has nontrivial spatial dependence even when $m = 1$. Apart from the assumptions that yield the comparison, the only hypotheses used for the Perron construction are (3.172).

As before, $S(t, t_0) : C_b^2(\mathbb{R}^d) \to C_b^2(\mathbb{R}^d)$ be the solution operator for local in time, spatially smooth solutions of

$$d\Phi = \sum_{i=1}^m H^i(D\Phi, x) \cdot dB_i \text{ in } \mathbb{R}^d \times (t_0 - h, t_0 + h) \quad \Phi(\cdot, t_0) = \phi. \qquad (3.174)$$

It is clear from the definition of stochastic viscosity subsolutions that the maximum of a finite number of subsolutions is also a subsolution, with a corresponding statement holding true for the minimum of a finite number of supersolutions. This observation to can be generalized to infinite families.

Lemma 11.1 *Given a family $\mathcal{F}$ of subsolutions (resp. supersolutions) of* (3.170), *let* $U(x,t) = \sup_{v \in \mathcal{F}} v(x,t)$ *(resp.* $\inf_{v \in \mathcal{F}} v(x,t)$). *If* $U^* < \infty$ *(resp.* $U_* > -\infty$), *then* U^* *(resp.* U_*) *is a subsolution (resp. supersolution) of* (3.170).

Proof I only a sketch of the proof of the subsolution property.

Let $\phi \in C_b^2(\mathbb{R}^d)$, $\psi \in C^1([0, T])$, $t_0 > 0$, and $h > 0$ be such that $S(\cdot, t_0)\phi \in C((t_0 - h, t_0 + h), C_b^2(\mathbb{R}^d))$, assume that $U^*(x,t) - S(t, t_0)\phi(x) - \psi(t)$ attains a strict local maximum at $(x_0, t_0) \in \mathbb{R}^d \times (t_0 - h, t_0 + h)$, and set $p = D\phi(x_0)$, $X = D^2\phi(x_0)$, and $a = \psi'(t_0)$. The goal is to show that

$$a \leq F(X, p).$$

The definition of upper-semicontinuous envelopes and arguments from the classical viscosity solution theory imply that there exist sequences $(x_n, t_n) \in \mathbb{R}^d \times (t_0 - h, t_0 + h)$ and $v_n \in \mathcal{F}$ such that $\lim_{n \to \infty}(x_n, t_n) = (x_0, t_0)$, $\lim_{n \to \infty} v_n(x_n, t_n) = U^*(x_0, t_0)$, and

$$v_n(x,t) - S(t, t_0)\phi(x) - \psi(t)$$

attains a local maximum at (x_n, t_0). Applying the definition of stochastic viscosity subsolutions and letting $n \to \infty$ completes the proof.

The second main step of the Perron construction is discussed next.

Lemma 11.2 *Suppose that w is a subsolution of (3.170), and that w_* fails to be a supersolution. Then there exists $(x_0, t_0) \in \mathbb{R}^d \times (0, T]$ such that, for all $\kappa > 0$, (3.170) has a subsolution w_κ such that*

$$w_\kappa \geq w, \quad \sup(w_\kappa - w) > 0, \quad and$$

$$w_\kappa = w \quad in \quad Q_T \setminus (B_\kappa(x_0) \times (t_0 - \kappa, t_0 + \kappa)).$$

Proof By assumption, there exist $\phi \in C_b^2(\mathbb{R}^d)$, $\psi \in C^1([0, T])$, $(x_0, t_0) \in \mathbb{R}^d \times (0, T]$, and $h \in (0, \kappa)$ such that $S(\cdot, t_0)\phi \in C((t_0 - h, t_0 + h), C_b^2(\mathbb{R}^d))$,

$$w_*(x, t) - S(t, t_0)\phi(x) - \psi(t)$$

attains a local minimum at (x_0, t_0), and

$$\psi'(t_0) - F(D^2\phi(x_0), D\phi(x_0)) < 0. \tag{3.175}$$

Assume, without loss of generality, that $x_0 = 0$, $\phi(0) = 0$, and $\psi(t_0) = 0$, set $X = D^2\phi(0)$, $p = D\phi(0)$, and $a = \psi'(t_0)$, fix $\gamma \in (0, 1)$, $r \in (0, \kappa)$, and $s \in (0, h)$, and choose $\hat\eta \in C_b^2(\mathbb{R}^d)$ and $h > 0$ so that

$$\hat\eta(x) = p \cdot x + \frac{1}{2}\langle Xx, x\rangle - \gamma|x|^2 \text{ in } B_r(x_0), \quad \hat\eta \leq \phi \text{ in } \mathbb{R}^d,$$

and

$$S(\cdot, t_0)\hat\eta \in C((t_0 - h, t_0 + h); C_b^2(\mathbb{R}^d)).$$

For $(x, t) \in \mathbb{R}^d \times (t_0 - h, t_0 + h)$ and $\delta > 0$, define

$$\widehat{w}(x, t) = w_*(0, t_0) + \delta + S(t, t_0)\hat\eta(x) + a(t - t_0) - \gamma(|t - t_0|^2 + \delta^2)^{1/2}.$$

In view of the strict inequality in (3.175), the continuity of the solution map $S(t, t_0)$ on $C_b^2(\mathbb{R}^d)$, and the continuity of F, if γ, r, s, and δ are sufficiently small, then $\widehat{w}$ satisfies the subsolution property in $B_r(0) \times (t_0 - s, t_0 + s)$.

The most important step in the proof is to show that, with all parameters sufficiently small, there exist $0 < r' < r$ and $0 < s' < s$ such that

$$w > \widehat{w} \quad \text{in } (B_r(0) \times (t_0 - s, t_0 + s)) \setminus \overline{B_{r'}(0) \times (t_0 - s', t_0 + s')}. \tag{3.176}$$

Achieving the inequality in (3.176) for points of the form (x, t_0) can be done using classical arguments. However, this is much more difficult for arbitrary $t \neq t_0$, because, in view of the definition of $\widehat{w}$, it is necessary to study the local in time, spatially smooth solution operator $S(t, t_0)$.

This difficulty is overcome by establishing a finite speed of propagation for such local in time, spatially smooth solutions. As has been discussed earlier in the notes, such a result cannot be true in general. Here it relies on access to the system of rough characteristics. Indeed, the domain of dependence result is proved by estimating the deviation of characteristics from their starting points, using tools from the theory of rough or stochastic differential equations.

Once (3.176) is established, define

$$
w_\kappa(x, t) = \begin{cases} \max(\widehat{w}(x, t), w(x, t)) & \text{for } (x, t) \in B_r(0) \times (t_0 - s, t_0 + s), \\ w(x, t) & \text{for } (x, t) \notin B_r(0) \times (t_0 - s, t_0 + s). \end{cases}
$$

Then $w_\kappa \geq w$, and $w_\kappa = w$ outside of $B_\kappa(0) \times (t_0 - \kappa, t_0 + \kappa)$. If (x_n, t_n) is such that $\lim_{n \to \infty}(x_n, t_n) = (0, t_0)$ and $\lim_{n \to \infty} w(x_n, t_n) = w_*(0, t_0)$, then

$$
\lim_{n \to \infty} \left(w(x_n, t_n) - \widehat{w}(x_n, t_n) \right) = -(1 - \gamma)\delta < 0,
$$

so that

$$
\sup_{B_\kappa(0) \times (t_0 - \kappa, t_0 + \kappa)} (w_\kappa - w) > 0.
$$

Finally, w_κ is a subsolution. This is evident outside of $B_r(0) \times (t_0 - s, t_0 + s)$, as well as in the interior of $B_r(0) \times (t_0 - s, t_0 + s)$, because there, w_κ is equal to the pointwise maximum of two subsolutions. It remains to verify the subsolution property on the boundary of $B_r(0) \times (t_0 - s, t_0 + s)$, and this follows because, in view of (3.176), $w_\kappa = w$ in a neighborhood of the boundary of $B_r(0) \times (t_0 - s, t_0 + s)$.

Proof (Proof of Theorem 11.1) The first step is to verify that u is well defined and bounded. This follows from the comparison principle, and the fact that, in view of the assumptions, it is possible to construct a subsolution and a supersolution with respectively initial datum $- \|u_0\|_\infty$ and $\|u_0\|_\infty$.

Fix $\varepsilon > 0$ and let $\phi^\varepsilon \in C_b^2(\mathbb{R}^d)$ be such that

$$
\phi^\varepsilon - \varepsilon \leq u_0 \leq \phi^\varepsilon + \varepsilon \quad \text{on } \mathbb{R}^d.
$$

It is possible to construct a subsolution and a supersolution $\underline{u}^\varepsilon$ and $\overline{u}^\varepsilon$ which are continuous in a neighborhood of $\mathbb{R}^d \times \{0\}$ and achieve respectively the initial datum $\phi^\varepsilon - \varepsilon$ and $\phi^\varepsilon + \varepsilon$. This can be done by using the solution operator $S(t_{k+1}, t_k)$ on successive, small intervals $[t_k, t_{k+1}]$ and the boundedness properties of F. Once again, see [99] for the details.

The comparison principle yields

$$
\underline{u}^\varepsilon \leq u_* \leq u \leq u^* \leq \overline{u}^\varepsilon \quad \text{in } Q_T,
$$

and, in view of the continuity of $\underline{u}^\varepsilon$ and $\overline{u}^\varepsilon$ near $\mathbb{R}^d \times \{0\}$,

$$\phi^\varepsilon - \varepsilon \leq u_*(\cdot, 0) \leq u^*(\cdot, 0) \leq \phi^\varepsilon + \varepsilon.$$

Since ε is arbitrary, it follows that $u(\cdot, 0) = u_0$ and $\lim_{(x,t) \to (x_0,0)} u(x, t) = u_0(x_0)$ for all $x_0 \in \mathbb{R}^d$.

Lemma 11.1 now implies that u^* is a subsolution of (3.170) with $u^*(x, 0) \leq u_0(x)$. The formula (3.173) for u then yields $u^* \leq u$, and, therefore, $u^* = u$. That is, u is itself upper-semicontinuous and a subsolution.

On the other hand, u_* is a supersolution. If this were not the case, then Lemma 11.2 would imply the existence of a subsolution $\tilde{u} \geq u$ and a neighborhood $N \subset \mathbb{R}^d \times (0, T]$ such that $\tilde{u} = u$ in $(\mathbb{R}^d \times [0, T]) \backslash N$ and $\sup_N (\tilde{u} - u) > 0$, contradicting the maximality of u.

The comparison principle gives $u^* \leq u_*$, and, as a consequence of the definition of semicontinuous envelopes, $u_* \leq u^*$. Therefore, $u = u_* = u^*$ is a solution of (3.170) with $u = u_0$ on $\mathbb{R}^d \times \{0\}$. The uniqueness of u follows from yet another application of the comparison principle.

3.12 Approximation Schemes, Convergence and Error Estimates

Here I discuss a general program for constructing convergent (numerical) approximation schemes for the pathwise viscosity solutions and obtain, for first-order equations, explicit error estimates.

The presentation focuses on the initial value problem

$$du = F(D^2u, Du)\, dt + \sum_{i=1}^{m} H^i(Du) \cdot dB_i \quad \text{in} \quad Q_T \quad u(\cdot, 0) = u_0, \qquad (3.177)$$

where $T > 0$ is a fixed finite horizon, $F \in C^{0,1}(\mathcal{S}^d \times \mathbb{R}^d)$ is degenerate elliptic, $H \in C^2(\mathbb{R}^d)$, $B = (B_1, \ldots, B_m) \in C([0, T]; \mathbb{R}^m)$, and $u_0 \in BUC(\mathbb{R}^d)$.

3.12.1 The Scheme Operator

Following the general methodology for constructing convergent schemes for "non-rough" viscosity solutions put forward by Barles and Souganidis [5], the approximations are constructed using a "scheme" operator, which, for $h > 0, 0 \leq s \leq t \leq T$, and $\zeta \in C([0, T]; \mathbb{R}^m)$, is a map $S_h(t, s; \zeta) : BUC(\mathbb{R}^d) \to BUC(\mathbb{R}^d)$.

Given a partition $\mathcal{P} = \{0 = t_0 < t_1 < \cdots < t_N = T\}$ of $[0, T]$ with mesh size $|\mathcal{P}|$ and a path $\zeta \in C_0([0, T]; \mathbb{R}^m)$, usually a piecewise linear approximation of B, the (approximating) function $\tilde{u}_h(\cdot; \zeta, \mathcal{P})$ is defined by

$$
\begin{cases}
\tilde{u}_h(\cdot, 0; \zeta, \mathcal{P}) := u_0 \quad \text{and} \\[2mm]
\tilde{u}_h(\cdot, t; \zeta, \mathcal{P}) := S_h(t, t_n; \zeta)\tilde{u}_h(\cdot, t_n; \zeta, \mathcal{P}) \\[2mm]
\text{for } n = 0, 1, \ldots, N - 1, \ t \in (t_n, t_{n+1}].
\end{cases}
\tag{3.178}
$$

The strategy is to choose families of approximating paths $\{B_h\}_{h>0}$ and partitions $\{\mathcal{P}_h\}_{h>0}$ satisfying

$$
\lim_{h \to 0^+} \|B_h - B\|_\infty = 0 = \lim_{h \to 0^+} |\mathcal{P}_h|,
\tag{3.179}
$$

in such a way that the function

$$
u_h(x, t) := \tilde{u}_h(x, t; B_h, \mathcal{P}_h)
\tag{3.180}
$$

is an efficient approximation of the solution of (3.177).

The main restriction on the scheme operator is that it has to be monotone, that is,

$$
\begin{cases}
\text{if } t_n \le t \le t_{n+1}, \, t_n, t_{n+1} \in \mathcal{P}_h, \text{ and } u, v \in BUC(\mathbb{R}^d) \text{ such that } u \le v \text{ in } \mathbb{R}^d, \\[2mm]
\quad \text{then} \\[2mm]
\qquad S_h(t, t_n; B_h)u \le S_h(t, t_n; B_h) \text{ in } \mathbb{R}^d.
\end{cases}
\tag{3.181}
$$

It will also be necessary for the scheme operator to commute with constants, that is, for all $u \in BUC(\mathbb{R}^d)$, $h > 0$, $0 \le s \le t < \infty$, $\zeta \in C_0([0, T], \mathbb{R}^m)$, and $k \in \mathbb{R}$,

$$
S_h(t, s; \zeta) \, (u + k) = S_h(t, s; \zeta)u + k.
\tag{3.182}
$$

Finally, the scheme operator must be "consistent" with the equation in some sense. This point, as well as the motivation for the above assumptions, are explained below.

3.12.2   The Method of Proof

I give here a brief sketch of the proof. All the details and concrete examples can be found in Seeger [96].

Assume for the moment that $\lim_{h \to 0} u_h = u$ locally uniformly for some $u \in BUC(Q_T)$. In fact, a rigorous proof involves studying the so-called half-relaxed limits of u_h, but I omit these cumbersome details.

The goal is to show that u is the unique pathwise solution of (3.177). To that end, suppose that

$$u(x, t) - \Phi(x, t) - \psi(t)$$

attains a strict maximum at $(y, s) \in \mathbb{R}^d \times I$, where $\psi \in C^1([0, T])$ and, for some small open interval $I \subset [0, T]$, $\Phi \in C(I; C^2(\mathbb{R}^d))$ is a local in time, smooth in space solution of

$$d\Phi = \sum_{i=1}^m H^i(D\Phi) \cdot dB_i \quad \text{in } \mathbb{R}^d \times I. \tag{3.183}$$

I will show that

$$\psi'(s) \leq F(D^2\Phi(y, s), D\Phi(y, s)),$$

which implies that u is a subsolution. The argument to show it is a supersolution is similar.

For $h > 0$, let Φ_h be the local in time, smooth in space solution of

$$\Phi_{h,t} = \sum_{i=1}^m H^i(D\Phi_h)\dot{B}_{i,h} \quad \text{in } \mathbb{R}^d \times I \quad \Phi_h(\cdot, s) = \Phi(\cdot, t_0). \tag{3.184}$$

Recall that such a solution can be shown to exist using the method of characteristics. The interval I may need to be shrunk, if necessary, but its length is uniform in h. Since $\lim_{h \to 0} B_h = B$ uniformly on $[0, T]$, it follows that, as $h \to 0$, Φ_h converges to Φ in $C(I; C^2(\mathbb{R}^d))$.

As a result, there exists $\{(y_h, s_h)\}_{h>0} \subset \mathbb{R}^d \times I$ such that $\lim_{h \to 0}(y_h, s_h) = (y, s)$ and

$$u_h(x, t) - \Phi_h(x, t) - \psi(t)$$

attains a local maximum at (y_h, s_h).

That $\lim_{h \to 0} |\mathcal{P}_h| = 0$ yields that, for h sufficiently small, there exist $n \in \mathbb{N}$ depending on h such that

$$t_n < s_h \leq t_{n+1} \quad \text{and} \quad t_n, t_{n+1} \in I.$$

It then follows that

$$u_h(\cdot, t_n) - \Phi_h(\cdot, t_n) - \psi(t_n) \leq u_h(y_h, s_h) - \Phi_h(y_h, s_h) - \psi(s_h),$$

or, after rearranging terms,

$$u_h(\cdot, t_n) \le u_h(y_h, s_h) + \Phi_h(\cdot, t_n) - \Phi_h(y_h, s_h) + \psi(t_n) - \psi(s_h). \qquad (3.185)$$

This is the place where the monotonicity (3.181) and the commutation with constants (3.182) of the scheme come into play. Applying $S_h(s_h, t_n; W_h)$ to both sides of (3.185), and evaluating the resulting expression at $x = y_h$ give

$$u_h(y_h, s_h) \le u_h(y_h, s_h) + S_h(s_h, t_n; B_h)\Phi_h(\cdot, t_n)(y_h) - \Phi_h(y_h, s_h) + \psi(t_n) - \psi(s_h),$$

whence

$$\frac{\psi(s_h) - \psi(t_n)}{s_h - t_n} \le \frac{S_h(s_h, t_n; B_h)\Phi_h(\cdot, t_n)(y_h) - \Phi_h(y_h, s_h)}{s_h - t_n}.$$

As $h \to 0$, the left-hand side converges to $\psi'(s)$. The construction of a convergent scheme then reduces to creating a scheme operator, partitions $\mathcal{P}_h$, and paths W_h satisfying (3.181) and (3.182), as well as the consistency requirement

$$\lim_{s,t \in I,\, t-s \to 0} \frac{S_h(t, s; B_h)\Phi_h(\cdot, s) - \Phi_h(\cdot, s)}{t - s} = F(D^2\Phi, D\Phi) \qquad (3.186)$$

whenever Φ and Φ_h are as in respectively (3.183) and (3.184).

3.12.3 The Main Examples

Presenting a full list of the types of schemes that may be constructed is beyond the scope of these notes. Here, I give a few specific examples that are representative of the general theory. More schemes and details can be found in [96].

Here I focus mainly on finite difference schemes. To simplify the presentation, assume $d = m = 1$, F and H are both smooth, and F depends only on u_{xx}, so that (3.177) becomes

$$du = F(u_{xx})\, dt + H(u_x) \cdot dB \text{ in } Q_T \quad u(\cdot, 0) = u_0, \qquad (3.187)$$

and, in the first-order case when $F \equiv 0$,

$$du = H(u_x) \cdot dB \text{ in } Q_T \quad u(\cdot, 0) = u_0. \qquad (3.188)$$

I present next a number of different partitions $\mathcal{P}_h$ and approximating paths B_h for which the program in the preceding subsection may be carried out. While technical, these are all made with the same idea in mind, namely, to ensure that the approximation B_h is "mild" enough with respect to the partition. In particular, for any consecutive points t_n and t_{n+1} of the partition $\mathcal{P}_h$ and for sufficiently small h, the ratio

$$\frac{|B_h(t_{n+1}) - B_h(t_n)|}{h}$$

should be less than some fixed constant. This is a special case of the well-known Courant-Lewy-Friedrichs (CFL) conditions required for the monotonicity of schemes in the "non-rough" setting.

For some $\varepsilon_h > 0$ to be determined, define

$$S_h(t, s; \zeta)u(x) = u(x) + H\left(\frac{u(x+h) - u(x-h)}{2h}\right)(\zeta(t) - \zeta(s))$$

$$+ \left[F\left(\frac{u(x+h) + u(x-h) - 2u(x)}{h^2}\right)\right. \tag{3.189}$$

$$\left. + \varepsilon_h\left(\frac{u(x+h) + u(x-h) - 2u(x)}{h^2}\right)\right](t - s).$$

The first result, which is qualitative in nature, applies to the simple setting above as follows.

Theorem 12.1 *Assume that, in addition to* (3.179), *B_h and $\mathcal{P}_h$ satisfy*

$$|\mathcal{P}_h| \leq \frac{h^2}{\|F'\|_\infty} \quad and \quad \varepsilon_h = h\|\dot{B}_h\| \xrightarrow{h\to 0} 0.$$

Then, as $h \to 0$, the function u_h defined by (3.180) *using the scheme operator* (3.189) *converges locally uniformly to the solution u of* (3.187).

The condition in Theorem 12.1 on the approximating path B_h can be satisfied in several different ways. For example, B_h could be a piecewise linear approximation of B of step-size $\eta_h > 0$, with $\lim_{h\to 0} \eta_h = 0$ in such a way that $\lim_{h\to 0} h\|\dot{B}_h\| = 0$.

By quantifying the method of proof in the previous subsection, it is possible to obtain explicit error estimates for finite difference approximations of the pathwise Hamilton-Jacobi equation (3.188). The results below are stated for the following scheme, which is defined, for some $\theta \in (0, 1]$, by

$$S_h(t, s; \zeta)u(x) = u(x) + H\left(\frac{u(x+h) - u(x-h)}{2h}\right)(\zeta(t) - \zeta(s))$$

$$+ \frac{\theta}{2}(u(x+h) + u(x-h) - 2u(x)); \tag{3.190}$$

note that this corresponds to choosing $\varepsilon_h = \frac{\theta h^2}{2(t-s)}$ in (3.189).

Assume that $\omega : [0, \infty) \to [0, \infty)$ is the modulus of continuity of the fixed continuous path B on $[0, T]$, define, for $h > 0$, ρ_h implicitly by

$$\lambda = \frac{(\rho_h)^{1/2}\omega((\rho_h)^{1/2})}{h} < \frac{\theta}{\|H'\|_\infty}, \tag{3.191}$$

and choose the partition $\mathcal{P}_h$ and path B_h so that

$$\begin{cases} \mathcal{P}_h = \{n\rho_h \wedge T\}_{n\in\mathbb{N}_0}, \ M_h := \lfloor(\rho_h)^{-1/2}\rfloor, \\ \text{and, for } k \in \mathbb{N}_0 \text{ and } t \in [kM_h\rho_h, (k+1)M_h\rho_h), \\ B_h(t) = B(kM_h\rho_h) \\ \quad + \left(\dfrac{B((k+1)M_h\rho_h) - B(kM_h\rho_h)}{M_h\rho_h}\right)(t - kM_h\rho_h). \end{cases} \tag{3.192}$$

Theorem 12.2 *There exists $C > 0$ depending only on L such that, if u_h is constructed using (3.180) and (3.190) with $\mathcal{P}_h$ and B_h as in (3.191) and (3.192), and u is the pathwise viscosity solution of (3.188), then*

$$\sup_{(x,t)\in\mathbb{R}^d\times[0,T]} |u_h(x, t) - u(x, t)| \le C(1 + T)\omega((\rho_h)^{1/2}).$$

If, for example, $B \in C^{0,\alpha}([0, T])$, then (3.191) means that $\rho_h = O(h^{2/(1+\alpha)})$, and the rate of convergence in Theorem 12.2 is $O(h^{\alpha/(1+\alpha)})$.

I describe next some examples in the case that B is a Brownian motion.

As a special case of Theorem 12.2, the approximating paths and partitions may be taken to satisfy (3.192) with ρ_h given by

$$\lambda = \frac{(\rho_h)^{3/4}|\log\rho_h|^{1/2}}{h} < \frac{\theta}{\|H'\|_\infty}, \tag{3.193}$$

in which case the scheme operator will be monotone almost surely for all h smaller than some (random) threshold $h_0 > 0$.

It is also possible to define the partitions and approximating paths using certain stopping times that ensure that the scheme is monotone almost surely for all $h > 0$. More details can be found in [96].

Theorem 12.3 *Suppose that B is a Brownian motion, and assume that $\mathcal{P}_h$ and B_h are as in (3.192) with ρ_h defined by (3.193). If u_h is constructed using (3.180) and (3.190), and u is the solution of (3.188), then there exists a deterministic constant $C > 0$ depending only on L and λ such that, with probability one,*

$$\limsup_{h\to 0}\ \sup_{(x,t)\in\mathbb{R}^d\times[0,T]} \frac{|u_h(x, t) - u(x, t)|}{h^{1/3}|\log h|^{1/3}} \le C(1 + T).$$

The final result presented here is about a scheme that converges in distribution in the space $BUC(\mathbb{R}^d \times [0, T])$ equipped with the topology of local uniform convergence.

Recall that, given random variables $(X_\delta)_{\delta>0}$ and X taking values in some topological space $\mathcal{X}$, it is said that X_δ converges, as $\delta \to 0$ in distribution (or in law) to X, if the law ν_δ of X_δ on $\mathcal{X}$ converges weakly to the law ν of X. That is, for any bounded continuous function $\phi : \mathcal{X} \to \mathbb{R}$,

$$\lim_{\delta \to 0} \int_{\mathcal{X}} \phi \, d\nu_\delta = \int_{\mathcal{X}} \phi \, d\nu.$$

Below, the paths B_h are taken to be appropriately scaled simple random walks, and, as a consequence, B_h converges in distribution to a Brownian motion B (see for instance Billingsley [8]). This corresponds above to $\mathcal{X} = C([0, T]; \mathbb{R}^m)$ and ν the Wiener measure on $\mathcal{X}$.

Let λ, ρ_h, B_h, and $\mathcal{P}_h$ be given, for some probability space $(\mathcal{A}, \mathcal{G}, \mathbf{P})$, by

$$\begin{cases} \lambda = \dfrac{(\rho_h)^{3/4}}{h} \leq \dfrac{\theta}{\|H'\|_\infty}, \quad M_h := \lfloor (\rho_h)^{-1/2} \rfloor, \\[2mm] \mathcal{P}_h := \{t_n\}_{n=0}^N = \{n\rho_h \wedge T\}_{n \in \mathbb{N}_0}, \\[2mm] \{\xi_n\}_{n=1}^\infty : \mathcal{A} \to \{-1, 1\} \text{ are independent,} \\[2mm] \mathbf{P}(\xi_n = 1) = \mathbf{P}(\xi_n = -1) = \dfrac{1}{2}, \\[2mm] B(0) = 0, \quad \text{and} \quad \text{for } k \in \mathbb{N}_0, \ t \in [kM_h\rho_h, (k+1)M_h\rho_h), \\[2mm] B_h(t) = B_h(kM_h\rho_h) + \dfrac{\xi_k}{\sqrt{M_h\rho_h}}(t - kM_h\rho_h). \end{cases} \qquad (3.194)$$

Theorem 12.4 *If u_h is constructed using (3.180) and (3.190) with B_h and $\mathcal{P}_h$ as in (3.194), and u is the solution of (3.188), then, as $h \to 0$, u_h converges to u in distribution.*

3.12.4 The Need to Regularize the Paths

A short discussion follows about the necessity to consider regularizations B_h of the continuous path B in all of the results above. To keep the presentation simple, I concentrate on the one-dimensional, pathwise Hamilton-Jacobi equation (3.188).

Consider the following naive attempt at constructing a scheme operator by setting

$$
\begin{aligned}
S_h(t, s)u(x) = u(x) + H &\left(\frac{u(x + h) - u(x - h)}{2h} \right) (B(t) - B(s)) \\
&+ \varepsilon_h \left(\frac{u(x + h) + u(x - h) - 2u(x)}{h^2} \right)(t - s).
\end{aligned}
\tag{3.195}
$$

A simple calculation reveals that $S_h(t, s)$ is monotone for $0 \le t - s \le \rho_h$, if ρ_h and ε_h are such that, for some $\theta \le 1$,

$$
\varepsilon_h = \frac{\theta h^2}{2(t - s)} \quad \text{and} \quad \lambda = \max_{|t-s| \le \rho_h} \frac{\mathrm{osc}(B, s, t)}{h} \le \lambda_0 = \frac{\theta}{\|H'\|_\infty}.
\tag{3.196}
$$

On the other hand, for any $s, t \in [0, T]$ with $|s - t|$ sufficiently small, spatially smooth solutions Φ of (3.188) have the expansion

$$
\begin{aligned}
\Phi(x, t) = \Phi(x, s) &+ H(\Phi_x(x, s))(B(t) - B(s)) \\
&+ H'(\Phi_x(x, s))^2 \Phi_{xx}(x, s)(B(t) - B(s))^2 + O(|B(t) - B(s)|^3).
\end{aligned}
\tag{3.197}
$$

It follows that, if $0 \le t - s \le \rho_h$, there exists $C > 0$ depending only on H such that

$$
\begin{cases}
\sup_\mathbb{R} |S_h(t, s)\Phi(\cdot, s) - \Phi(\cdot, t)| \\
\le C \sup_{r \in [s,t]} \|D^2\Phi(\cdot, r)\|_\infty \left(|B(t) - B(s)|^2 + h^2 \right) \\
\le C \sup_{r \in [s,t]} \|D^2\Phi(\cdot, r)\|_\infty (1 + \lambda_0^2)h^2.
\end{cases}
\tag{3.198}
$$

Therefore, in order for the scheme to have a chance of converging, ρ_h should satisfy

$$
\lim_{h \to 0} \frac{h^2}{\rho_h} = 0.
\tag{3.199}
$$

Both (3.196) and (3.199) can be achieved when B is continuously differentiable or merely Lipschitz continuous by setting

$$
\rho_h = \lambda h \|\dot{B}\|_\infty^{-1}.
$$

More generally, if $B \in C^{0,\alpha}([0, T])$ with $\alpha > \frac{1}{2}$ and

$$
(\rho_h)^\alpha = \frac{\lambda h}{[W]_{\alpha, T}},
\tag{3.200}
$$

then both (3.196) and (3.199) are satisfied, since

$$\frac{h^2}{\rho_h} = \left(\frac{[B]_{\alpha,T} h^{2\alpha-1}}{\lambda}\right)^{1/\alpha} \xrightarrow{h\to 0} 0.$$

However, this approach fails as soon as the quadratic variation path

$$\langle B \rangle_T := \lim_{|\mathcal{P}|\to 0} \sum_{n=0}^{N-1} |B(t_{n+1}) - B(t_n)|^2$$

is non-zero, as (3.196) and (3.199) together imply that $\langle B \rangle_T = 0$. This rules out, for instance, the case where B is the sample path of a Brownian motion, or, more generally, any nontrivial semimartingale.

Motivated by the theory of rough differential equations, it is natural to explore whether the scheme operator (3.195) can be somehow altered to refine the estimate in (3.198), potentially allowing (3.199) to be relaxed and ρ_h to converge more quickly to zero as $h \to 0^+$.

More precisely, the next term in the expansion (3.197) suggests taking $B \in C^{0,\alpha}([0, T]; \mathbb{R}^m)$ with $\alpha > \frac{1}{3}$, or, more generally, B with p-variation with $p < 3$, and defining

$$
\begin{aligned}
S_h(t, s)u(x) = u(x) &+ H\left(\frac{u(x + h) - u(x - h)}{2h}\right)(B(t) - B(s)) \\
&+ \frac{1}{2}H'\left(\frac{u(x + h) - u(x - h)}{2h}\right)^2 \\
&\times \left(\frac{u(x + h) + u(x - h) - 2u(x)}{h^2}\right)(B(t) - B(s))^2 \\
&+ \frac{\theta}{2}\left(u(x + h) + u(x - h) - 2u(x)\right).
\end{aligned}
\tag{3.201}
$$

As can easily be checked, (3.201) is monotone as long as (3.196) holds, and

$$\text{Lip}(u) \leq L, \quad \theta + \|H'\|_\infty \lambda^2 \leq 1, \quad \text{and} \quad \lambda \leq \frac{\theta}{\|H'\|_\infty (1 + 2L\|H''\|_\infty)}.$$

The error in (3.198) would then be of order $h^2 + |B(t) - B(s)|^3$, which again leads to a requirement like (3.199). This seems to indicate that it is necessary to incorporate higher order corrections in (3.201) to deal with the second-order spatial derivatives of u. However, this will disrupt, in general, the monotonicity of the scheme, since it will no longer be possible to use discrete maximum principle techniques.

For this reason, it is more convenient to concentrate on the more effective strategy of regularizing the path B. If $\{B_h\}_{h>0}$ is a family of smooth paths converging uniformly, as $h \to 0$, to B, then $\langle B_h \rangle_T = 0$ for each fixed $h > 0$, and therefore, B_h and ρ_h can be chosen so that (3.196) and (3.199) hold for B_h rather than W.

3.13 Homogenization

I present a variety of results regarding the asymptotic properties, for small $\varepsilon > 0$, of equations of the form

$$u_t^\varepsilon + \sum_{i=1}^{m} H^i(Du^\varepsilon, x/\varepsilon)\dot{\zeta}_i^\varepsilon = 0 \text{ in } Q_\infty \quad u^\varepsilon(\cdot, 0) = u_0. \tag{3.202}$$

Many proofs and details are omitted here, and can be found in Seeger [98].

Each Hamiltonian H^i in (3.202) is assumed to have some averaging properties in the variable $y = x/\varepsilon$. The paths $\zeta^\varepsilon = (\zeta_1^\varepsilon, \cdots, \zeta_m^\varepsilon)$, which converge locally uniformly to some limiting path $\zeta \in C_0([0, \infty); \mathbb{R}^m)$, will be assumed to be piecewise C^1, although I present some results where they are only continuous.

One motivation for considering such problems is to study general equations of the form

$$u_t^\varepsilon + \frac{1}{\varepsilon^\gamma} H\left(Du^\varepsilon, \frac{x}{\varepsilon}, \frac{t}{\varepsilon^{2\gamma}}\right) = 0 \text{ in } Q_\infty \quad u^\varepsilon(\cdot, 0) = u_0. \tag{3.203}$$

In addition to the averaging dependence on space, the Hamiltonian H is assumed to have zero expectation, so that, on average, u^ε is close to its initial value u_0. The dependence on time, meanwhile, is assumed to be "mixing" with a certain rate, so that, with the scaling of the central limit theorem, $\varepsilon^{-\gamma} H(\cdot, \cdot, t\varepsilon^{-2\gamma})$ will resemble, as $\varepsilon \to 0$, to white noise in time.

When $\gamma = 1$, (3.203) arises naturally as a scaled version of

$$u_t + H(Du, x, t) = 0 \text{ in } Q_\infty \quad u(\cdot, 0) = \varepsilon^{-1} u_0(\varepsilon \cdot), \tag{3.204}$$

with u and u^ε related by $u^\varepsilon(x, t) = \varepsilon u(x/\varepsilon, t/\varepsilon^2)$.

Studying the $\varepsilon \to 0$ limit of u^ε then amounts to understanding the averaged large space, long time behavior of solutions of (3.204) with large, slowly-varying initial data.

Although it is of interest to examine (3.203) for different values of γ, it turns out that the nature of the limiting behavior does not change for different values of γ. Hence, from a practical point of view, ε and $\delta = \varepsilon^\gamma$ can be viewed as small, independent parameters. It should be, however, noted that for technical reasons, some results can only be proved under a mildness assumption on the approximate white noise dependence, which translates to a smallness condition on γ.

The Hamiltonians considered in (3.203) have the form

$$H(p, y, t) = \sum_{i=1}^{m} H^i(p, y)\xi_i(t),$$ (3.205)

where the random fields $\xi_i : [0, \infty) \to \mathbb{R}$ are defined on a probability space $(\Omega, \mathcal{F}, \mathbb{P})$ and are assumed to be mixing with rate ρ as explained below.

For $0 \leq s \leq t \leq \infty$, consider the sigma algebras $\mathcal{F}_{s,t}^i \subset \mathcal{F}$ generated by $\{\xi_i(r)\}_{r \in [s,t]}$. The mixing rate is then defined by

$$\rho(t) = \max_{i=1,2,\dots,m} \sup_{s \geq 0} \sup_{A \in \mathcal{F}_{s+t,\infty}^i} \sup_{B \in \mathcal{F}_{0,s}^i} |\mathbb{P}(A \mid B) - \mathbb{P}(A)|.$$ (3.206)

The quantitative mixing assumptions for the ξ^i are that

$$\begin{cases} t \mapsto \xi_i(t) \text{ is stationary,} \quad \rho(t) \xrightarrow{t \to \infty} 0, \quad \int_0^\infty \rho(t)^{1/2}\, dt < \infty, \\ \mathbb{E}[\xi_i(0)] = 0, \text{ and } \mathbb{E}[\xi_i(0)^2] = 1. \end{cases}$$ (3.207)

Above stationarity means that

$$(\xi(s_1), \xi(s_2), \dots, \xi(s_M)) \quad \text{and} \quad (\xi(s_1 + t), \xi(s_2 + t), \dots, \xi(s_M + t))$$

have the same joint distribution for any choice of $s_1, s_2, \dots, s_M \in [0, \infty)$ and $t \geq -\min_j s_j$.

It follows from the ergodic theorem, the stationarity and the centering assumptions that

$$\lim_{\delta \to 0} \delta \int_0^{\frac{t}{\delta}} \xi_i(s)\, ds = 0.$$

The properties of the long time fluctuations of $\zeta = \int_0^t \xi(s)ds$ around 0 can be studied using the central limit theorem scaling. Indeed setting $\zeta_i^\delta(t) = \delta\zeta_i(t/\delta^2)$, it is well-known that, as $\delta \to 0$, ζ_i^δ converges in distribution and locally uniformly to a standard Brownian motion. Indeed, with $\delta = \varepsilon^\gamma$, (3.203) is then a specific form of (3.202).

3.13.1   The Difficulties and General Strategy

Here I discuss some of the difficulties in the study of the $\varepsilon \to 0$ behavior of (3.202) and the strategies that can be used to overcome them. To keep things simple, I only consider Hamiltonians that are periodic in space.

The starting (formal) assumption is that the noise is "mild" enough to allow for averaging behavior in space, and therefore, u^ε is closely approximated by a solution $\overline{u}^\varepsilon$ of an equation of the form $\overline{u}_t^\varepsilon + \overline{H}^\varepsilon(D\overline{u}^\varepsilon, t) = 0$.

More precisely, following the standard strategy of the homogenization theory, it is assumed that there exists some auxiliary function $v : \mathbb{T}^d \times [0, \infty) \to \mathbb{R}$, so that u^ε has the formal expansion

$$u^\varepsilon(x, t) \approx \overline{u}^\varepsilon(x, t) + \varepsilon v(x/\varepsilon, t).$$

An asymptotic analysis yields that, for fixed $p \in \mathbb{R}^d$ (here, $p = D\overline{u}^\varepsilon(x, t)$ and $y = \frac{x}{\varepsilon}$), v solves the so called "cell problem"

$$\sum_{i=1}^m H^i(D_y v + p, y)\xi_i = \overline{H}(p, \xi) \text{ in } . \mathbb{R}^d, \tag{3.208}$$

where the fixed parameter $\xi \in \mathbb{R}^m$ stands in place of the mild white noise $\varepsilon^{-\gamma}\xi(t/\varepsilon^{2\gamma})$.

It is standard the theory of periodic homogenization of Hamilton-Jacobi equations that, under the right conditions, there is a unique constant $\overline{H}(p, \xi)$ for which (3.208) has periodic solutions, which are called "correctors."

Taking this fact for granted for now and always arguing formally yields that u^ε will be closely approximated by $\overline{u}^\varepsilon$ which solves

$$\overline{u}_t^\varepsilon + \frac{1}{\varepsilon^\gamma}\overline{H}\left(D\overline{u}^\varepsilon, \xi\left(\frac{t}{\varepsilon^{2\gamma}}\right)\right) = 0 \text{ in } Q_\infty \quad \overline{u}^\varepsilon(\cdot, 0) = u_0. \tag{3.209}$$

Note that, in deriving (3.209), it was used that $\xi \mapsto \overline{H}(\cdot, \xi)$ is positively homogenous, which follows from multiplying (3.208) by a positive constant and using the uniqueness of the right-hand side.

If

$$\mathbb{E}\left[\overline{H}(p, \xi(t))\right] = 0 \text{ for all } p \in \mathbb{R}^d, \tag{3.210}$$

then the solution of (3.209) with $u_0(x) = \langle p_0, x \rangle$, which is given by

$$\overline{u}^\varepsilon(x, t) = \langle p_0, x \rangle - \frac{1}{\varepsilon^{2\gamma}}\int_0^t \overline{H}\left(p_0, \xi\left(\frac{s}{\varepsilon^{2\gamma}}\right)\right) ds,$$

converges, as $\varepsilon \to 0$ and in distribution, to $p_0 \cdot x + \sigma(p_0) B(t)$, where B is a standard Brownian motion and

$$\sigma(p_0) = \left(\mathbb{E}\left[\overline{H}(p_0, \xi(0))^2 \right] \right)^{1/2}.$$

Due, however, to the nonlinearity of the map $\xi \mapsto \overline{H}(\cdot, \xi)$ and the difficulties associated with the "rough" pathwise solutions, it is not clear how to study the (3.209) for an arbitrary $u_0 \in UC(\mathbb{R}^d)$. It turns out that the answers are subtle, and, in the multiple path case considered below, depend strongly on the nature of the mixing field ξ.

When $m = 1$, the characterization of $\overline{H}(p, \xi)$ reduces to the study of the two Hamiltonians

$$\overline{H}(p) = \overline{H}(p, 1) \quad \text{and} \quad \overline{(-H)}(p) = \overline{H}(p, -1).$$

Then (3.209) takes the form

$$\begin{cases} \overline{u}_t^\varepsilon + \frac{1}{\varepsilon^\gamma} \overline{H}^1 (D\overline{u}^\varepsilon) \xi \left(\frac{t}{\varepsilon^{2\gamma}} \right) + \frac{1}{\varepsilon^\gamma} \overline{H}^2 (D\overline{u}^\varepsilon) \left| \xi \left(\frac{t}{\varepsilon^{2\gamma}} \right) \right| = 0 \text{ in } Q_T \\ \overline{u}^\varepsilon(\cdot, 0) = u_0, \end{cases} \tag{3.211}$$

where

$$\overline{H}^1(p) = \frac{\overline{H}(p) - \overline{(-H)}(p)}{2} \quad \text{and} \quad \overline{H}^2(p) = \frac{\overline{H}(p) + \overline{(-H)}(p)}{2}.$$

Note that $\overline{H}^2 = 0$ if and only if

$$\overline{(-H)} = -\overline{H}, \tag{3.212}$$

and, moreover, that (3.210) is equivalent to (3.212) when $m = 1$.

Since (3.208) is interpreted in the viscosity solution sense, it is not possible to multiply the equation by -1, and so (3.212) is not only not obvious, but actually false in general.

Indeed, assume that, for some $p_0 \in \mathbb{R}^d$, $\overline{(-H)}(p_0) \neq -\overline{H}(p_0)$. Then $\overline{u}^\varepsilon$ with $u_0(x) = \langle p_0, x \rangle$ is given by

$$\overline{u}^\varepsilon(x, t) = \langle p_0, x \rangle - \varepsilon^\gamma \frac{\overline{H}(p_0) - \overline{(-H)}(p_0)}{2} \zeta \left(\frac{t}{\varepsilon^{2\gamma}} \right)$$

$$- \varepsilon^\gamma \frac{\overline{H}(p_0) + \overline{(-H)}(p_0)}{2} \int_0^{t/\varepsilon^{2\gamma}} \left| \xi \left(\frac{s}{\varepsilon^{2\gamma}} \right) \right| ds,$$

and, hence,

$$\varepsilon^\gamma \overline{u}^\varepsilon(x,t) \xrightarrow{\varepsilon \to 0} -\frac{\overline{H}(p_0) + \overline{(-H)}(p_0)}{2} \mathbb{E}\,|\xi(0)|\,t \quad \text{in distribution.}$$

On the other hand, if (3.212) holds, then (3.209) becomes

$$\overline{u}_t^\varepsilon + \frac{1}{\varepsilon^\gamma}\overline{H}(D\overline{u}^\varepsilon)\dot{\xi}\left(\frac{t}{\varepsilon^{2\gamma}}\right) = 0 \text{ in } Q_T \quad \overline{u}^\varepsilon(\cdot, 0) = u_0, \tag{3.213}$$

and the determination of whether or not $\overline{u}^\varepsilon$ has a limit depends on the properties of the effective Hamiltonian $\overline{H}$, and, in particular, whether or not it is the difference of two convex functions.

3.13.2 The Single-Noise Case

I state next some results about

$$u_t^\varepsilon + \frac{1}{\varepsilon^\gamma}H\left(Du^\varepsilon, \frac{x}{\varepsilon}\right)\dot{\xi}\left(\frac{t}{\varepsilon^{2\gamma}}\right) = 0 \text{ in } Q_T \quad u^\varepsilon(\cdot, 0) = u_0. \tag{3.214}$$

As suggested in the previous subsection, the fact that there is only one source of noise simplifies the structure of the problem. Consequently, the results are more comprehensive than in the multiple-path setting.

It is assumed that

$$H \in C(\mathbb{R}^d \times \mathbb{R}^d) \text{ is convex and coercive in the gradient variable.} \tag{3.215}$$

The convexity assumption is important for two reasons. It guarantees that the consistency condition (3.212) holds, and it also implies strong path-stability estimates for the solutions. The latter were already alluded to earlier in the notes, in the section on the comparison principle for equations with convex, spatially-dependent Hamiltonians.

Regarding the spatial environment, the results are general enough to allow for a variety of different assumptions. Here, I list two well-studied examples.

The first possible self-averaging assumption is that

$$y \mapsto H(p, y) \text{ is } \mathbb{Z}^d\text{-periodic.} \tag{3.216}$$

The periodic homogenization of (time-homogenous) Hamilton-Jacobi equations has a vast literature going back to Lions et al. [63] and Evans [24, 25].

Another type of averaging dependence, which in general is more physically relevant, is stationary-ergodicity. In this setting, the Hamiltonians $H = H(p, x, \omega)$

are defined on a probability space $(\Omega, \mathbf{F})$ that is independent of the random field ξ and is equipped with a group of translation operators $T_z : \Omega \to \Omega$ such that $H(\cdot, T_z y) = H(\cdot, y + z)$. It is assumed that $\{T_z\}_{z \in \mathbb{R}^d}$ is stationary and ergodic, that is,

$$\begin{cases} \mathbf{P} = \mathbf{P} \circ T_z \text{ for all } z \in \mathbb{R}^d \text{ , and} \\ \text{if } E \in \mathbf{F} \text{ and } T_z E = E \text{ for all } z \in \mathbb{R}^d, \text{ then } \mathbf{P}[E] = 1 \text{ or } \mathbf{P}[E] = 0. \end{cases} \tag{3.217}$$

In the time-inhomogenous setting, this homogenization problem was studied by Souganidis [103] and Rezakhanlou and Tarver [93].

The first result is stated next.

Theorem 13.1 *There exists a Brownian motion* $B : [0, \infty) \to \mathbb{R}$ *such that, as* $\varepsilon \to 0$, $(u^\varepsilon, \zeta^\varepsilon)$ *converges in distribution to* $(\overline{u}, B)$ *in* $BUC(\mathbb{R}^d \times [0, \infty)) \times C([0, \infty))$, *where* $\overline{u}$ *is the pathwise viscosity solution of*

$$d\overline{u} + \overline{H}(D\overline{u}) \cdot dB = 0 \text{ in } Q_\infty \quad \overline{u}(\cdot, 0) = u_0. \tag{3.218}$$

Since $\delta B(t/\delta^2)$ equals $B(t)$ in distribution, it is also an interesting question to study the limiting behavior of

$$du^\varepsilon + H(Du^\varepsilon, x/\varepsilon) \cdot dB = 0 \text{ in } Q_\infty \quad u^\varepsilon(\cdot, 0) = u_0. \tag{3.219}$$

Theorem 13.2 *In addition to the hypotheses of Theorem 13.1, assume that the comparison principle holds for* (3.219). *Then, with probability one, as* $\varepsilon \to 0$, *the solution* u^ε *of* (3.219) *converges locally uniformly to the solution of* (3.218).

The final remark is that the theorems above can be applied to a variety of other settings like, for instance, the homogenization of

$$u_t^\varepsilon + H\left(Du^\varepsilon, x, \frac{x}{\varepsilon}\right) \dot{\zeta}^\varepsilon(t) = 0 \text{ in } Q_\infty \quad u^\varepsilon(\cdot, 0) = u_0$$

with $(\zeta^\varepsilon)_{\varepsilon > 0}$ any collection of paths converging locally uniformly and almost surely (or in distribution) to a Brownian motion or other stochastic process, and with the dependence of H on the fast variable being, for instance, periodic, quasi-periodic, or stationary-ergodic.

3.13.3 The Multiple-Noise Case

Since in this setting the results so far are less general and quite technical, I only present an overview here. Details and more results can be found in a forthcoming work of Seeger [97].

The problem is the behavior of equations like

$$u_t^\varepsilon + \frac{1}{\varepsilon^\gamma} \sum_{i=0}^m H^i(Du^\varepsilon, x/\varepsilon)\xi_i\left(\frac{t}{\varepsilon^{2\gamma}}\right) = 0 \text{ in } Q_\infty \quad u^\varepsilon(\cdot, 0) = u_0, \tag{3.220}$$

where, for each $i = 0, \ldots, m$, ξ^i is a mixing field satisfying (3.207). More assumptions on the Hamiltonians and the paths will need to be made later.

To simplify the presentation, here I only consider the periodic setting (3.216). It turns out that, under appropriate conditions on the H^i's which are made more specific below, for every $p \in \mathbb{R}^d$ and $\xi \in \mathbb{R}^m$, there exists a unique constant $\overline{H}(p, \xi)$ such that the cell problem

$$\sum_{i=1}^m H^i(p + D_y v, y)\xi^i = \overline{H}(p, \xi) \tag{3.221}$$

admits periodic solutions $v : \mathbb{T}^d \to \mathbb{R}$. Moreover, $\xi \mapsto \overline{H}(p, \xi)$ is positively homogenous, and

$$\mathbb{E}[\overline{H}(p, \xi(0)] = 0 \text{ for all } p \in \mathbb{R}^d. \tag{3.222}$$

Using error estimates for the theory of periodic homogenization of Hamilton-Jacobi equations, it is possible to show that u^ε is closely approximated by the solution $\overline{u}^\varepsilon$ of

$$\overline{u}_t^\varepsilon + \frac{1}{\varepsilon^\gamma} \overline{H}\left(D\overline{u}^\varepsilon, \xi\left(\frac{t}{\varepsilon^{2\gamma}}\right)\right) = 0 \text{ in } Q_\infty \quad \overline{u}^\varepsilon(\cdot, 0) = u_0. \tag{3.223}$$

The limiting behavior of (3.223) is well understood if $u_0(x) = \langle p_0, x \rangle$ for some fixed $p_0 \in \mathbb{R}^d$. Indeed, in view of the mixing properties of ξ and the centering property (3.222), there exists a Brownian motion B such that, as $\varepsilon \to 0$, $\overline{u}^\varepsilon$ converges locally uniformly in distribution to

$$\langle p_0, x \rangle + \mathbb{E}\left[\overline{H}(p_0, \xi(0))^2\right]^{1/2} B(t).$$

I comment next about the limit of $\overline{u}^\varepsilon$ for arbitrary initial data u_0. The goal is to show that, under assumptions on the Hamiltonians and mixing fields, there exists $M \geq 1$ and, for each $j = 1, \ldots, M$, an effective Hamiltonian $\overline{H}^j : \mathbb{R}^d \to \mathbb{R}$ which is the difference of two convex functions, and a Brownian motion B^j such that, as $\varepsilon \to 0$ and in distribution, $\overline{u}^\varepsilon$ and, therefore, u^ε converges in $BUC(Q_T)$ to the pathwise viscosity solution $\overline{u}$ of

$$d\overline{u} + \sum_{j=1}^M \overline{H}^j(D\overline{u}) \cdot dB_j = 0 \text{ in } Q_\infty \quad \overline{u} = u_0. \tag{3.224}$$

Although at first glance, the nature of the problem is similar to the single path case, there are some fundamental differences. Most importantly, the deterministic effective Hamiltonians $\{\overline{H}^j\}_{j=1}^M$, and even the number M, depend on the particular law of the mixing field ξ.

Next I introduce some further assumptions that give rise to a rich class of examples and results.

As far as the Hamiltonians $(H_i, \ldots, H_m)$ are concerned, it is assumed that

$$
\begin{cases}
H^i \in C^{0,1}(\mathbb{R}^d \times \mathbb{T}^d), \\[2ex]
p \mapsto H^1(p, \cdot) + \displaystyle\sum_{i=2}^m H^i(p, \cdot)\xi_i \text{ is convex} \\[2ex]
\text{for all } \xi_2, \ldots, \xi_m \in \{-1, 1\}, \text{ and} \\[2ex]
\displaystyle\lim_{|p| \to +\infty} \inf_{y \in \mathbb{T}^d} \left(H^1(p, y) - \sum_{i=2}^m \left| H^i(p, y) \right| \right) = +\infty.
\end{cases}
\tag{3.225}
$$

As a consequence, the cell problem (3.221) is solvable for all $p \in \mathbb{R}^d$ and $\xi \in \{-1, 1\}^m$, and furthermore, $p \mapsto \overline{H}(p, 1, \xi)$ is convex and $\xi \mapsto \overline{H}(p, \xi)$ is homogenous, that is, for all $\lambda \in \mathbb{R}$ and $\xi \in \{-1, 1\}^m$,

$$
\overline{H}(\cdot, \lambda\xi) = \lambda\overline{H}(\cdot, \xi).
\tag{3.226}
$$

The mixing fields are assumed to be, for $i = 1, \ldots, m$, of the form

$$
\begin{cases}
\xi_i = \displaystyle\sum_{k=0}^\infty X_k^i \mathbf{1}_{(k, k+1)} \quad \text{where} \\[2ex]
\left(X_k^i \right)_{i=1,2,\ldots,m, \ k=0,1,\ldots}
\end{cases}
\tag{3.227}
$$

are independent Rademacher random variables.

In particular, if

$$
\xi_i^\varepsilon(t) = \frac{1}{\varepsilon^\gamma} \xi_i(t/\varepsilon^{2\gamma}) \quad \text{and} \quad \zeta_i^\varepsilon(t) = \int_0^t \xi^\varepsilon(s)_i \, ds,
\tag{3.228}
$$

then each $\zeta^{i,\varepsilon}$ is a scaled, linearly-interpolated, simple random walk on $\mathbb{Z}$, and there exists an m-dimensional Brownian motion $(B_1, \ldots, B_m)$, such that, in distribution,

$$
(\zeta^{1,\varepsilon}, \zeta^{2,\varepsilon}, \ldots, \zeta^{m,\varepsilon}) \xrightarrow{\varepsilon \to 0} (B_1, \ldots, B_m) \text{ in } C([0, \infty); \mathbb{R}^m).
$$

Consider the sets of indices

$$
\begin{cases}
\mathcal{A}^m := \{\mathbf{j} = (j_1, \ldots, j_l) : j_i \in \{1, \ldots, m\}, \ j_1 < \cdots < j_l\} \\[4pt]
\text{with } l = |\mathbf{j}| = |(j_1, j_2, \ldots, j_l)| \\[4pt]
\mathcal{A}^m_0 := \{\mathbf{j} \in \mathcal{A}^m : |\mathbf{j}| \text{ is odd}\},
\end{cases}
$$

noting that $\#\mathcal{A}^m = 2^m - 1$ and $\#\mathcal{A}^m_0 = 2^{m-1}$.

For any $\mathbf{j} = (j_1, j_2, \ldots, j_l) \in \mathcal{A}^m$, define

$$
\begin{cases}
\xi_{\mathbf{j}} := \xi_{j_1} \cdots \xi_{j_l} \ \text{ for } \ \xi = (\xi_1, \ldots, \xi_m) \in \{-1, 1\}^m, \\[6pt]
\overline{H}^{\mathbf{j}}(p) := \displaystyle\sum_{\xi \in \{-1,1\}^m} 2^{-m} \overline{H}(p, \xi) \xi_{\mathbf{j}}, \\[10pt]
X^{\mathbf{j}}_k := X^{j_1}_k X^{j_2}_k \cdots X^{j_l}_k, \\[8pt]
\zeta_{\mathbf{j}}(0) := 0, \quad \dot{\zeta}_{\mathbf{j}} = \displaystyle\sum_{k=0}^{\infty} X^{\mathbf{j}}_k \mathbf{1}_{(k, k+1)}, \quad \text{and} \quad \zeta^{\varepsilon}_{\mathbf{j}}(t) = \varepsilon^{\gamma} \zeta_{\mathbf{j}}(t/\varepsilon^{2\gamma}),
\end{cases}
\tag{3.229}
$$

and observe that, for each $\mathbf{j} \in \mathcal{A}^m_0$, $\overline{H}^{\mathbf{j}}$ is a difference of convex functions. Note also that, if $|\mathbf{j}|$ is even, then the homogeneity property (3.226) implies that $\overline{H}^{\mathbf{j}} = 0$.

The following is true.

Theorem 13.3 *Assume that $\gamma \in (0, 1/6)$, $u_0 \in C^{0,1}(\mathbb{R}^d)$, (3.225), and (3.227), and let u^{ε} be the solution of (3.220). Then there exist 2^{m-1} independent Brownian motions $\{B^{\mathbf{j}}\}_{\mathbf{j} \in \mathcal{A}^m_0}$, such that, in distribution,*

$$
\left(u^{\varepsilon}, \{\zeta^{\mathbf{j},\varepsilon}\}_{\mathbf{j} \in \mathcal{A}^m_0}\right) \xrightarrow{\ \varepsilon \to 0\ } \left(\overline{u}, \{B^{\mathbf{j}}\}_{\mathbf{j} \in \mathcal{A}^m_0}\right) \ \text{in} \ BUC(Q_T) \times C\left([0, T]; \mathbb{R}^{2^{m-1}}\right),
$$

where $\overline{u}$ is the stochastic viscosity solution of

$$
d\overline{u} + \sum_{\mathbf{j} \in \mathcal{A}^m_0} \overline{H}^{\mathbf{j}}(D\overline{u}) \cdot dB_{\mathbf{j}} = 0 \text{ in } Q_{\infty} \quad \overline{u}(\cdot, 0) = u_0.
\tag{3.230}
$$

The result relies on the fact that, in view of the assumptions on the mixing fields ξ_i, which take their values only in $\{-1, 1\}$, the general effective Hamiltonian $\overline{H}(p, \xi)$ can be decomposed using a combinatorial argument.

As already mentioned, the above theorem covers only some of the possible homogenization problems that can be studied in the multiple-noise case. In particular, it is shown in [97] that the limiting equation depends on the law of the mixing field ξ. This is in stark contrast to the single-noise case, where the limiting equation is independent of the mild-noise approximation.

3.14   Stochastically Perturbed Reaction-Diffusion Equations and Front Propagation

I discuss here a result of Lions and Souganidis [69] about the onset of fronts in the long time and large space asymptotics of bistable reaction-diffusion equations which are additively perturbed by small relatively smooth (mild) stochastic in time forcing. The prototype problem is the so called stochastic Allen-Cahn equation. The interfaces evolve with curvature dependent normal velocity which is additively perturbed by time white noise. No regularity assumptions are made on the fronts. The results can be extended to more complicated equations with anisotropic diffusion, drift and reaction which may be periodically oscillatory in space. To keep the ideas simple, in this section I concentrate on the classical Allen-Cahn equation.

The goal is to study the behavior, as $\varepsilon \to 0$, of the parabolically rescaled Allen-Cahn equation

$$u_t^\varepsilon - \Delta u^\varepsilon + \frac{1}{\varepsilon^2}(f(u^\varepsilon) - \varepsilon \dot{B}^\varepsilon(t, \omega)) = 0 \text{ in } Q_\infty \quad u^\varepsilon(\cdot, 0) = u_0^\varepsilon, \qquad (3.231)$$

where, $f \in C^2(\mathbb{R}^d; \mathbb{R})$ is such that

$$\begin{cases} f(\pm 1) = f(0) = 0, \ f'(\pm 1) > 0, \ f'(0) < 0 \\ f > 0 \text{ in } (-1, 0), \ f < 0 \text{ in } (0, 1), \text{ and } \int_{-1}^{+1} f(u)du = 0, \end{cases} \qquad (3.232)$$

that is, f is the derivative of a double well potential with wells of equal depth at, for definiteness, ± 1 and in between maximum at 0,

$$B^\varepsilon(\cdot, \omega) \in C^2([0, \infty); \mathbb{R}) \text{ is an a.s. mild approximation of } B(\cdot, \omega), \qquad (3.233)$$

that is, a.s. in ω and locally uniformly $[0, \infty)$,

$$\lim_{\varepsilon \to 0} B^\varepsilon(t, \omega) = B, \ B^\varepsilon(0, \omega) = 0, \text{ and } \lim_{\varepsilon \to 0} \varepsilon |\ddot{B}^\varepsilon(t, \omega)| = 0, \qquad (3.234)$$

and there exists an open $\mathcal{O}_0 \subset \mathbb{R}^d$ such that

$$\begin{cases} \mathcal{O}_0 = \{x \in \mathbb{R}^d : u_0^\varepsilon(x) > 0\}, \ \mathbb{R}^d \setminus \overline{\mathcal{O}_0} = \{x \in \mathbb{R}^d : u_0^\varepsilon(x) < 0\}, \\ \text{and} \\ \Gamma_0 = \partial \mathcal{O}_0 = \partial(\mathbb{R}^d \setminus \overline{\mathcal{O}_0}) = \{x \in \mathbb{R}^d : u_0^\varepsilon(x) = 0\}. \end{cases} \qquad (3.235)$$

Although it is not stated explicitly, it assumed that there exists an underlying probability space, but, for ease of the notation, we omit the dependence on ω unless necessary.

Here are two classical examples of mild approximations. The first is the convolution $B^\varepsilon(t) = B \star \rho^\varepsilon(t)$, where $\rho^\varepsilon(t) = \varepsilon^{-\gamma}\rho(\varepsilon^{-\gamma}t)$ with $\rho \in C^\infty$, even and compactly supported in $(-1, 1)$, $\int \rho(t)dt = 1$ and $\gamma \in (0, 1/2)$. The second is $\dot{B}^\varepsilon(t) = \varepsilon^{-\gamma}\xi(\varepsilon^{-2\gamma}t)$, where $\xi(t)$ is a stationary, strongly mixing, mean zero stochastic process such that $\max(|\xi|, |\dot{\xi}|) \leq M$ and $\gamma \in (0, 1/3)$. I refer to [51] for a discussion.

Next I use the notion of stochastic viscosity solutions and the level set approach to describe the generalized evolution (past singularities) of a set with normal velocity

$$V = -\text{tr}[Dn]\, dt + d\zeta, \qquad (3.236)$$

for some a continuous path $\zeta \in C_0([0, \infty); \mathbb{R})$. Here n is the external normal to the front and, hence, $\text{tr}[Dn]$ is the mean curvature.

Given a triplet $(\mathcal{O}_0, \Gamma_0, \mathbb{R}^d \setminus \overline{\mathcal{O}_0})$ with $\mathcal{O}_0 \subset \mathbb{R}^d$ open, we say that the sets $(\Gamma_t)_{t>0}$ move with normal velocity (3.236), if, for each $t > 0$, there exists a triplet $(\mathcal{O}_t, \Gamma_t, \mathbb{R}^d \setminus \overline{\mathcal{O}_t})$, with $\mathcal{O}_t \subset \mathbb{R}^d$ open, such that

$$\begin{cases} \mathcal{O}_t = \{x \in \mathbb{R}^d : w(x, t) > 0\}, \ \mathbb{R}^d \setminus \overline{\mathcal{O}_t} = \{x \in \mathbb{R}^d : w(x, t) < 0\}, \\ \text{and} \\ \Gamma_t = \{x \in \mathbb{R}^d : w(x, t) = 0\}, \end{cases}$$

$$(3.237)$$

where $w \in \text{BUC}(\mathbb{R}^d \times [0, \infty))$ is the unique stochastic (pathwise) solution of the level-set initial value pde

$$dw = (I - \widehat{Dw} \otimes \widehat{Dw}) : D^2 w - |Dw| \cdot d\zeta \text{ in } Q_\infty \quad w(\cdot, 0) = w_0, \qquad (3.238)$$

with $\hat{p} := p/|p|$ and $w_0 \in \text{BUC}(\mathbb{R}^d)$ such that

$$\begin{cases} \mathcal{O}_0 = \{x \in \mathbb{R}^d : w_0(x) > 0\}, \ \mathbb{R}^d \setminus \overline{\mathcal{O}_0} = \{x \in \mathbb{R}^d : w_0(x) < 0\}, \\ \text{and} \\ \Gamma_0 = \{x \in \mathbb{R}^d : w_0(x) = 0\}. \end{cases}$$

$$(3.239)$$

The properties of (3.238) are used here to adapt the approach introduced in Evans et al. [27], Barles et al. [7], and Barles and Souganidis [6] to study the onset of moving fronts in the asymptotic limit of reaction-diffusion equations and interacting particle systems with long range interactions. This methodology allows to prove global in time asymptotic results and is not restricted to smoothly evolving fronts.

The main result of the paper is stated next.

Theorem 14.1 *Assume* (3.232)–(3.235), *and let* u^ε *be the solution of* (3.231). *There exists* $\alpha_0 \in \mathbb{R}$ *such that, if* w *is the solution of* (3.238) *with* w_0 *satisfying* (3.239) *and $\zeta \equiv \alpha_0 B$, where B is a standard Brownian path, then, as $\varepsilon \to 0$, a.s. in ω and locally uniformly in (x, t), $u^\varepsilon \to 1$ in $\{(x, t) \in \mathbb{R}^d \times (0, \infty) : w(x, t) > 0\}$ and $u^\varepsilon \to -1$ in $\{(x, t) \in \mathbb{R}^d \times (0, \infty) : w(x, t) < 0\}$, that is, $u^\varepsilon \to 1$ (resp. $u^\varepsilon \to -1$) inside (resp. outside) a front moving with normal velocity $V = -tr[Dn]\, dt + \alpha_0 dB$.*

Theorem 14.1 provides a complete characterization of the asymptotic behavior of the Allen-Cahn equation perturbed by mild approximations of the time white noise. The result holds in all dimensions, it is global in time and does not require any regularity assumptions on the moving interface.

In [32] Funaki studied the asymptotics of (3.231) when $d = 2$ assuming that the initial set is a smooth curve bounding a convex set. Under these assumptions the evolving curve remains smooth and (3.295) reduces to a stochastic differential equation in the arc length variable. Under the assumption that the evolving set is smooth, which is true if the initial set is smooth and for small time, a similar result was announced recently by Alfaro et al. [1]. Assuming convexity at $t = 0$, Yip [106] showed a similar result for all times using a variational approach. There have also been several other attempts to study the asymptotics of (3.231) in the graph-like setting and always for small time.

Reaction-diffusion equations perturbed additively by white noise arise naturally in the study of hydrodynamic limits of interacting particles. The relationship between the long time, large space behavior of the Allen-Cahn perturbed additively by space-time white noise and fronts moving by additively perturbed mean curvature was conjectured by Ohta et al. [84]. Funaki [31] obtained results in this direction when $d = 1$ where there is no curvature effect. A recent observation of Lions and Souganidis [70] shows that the general conjecture cannot be correct. Indeed, it is shown in [70] that the formally conjectured interfaces, which should move by mean curvature additively perturbed with space-time white noise, are not well defined.

From the phenomelogical point of view, problems like (3.231) arise naturally in the phase-field theory when modeling double-well potentials with depths (stochastically) oscillating in space-time around a common one. This leads to stable equilibria that are only formally close to ± 1. As a matter of fact, the locations of the equilibria may diverge due to the strong effect of the white noise.

The history and literature about the asymptotics of (3.231) with or without additive continuous perturbations is rather long. I refer to [6] for an extensive review as well as references.

An important tool in the study of evolving fronts is the signed distance function to the front which is defined as

$$
\rho(x, t) = \begin{cases} \rho(x, \{x \in \mathbb{R}^d : w(x, t) \leq 0\}), \\ -\rho(x, \{x \in \mathbb{R}^d : w(x, t) \geq 0\}), \end{cases} \tag{3.240}
$$

where $\rho(x, A)$ is the usual distance between a point x and a set A.

When there is no interior, that is,

$$\partial\{x \in \mathbb{R}^d : w(x,t) < 0\} = \partial\{x \in \mathbb{R}^d : w(x,t) > 0\},$$

then

$$\rho(x,t) = \begin{cases} \rho(x, \Gamma_t) \text{ if } w(x,t) > 0, \\ -\rho(x, \Gamma_t) \text{ if } w(x,t) < 0. \end{cases}$$

The next claim is a direct consequence of the stability properties of the pathwise solutions and the fact that a nondecreasing function of the solution is also a solution. When ζ is a smooth path, the claim below is established in [7]. The result for the general path follows by the stability of the pathwise viscosity solutions with respect to the local uniform convergence of the paths.

Theorem 14.2 *Let $w \in BUC(\mathbb{R}^d \times [0, \infty))$ be the solution of (3.295) and ρ the signed distance function defined by (3.240). Then $\underline{\rho} = \min(\rho, 0)$ and $\overline{\rho} = \max(\rho, 0)$ satisfy respectively*

$$d\underline{\rho} \leq \left[\left(I - \frac{D\underline{\rho} \otimes D\underline{\rho}}{|D\underline{\rho}|^2} \right) : D^2\underline{\rho} \right] dt + |D\underline{\rho}| \circ d\zeta \leq \text{ in } Q_\infty, \tag{3.241}$$

and

$$d\overline{\rho} \geq \left[\left(I - \frac{D\overline{\rho} \otimes D\overline{\rho}}{|D\overline{\rho}|^2} \right) : D^2\overline{\rho} \right] dt + |D\overline{\rho}| \circ d\zeta \geq 0 \text{ in } Q_\infty. \tag{3.242}$$

In addition,

$$-(D^2\underline{\rho}D\underline{\rho}, \underline{\rho}) \leq 0 \text{ and } d\underline{\rho} \leq \Delta\underline{\rho} - d\zeta \text{ in } \{\rho < 0\}, \tag{3.243}$$

and

$$-(D^2\overline{\rho}D\overline{\rho}, \overline{\rho}) \geq 0 \text{ and } d\overline{\rho} \geq \Delta\overline{\rho} - d\zeta \text{ in } \{\rho > 0\}. \tag{3.244}$$

Following the arguments of [7], it is possible to construct global in time subsolutions and supersolutions of (3.231) which do not rely on the regularity of the evolving fronts. In view of the stabilities of the solutions, it is then possible to conclude.

An important ingredient of the argument is the existence and properties of traveling wave solutions of (3.231) and small additive perturbations of it, which we describe next.

It is well known (see, for example, [7] for a long list of references) that, if f satisfies (3.232), then for every sufficiently small b, there exists a unique strictly increasing traveling wave solution $q = q(x, b)$ and a unique speed $c = c(b)$ of

$$cq_\xi + q_{\xi\xi} = f(q) - b \text{ in } \mathbb{R} \quad q(\pm\infty, a) = h_\pm(b) \quad q(0, a) = h_0(b), \qquad (3.245)$$

where $h_-(b) < h_0(b) < h_+(b))$ are the three solutions of the algebraic equation $f(u) = b$. Moreover, as $b \to 0$,

$$h_\pm(b) \to \pm 1 \text{ and } h_0(b) \to 0. \qquad (3.246)$$

The results needed here are summarized in the next lemma. For a sketch of its proof I refer to [7] and the references therein. In what follows, q_ξ and $q_{\xi\xi}$ denote first and second derivatives of q in ξ and q_b the derivative with respect to b.

Lemma 14.1 *Assume (3.232). There exist $b_0 > 0, C > 0, \lambda > 0$ such that, for all $|b| < b_0$, there exist a unique $c(b) \in \mathbb{R}$, a unique strictly increasing $q(\cdot, b) : \mathbb{R} \to \mathbb{R}$ satisfying (7.3), (3.246) and $\alpha_0 \in \mathbb{R}$ such that*

$$\begin{cases} 0 < h_+(b) - q(\xi; b) \le Ce^{-\lambda|\xi|} \text{ if } \xi \ge 0 \\ \text{and } 0 < q(\xi; b) - h_-(b) \le Ce^{-\lambda|\xi|} \text{ if } \xi \le 0, \end{cases} \qquad (3.247)$$

$$0 < q_\xi(\xi; b) \le Ce^{-\lambda|\xi|}, \ |q_{\xi\xi}(\xi; b)| \le Ce^{-\lambda|\xi|} \text{ and } |q_b| \le C, \qquad (3.248)$$

$$\begin{cases} c(b) = -\dfrac{h_+(b) - h_-(b)}{\displaystyle\int_{-\infty}^{\infty} q_\xi(\xi; b)^2 d\xi}, \\ -\alpha_0 = -\dfrac{dc}{db}(0) = \dfrac{2}{\displaystyle\int_{-1}^{1} q_\xi^2(\xi, 0)d\xi} \\ \left|\dfrac{c(b)}{b} + \alpha_0\right| \le C|b|. \end{cases} \qquad (3.249)$$

In the proof of Theorem 14.1 we work with $b = \varepsilon\dot{B}^\varepsilon(t) - \varepsilon a$ for $a \in (-1, 1)$; note that, in view of (3.234), for ε sufficiently small, $|b| < b_0$. To ease the notation, I write

$$q^\varepsilon(\xi, t, a) = q(\xi, \varepsilon(\dot{B}^\varepsilon(t) - a)) \text{ and } c^\varepsilon(a) = c(\varepsilon(\dot{B}^\varepsilon(t) - a)),$$

and I summarize in the next lemma, without a proof, the key properties of q^ε and c^ε that we need later.

Lemma 14.2 *Assume the hypotheses of Lemma 14.1 and* (3.234). *Then, there exists* $C > 0$ *such that*

$$\begin{cases} \lim_{\varepsilon \to} \varepsilon |q_t^{\varepsilon}(\xi, t, a)| = 0 \text{ uniformly on } \xi \\ \text{and a and locally uniformly in } t \in [0, \infty), \end{cases} \tag{3.250}$$

$$\frac{1}{\varepsilon} q_{\xi}^{\varepsilon}(\xi, t, a) + \frac{1}{\varepsilon^2} |q_{\xi\xi}^{\varepsilon}\xi, t, a)| \leq C e^{-C\eta/\varepsilon} \text{ for all } |\xi| \geq \eta \text{ and all } \eta > 0, \tag{3.251}$$

$$q_{\xi}^{\varepsilon} \geq 0 \text{ and } q_a^{\varepsilon} \geq 0 \text{ for all } t \geq 0 \text{ and } \varepsilon, |a| \text{ sufficiently small}, \tag{3.252}$$

and

$$|\frac{c^{\varepsilon}}{\varepsilon} + \alpha_0 \varepsilon(\dot{B}^{\varepsilon}(t) - a)| = o(1) \text{ uniformly for bounded } t \text{ and } a. \tag{3.253}$$

Theorem 14.1 is proved assuming that u_0^{ε} in (3.231) is well prepared, that is, has the form

$$u_0^{\varepsilon}(x) = q^{\varepsilon}(\frac{\rho(x)}{\varepsilon}, 0), \tag{3.254}$$

where ρ is the signed distance function to Γ_0 and $q(\cdot, 0)$ is the standing wave solution of (7.3).

Going from (3.254) to a general u_0^{ε} as in the statement of the theorem is standard in the theory of front propagation. It amounts to showing that, in a conveniently small time interval, u^{ε} can be "sandwiched" between functions like the ones in (3.254). Since this is only technical, I omit the details and I refer to [6] for the details.

The proof of the result is a refinement of the analogous results of [27] and [7]. It is based on using two approximate flows, which evolve with normal velocity $V = -\text{tr}[Dn] + \alpha_0(\dot{B}^{\varepsilon}(t) - \varepsilon a)$, to construct a subsolution and supersolution (3.231). Since the arguments are similar, here we show the details only for the supersolution construction.

For fixed $\delta, a > 0$ to be chosen below and any $T > 0$, consider the solution $w^{a,\delta,\varepsilon}$ of

$$\begin{cases} w_t^{a,\delta,\varepsilon} - \left(I - \widehat{Dw^{a,\delta,\varepsilon}} \otimes \widehat{Dw^{a,\delta,\varepsilon}}\right) : D^2 w^{a,\delta,\varepsilon} \\ \qquad\qquad + \alpha_0(\dot{B}^{\varepsilon} - a)|Dw^{a,\delta,\varepsilon}| = 0 \text{ in } Q_T, \\ w^{a,\delta,\varepsilon}(\cdot, 0) = \rho + \delta. \end{cases} \tag{3.255}$$

Let $\rho^{a,\delta,\varepsilon}$ be the signed distance from $\{w^{a,\delta,\varepsilon} = 0\}$. It follows from Theorem 14.2 (see also Theorem 3.1 in [7]) that

$$\rho^{a,\delta,\varepsilon} - \Delta\rho^{a,\delta,\varepsilon} - \alpha_0(\dot{B}^\varepsilon - a) \geq 0 \text{ in } \{\rho^{a,\delta,\varepsilon}\ 0\}. \qquad (3.256)$$

Following the proof of Lemma 3.1 of [27], define

$$W^{a,\delta,\varepsilon} = \eta_\delta(\rho^{a,\delta,\varepsilon}), \qquad (3.257)$$

where $\eta_\delta : \mathbb{R} \to \mathbb{R}$ is smooth and such that, for some $C > 0$ independent of δ,

$$\begin{cases} \eta_\delta \equiv -\delta \text{ in } (-\infty, \delta/4], \quad \eta_\delta \leq -\delta/2 \text{ in } (-\infty, \delta/2], \\ \eta_\delta(z) = z - \delta \text{ in } [\delta/2, \infty), \ 0 \leq \eta'_\delta \leq C \text{ and } |\eta''_\delta| \leq C\delta^{-1} \text{ on } \mathbb{R}. \end{cases} \qquad (3.258)$$

Let $T^\star$ be the extinction time of $\{w^{a,\delta,\varepsilon} = 0\}$. A straightforward modification of Lemma 3.1 of [27] leads to the following claim.

Lemma 14.3 *There exists a constant $C > 0$, which is independent of ε, δ and a, such that*

$$W_t^{a,\delta,\varepsilon} - \Delta W^{a,\delta,\varepsilon} - \alpha_0(\dot{B}^\varepsilon - a)|DW^{a,\delta,\varepsilon}| \geq -\frac{C}{\delta} \text{ in } \mathbb{R}^d \times [0, T^\star], \qquad (3.259)$$

$$W_t^{a,\delta,\varepsilon} - \Delta W^{a,\delta,\varepsilon} - \alpha_0(\dot{B}^\varepsilon - a) \geq 0 \text{ in } \{\rho^{a,\delta,\varepsilon} > \delta/2\}, \qquad (3.260)$$

and

$$|DW^{a,\delta,\varepsilon}| = 1 \text{ in } \{\rho^{a,\delta,\varepsilon} > \delta/2\}. \qquad (3.261)$$

Finally, set

$$U^{a,\delta,\varepsilon}(x, t) = q^\varepsilon\left(\frac{W^{a,\delta,\varepsilon}(x, t)}{\varepsilon}, t, a\right) \text{ on } \mathbb{R}^d \times [0, \infty). \qquad (3.262)$$

Proposition 14.1 *Assume (3.232), (3.234) and (3.235). Then, for every $a \in (0, 1)$, U^ε is a supersolution of (3.231) if $\varepsilon \leq \varepsilon_0 = \varepsilon_0(\delta, a)$ and $\delta \leq \delta_0 = \delta_0(a)$.*

Proof Since the arguments are similar to the ones used to prove the analogous result (Proposition 10.2) in [7], here I only sketch the argument. Note that since everything takes place at the $\varepsilon > 0$ level, there is no reason to be concerned about anything "rough". Below, for simplicity, I argue as if $w^{\varepsilon,\delta,a}$ had actual derivatives, and is left up to the reader to argue in the viscosity sense. Note that, throughout the proof, o(1) stands for a function such that $\lim_{\varepsilon\to 0} o(1) = 0$. Finally, throughout the proof q^ε and its derivatives are evaluated at $(W^{a,\delta,\varepsilon}/\varepsilon, t, a)$.

Using the equation satisfied by q^ε gives

$$
U_t^{a,\delta,\varepsilon} - \Delta U^{a,\delta,\varepsilon} + \frac{1}{\varepsilon^2}[f(U^{a,\delta,\varepsilon}) - \varepsilon \dot{B}(t))
$$
$$
= J^\varepsilon - \frac{1}{\varepsilon^2} q_{\xi\xi}^\varepsilon (|DW^{a,\delta,\varepsilon}|^2 - 1) \tag{3.263}
$$
$$
+ \frac{1}{\varepsilon} q_\xi^\varepsilon (DW_t^{a,\delta,\varepsilon} - \Delta W^{a,\delta,\varepsilon} + \frac{c^\varepsilon}{\varepsilon}) + \frac{a}{\varepsilon},
$$

and

$$
J^\varepsilon(x,t) = q_b \left(\frac{W^{a,\delta,\varepsilon}(x,t)}{\varepsilon}, \varepsilon \dot{B}^\varepsilon(t) - \varepsilon a \right) \varepsilon \ddot{B}^\varepsilon(t). \tag{3.264}
$$

In view of its definition, it is immediate that $|DW^{a,\delta,\varepsilon}| \le C$ with C as in (14.3), while it follows from Lemma 14.1 that, as $\varepsilon \to 0$ and uniformly in (x,t,δ,a)

$$
J^\varepsilon = \frac{o(1)}{\varepsilon}. \tag{3.265}
$$

Three different cases, which depend on the relationship between $\rho^{a,\delta,\varepsilon}$ and δ, need to be considered.

If $\delta/2 < \rho^{a,\delta,\varepsilon} < 2\delta$, then (3.260), (3.261), (3.253) and the form of η_δ allow to rewrite (3.263) as

$$
U_t^{a,\delta,\varepsilon} - \Delta U^{a,\delta,\varepsilon} + \frac{1}{\varepsilon^2}[f(U^{a,\delta,\varepsilon}) - \varepsilon \dot{B}^{(}t))
$$
$$
\ge -\frac{1}{\varepsilon}\left[q_\xi^\varepsilon \left(\frac{c^\varepsilon}{\varepsilon} + \alpha_0(\varepsilon \dot{B}^\varepsilon - \varepsilon a) \right) + a + o(1) \right] \tag{3.266}
$$
$$
\ge -\frac{1}{\varepsilon}\left[q_\xi^\varepsilon o(1) + a + o(1) \right].
$$

It easily now follows that the right side of (3.266) is positive, if ε and δ are small.

If $d^{a,\delta,\varepsilon} \le \delta/2$, the choice of η_δ implies that $W^{a,\delta,\varepsilon} \le -\delta/2$. Hence, (3.251) yields that, for some $C > 0$,

$$
\frac{1}{\varepsilon} q_\xi^\varepsilon + \frac{1}{\varepsilon^2}|q_{\xi\xi}^\varepsilon| \le C e^{-C\delta/\varepsilon}.
$$

Then $|DW^{a,\delta,\varepsilon}| \le C$ and (3.259) and (3.260) in (3.266) give

$$
U_t^{a,\delta,\varepsilon} - \Delta U^{a,\delta,\varepsilon} + \frac{1}{\varepsilon^2}[f(U^{a,\delta,\varepsilon}) - \varepsilon \dot{B}(t)] \le -C(\frac{1}{\delta} + 1)e^{-C\delta/\varepsilon} + o(1) + \frac{a}{\varepsilon};
$$

note that, for ε small enough the right hand side of the inequality above is positive.

Finally, if $\rho^{a,\delta,\varepsilon} > \delta$, it is possible to conclude as in the previous case using (3.260) and (3.251).

The proof of the main result is sketched next.

Proof (The Proof of Theorem 14.1) Fix $(x_0, t_0) \in \mathbb{R}^d \times [0, T^\star)$ such that $w(x_0, t_0) = -\beta < 0$. The stability of the pathwise solutions yields that, in the limit $\varepsilon \to 0$, $\delta \to 0$ and $a \to 0$ and uniformly in (x, t), $w^{a,\delta,\varepsilon} \to w$. Thus, for sufficiently small ε, δ and a,

$$w^{a,\delta,\varepsilon}(x_0, t_0) < -\frac{\beta}{2} < 0. \tag{3.267}$$

Then $U^{a,\delta,\varepsilon}$, which is defined in (3.261), is a supersolution of (3.231) for sufficiently small ε and also satisfies, in view of (3.252),

$$U^{a,\delta,\varepsilon}(x, 0) \geq q^\varepsilon\left(\frac{\rho(x)}{\varepsilon}, 0\right) \text{ on } \mathbb{R}^d,$$

since

$$w^{a,\delta,\varepsilon}(x, 0) = \eta_\delta(\rho(x) + \delta) \geq \rho(x).$$

The comparison of viscosity solutions of (3.231) then gives

$$u^\varepsilon \leq U^{a,\delta,\varepsilon} \text{ in } \mathbb{R}^d \times [0, T^\star).$$

Recall that, in view of (3.267), $\rho^{a,\delta,\varepsilon}(x_0, t_0) < 0$, and, hence,

$$\limsup_{\varepsilon \to 0} u^\varepsilon(x_0, t_0) \leq \limsup_{\varepsilon \to 0} U^{a,\delta,\varepsilon}(x_0, t_0) = -1.$$

For the reverse inequality, observe that $\hat{U}(x, t) = -1 - \gamma$ is a subsolution of (3.231) if ε and $\gamma > 0$ are chosen sufficiently small as can be seen easily from

$$\hat{U}_t - \Delta\hat{U} + \frac{1}{\varepsilon^2}(f(\hat{U}) + \varepsilon\dot{B}^\varepsilon) \leq C + \frac{1}{\varepsilon^2}[-\gamma f'(-1) + o(1)].$$

The maximum principle then gives, for all (x, t) and sufficiently small $\gamma > 0$,

$$\liminf_{\varepsilon \to 0} u^\varepsilon(x_0, t_0) \geq -1 - \gamma.$$

The conclusion now follows after letting $\gamma \to 0$.

Finally note that a simple modification of the argument above yields the local uniform convergence of u^ε to -1 in compact subsets of $\{w < 0\}$.

3.15 Pathwise Entropy/Kinetic Solutions for Scalar Conservation Laws with Multiplicative Rough Time Signals

3.15.1 Introduction

Ideas similar to the ones described up to the previous sections were used by Lions et al. [65, 66], Gess and Souganidis [38–40] and Gess et al. [41] to study pathwise entropy/kinetic solutions for scalar conservation laws with multiplicative rough time signals as well as their long time behavior, the existence of invariant measures and the convergence of general relaxation schemes with error estimates.

To keep the ideas simple the presentation here is about the simplest possible case, that is the spatially homogeneous initial value problem

$$du + \sum_{i=1}^{d} A^i(u)_{x_i} \cdot dB_i = 0 \text{ in } Q_T \quad u_0(\cdot, 0) = u_0, \tag{3.268}$$

with

$$\mathbf{A} = (A_1, \ldots, A_d) \in C^2(\mathbb{R}; \mathbb{R}^d) \tag{3.269}$$

and merely continuous paths

$$\mathbf{B} = (B_1, \ldots, B_d) \in C([0, \infty); \mathbb{R}^d). \tag{3.270}$$

If, instead of (3.270), $\mathbf{B} \in C^1([0, \infty); \mathbb{R}^d)$, (3.268) is a "classical" problem with a well known theory; see, for example, the books by Dafermos [19] and Serre [101]. The solution can develop singularities in the form of shocks (discontinuities). Hence it is necessary to consider entropy solutions which, although not regular, satisfy the L^1-contraction property established by Kruzkov [47].

Solutions of deterministic non-degenerate conservation laws have remarkable regularizing effects in Sobolev spaces of low order. It is an interesting question to see if they are still true in the present case. This is certainly possible with different exponents as shown in [66] and [39].

Contrary to the Hamilton-Jacobi equation, the approach put forward for (3.268) does not work for conservation laws with semilinear rough path dependence like

$$du + \sum_{i=1}^{d} (A^i(u))_{x_i} dt = \Phi(u) \cdot d\tilde{\mathbf{B}} \text{ in } Q_T \quad u(\cdot, 0) = u_0, \tag{3.271}$$

for $\mathbf{\Phi} = (\Phi_1, \ldots, \Phi_m) \in C^2(\mathbb{R}; \mathbb{R}^m)$ and an m-dimensional path $\tilde{\mathbf{B}} = (\tilde{B}_1, \ldots, \tilde{B}_m)$.

Semilinear stochastic conservation laws in Itô's form like

$$du + \sum_{i=1}^{d} (A^i(u))_{x_i} dt = \Phi(u) d\tilde{\mathbf{B}} \quad \text{in } Q_T \tag{3.272}$$

have been studied by Debussche and Vovelle [20–22], Feng and Nualart [28], Chen et al. [11], and Hofmanova [43, 44].

It turns out that pathwise solutions are natural in problems with nonlinear dependence. Indeed, let u, v be solutions of the simple one dimensional problems

$$du + A(u)_x \cdot dB = 0 \quad \text{and} \quad dv + A(v)_x \cdot dB = 0.$$

Then

$$d(u - v) + (A(u) - A(v))_x \cdot dB = 0.$$

Multiplying by the $\text{sign}(u - v)$ and integrating over $\mathbb{R}$ formally leads to

$$d \int_{\mathbb{R}} |u - v| dx + \int_{\mathbb{R}} (\text{sign}(u - v)(A(u) - A(v)))_x \cdot dB = 0$$

and, hence,

$$d \int_{\mathbb{R}} |u - v| dx = 0.$$

On the other hand, if $du = \Phi(u) \cdot dB$ and $dv = \Phi(v) \cdot dB$, then the previous argument cannot be used since the term $\int_R \text{sign}(u - v)(\Phi(u) - \Phi(v)) \cdot dB$ is neither 0 nor has a sign. More about this is presented in the last subsection.

3.15.2   The Kinetic Theory When B Is Smooth

To make the connection with the "non rough" theory, assume that $\mathbf{B} \in C^1((0, \infty); \mathbb{R}^d)$, in which case du stands for the usual derivative and $\cdot$ is the usual multiplication and, hence, should be ignored.

The entropy inequality (see [19, 101]), which guarantees the uniqueness of the weak solutions, is that

$$dS(u) + \sum_{i=1}^{d} (A^{i,S}(u))_{x_i} \cdot dB_i \leq 0 \quad \text{in } Q_T \quad S(u(\cdot, 0)) = S(u_0), \tag{3.273}$$

for all C^2 -convex functions S and fluxes $\mathbf{A}^S$ defined by

$$\left(\mathbf{A}^S(u)\right)' = \mathbf{a}(u)S'(u) \quad \text{with} \quad \mathbf{a} = \mathbf{A}'.$$

It is by now well established that the simplest way to handle conservation laws is through their kinetic formulation developed through a series of papers—see Perthame and Tadmor [92], Lions et al. [62], Perthame [89, 90], and Lions et al. [64]. The basic idea is to write a linear equation on the nonlinear function

$$\chi(x, \xi, t) = \chi(u(x, t), \xi) = \begin{cases} +1 & \text{if} \quad 0 \leq \xi \leq u(x, t), \\ -1 & \text{if} \quad u(x, t) \leq \xi \leq 0, \\ 0 & \text{otherwise.} \end{cases} \quad (3.274)$$

The kinetic formulation states that using the entropy inequalities (3.273) for all convex entropies S is equivalent to χ solving, in the sense of distributions,

$$d\chi + \sum_{i=1}^{d} A^i(\xi)\partial_{x_i}\chi \cdot dB_i = \partial_\xi m dt \text{ in } \mathbb{R}^d \times \mathbb{R} \times (0, \infty) \quad \chi(x, \xi, 0)$$
$$= \chi(u_0(x), \xi), \quad (3.275)$$

where

$$m \text{ is a nonnegative bounded measure in } \mathbb{R}^d \times \mathbb{R} \times (0, \infty). \quad (3.276)$$

At least formally, one direction of this equivalence can be seen easily. Indeed since, for all $(x, t) \in \mathbb{R}^d \times (0, \infty)$,

$$S(u(x, t)) - S(0) = \int S'(\xi)\chi(u(x, t), \xi)d\xi,$$

multiplying (3.275) by $S'(\xi)$ and integrating in ξ leads to (3.273).

The next proposition, which is stated without proof, summarizes the basic estimates of the kinetic theory, which hold for smooth paths and are independent of the regularity of the paths. They are the $L^p(Q_T)$ and $BV(Q_T)$ bounds (for all $T > 0$) for the solutions, as well as the bounds on the kinetic defect measures m, which imply that the latter are weakly continuous in ξ as measures on Q_T.

Proposition 15.1 *Assume (3.269). The entropy solutions to (3.268) satisfy, for all* $t > 0$,

$$\|u(\cdot, t)\|_{L^p(\mathbb{R}^d)} \leq \|u_0\|_{L^p(\mathbb{R}^d)} \quad \text{for all} \quad p \in [1, \infty], \tag{3.277}$$

$$\|Du(\cdot, t)\|_{L^1(\mathbb{R}^d)} \leq \|Du_0\|_{L^1(\mathbb{R}^d)}, \tag{3.278}$$

$$\{\xi \in \mathbb{R} : |\chi(x, \xi, t) > 0\} \subset [-|u(x, t)|, |u(x, t)|] \quad \text{for all} \ (x, t) \in \mathbb{R} \times (0, \infty), \tag{3.279}$$

$$\int_0^\infty \int_{\mathbb{R}^d} \int_{\mathbb{R}} m(x, \xi, t) dx d\xi dt \leq \frac{1}{2} \|u_0\|_{L^2(\mathbb{R}^d)}^2, \tag{3.280}$$

$$\int_0^\infty \int_{\mathbb{R}^d} m(x, \xi, t) dx \, dt \leq \|u_0\|_{L^1(\mathbb{R}^d)} \quad \text{for all} \quad \xi \in \mathbb{R}, \tag{3.281}$$

and, for all smooth test functions ψ,

$$\frac{d}{d\xi} \int_0^\infty \int_{\mathbb{R}^d} \psi(x, t) m(x, \xi, t) dx \, dt \tag{3.282}$$

$$\leq \left[\|D_{x,t}\psi\|_{L^\infty(\mathbb{R}^{d+1})} + \|\psi(\cdot, 0)\|_{L^\infty(\mathbb{R}^d)} \right] \|u^0\|_{L^1(\mathbb{R}^d)}.$$

The next observation is the backbone of the theory of pathwise entropy/kinetic solutions. The reader will recognize ideas described already in the earlier parts of these notes.

Since the flux in (3.268) is independent of x, it is possible to use the characteristics associated with (3.275) to derive an identity which is equivalent to solving (3.275) in the sense of distributions. Indeed, choose

$$\rho_0 \in C^\infty(\mathbb{R}^d) \quad \text{such that} \quad \rho_0 \geq 0 \quad \text{and} \quad \int_{\mathbb{R}^d} \rho_0(x) dx = 1, \tag{3.283}$$

and observe that

$$\rho(y, x, \xi, t) = \rho_0\big(y - x + \mathbf{a}(\xi)\mathbf{B}(t)\big), \tag{3.284}$$

where

$$\mathbf{a}(\xi)\mathbf{B}(t) := (a_1(\xi) B_1(t), a_2(\xi) B_2(t), \dots, a_N(\xi) B_N(t)), \tag{3.285}$$

solves the linear transport equation (recall that in this subsection it is assumed that
B is smooth)

$$d\rho + \sum_{i=1}^{d} A^i(\xi)\partial_{x_i}\rho \cdot dB_i = 0 \text{ in } \mathbb{R}^d \times \mathbb{R} \times (0, \infty),$$

and, hence,

$$d(\rho(y, x, \xi, t)\chi(x, \xi, t)) + \sum_{i=1}^{d} A^i(\xi)\partial_{x_i}(\rho(y, x, \xi, t)\chi(x, \xi, t)) \cdot dB_i$$

$$= \rho(y, x, \xi, t)\partial_\xi m(x, \xi, t)dt.$$ (3.286)

Integrating (3.286) with respect to x (recall that ρ_0 has compact support) yields that,
in the sense of distributions in $\mathbb{R} \times (0, \infty)$,

$$\frac{d}{dt} \int_{\mathbb{R}^d} \chi(x, \xi, t)\rho(y, x, \xi, t)dx = \int_{\mathbb{R}^d} \rho(y, x, \xi, t)\partial_\xi m(x, \xi, t)dx.$$ (3.287)

Observe that, although the regularity of the path was used to derive (3.287),
the actual conclusion does not need it. In particular, (3.287) holds for paths
which are only continuous. Moreover, (3.287) is basically equivalent to the kinetic
formulation, if the measure m satisfies (3.276).

Finally, note that (3.287) makes sense only after integrating with respect to ξ
against a test function. This requires that $\mathbf{a}' \in C^1(\mathbb{R}; \mathbb{R}^d)$ as long as we only use
that m is a measure. Indeed, integrating against a test function Ψ yields

$$\int_{\mathbb{R}^{d+1}} \Psi(\xi)\rho(y, x, \xi, t)\partial_\xi m(x, \xi, t) \, dxd\xi$$

$$= -\int_{\mathbb{R}^{d+1}} \Psi'(\xi)\rho(y, x, \xi, t) \, m(x, \xi, t) \, dxd\xi$$

$$+ \int_{\mathbb{R}^{d+1}} \Psi(\xi)(\sum_{i=1}^{d} \partial_{x_i}\rho(y, x, \xi, t)(a^i)'(\xi)B_i(t)) \, m(x, \xi, t) \, dxd\xi$$

and all the terms make sense as continuous functions tested against a measure.

Some (new) estimates and identities, needed for the proof of the main results of
this section and derived from (3.287), are stated next. Here δ denotes the Dirac mass
at the origin.

Proposition 15.2 *Assume (3.269) and $u_0 \in (L^1 \cap L^\infty \cap BV)(\mathbb{R}^d)$. Then, for all
$t > 0$,*

$$\frac{d}{dt} \int_{\mathbb{R}^{d+1}} |\chi(x, \xi, t)|dx \, d\xi = -2\int_{\mathbb{R}^d} m(x, 0, t)dx,$$ (3.288)

and

$$\int_{\mathbb{R}^{d+1}} \int_{\mathbb{R}^{2d}} \delta(\xi - u(z,t))\, \rho(y,z,\xi,t)\rho(y,x,\xi,t)\, m(t,x,\xi) dxdydzd\xi$$

$$= \tfrac{1}{2}\frac{d}{dt} \int_{\mathbb{R}^{d+1}} \left[\left(\int_{\mathbb{R}^d} \chi(x,\xi,t)\rho(y,x,\xi,t)dx\right)^2 - |\chi(y,\xi,t)|\right]dyd\xi.$$

$$(3.289)$$

Proof The first identity is classical and is obtained from multiplying (3.268) by sign(ξ) and using that the fact that sign(ξ)$\chi(x,\xi,t) = |\chi(x,\xi,t)|$. Notice that taking the value $\xi = 0$ in m is allowed by the Lipschitz regularity in Proposition 15.1.

The proof of (3.289) uses the regularization kernel along the characteristics (3.284). Indeed, (3.287) and the facts that $\chi_\xi(z,\xi,t) = \delta(\xi) - \delta(\xi - u(z,t))$ and, for all $\xi \in \mathbb{R}$,

$$\int_{\mathbb{R}^d} \int_{\mathbb{R}^{2d}} \chi(z,\xi,t)[D_y\rho(y,z,\xi,t)\rho(y,x,\xi,t)$$

$$+ \rho(y,z,\xi,t)D_y\rho(y,x,\xi,t)]\, m(t,x,\xi)dzdxdy = 0,$$

which follows from the observation that the integrand is an exact derivative with respect to y.

$$\frac{1}{2}\frac{d}{dt} \int_{\mathbb{R}^{d+1}} \left(\int_{\mathbb{R}^d} \chi(x,\xi,t)\rho(y,x,\xi,t)dx\right)^2 dyd\xi$$

$$= \int_{\mathbb{R}^{d+1}} \left[\int_{\mathbb{R}^d} \chi(z,\xi,t)\rho(y,z,\xi,t)dz \int_{\mathbb{R}^d} \rho(y,x,\xi,t)\partial_\xi m(x,\xi,t)\, dx\right] dyd\xi$$

$$= -\int_{\mathbb{R}^{d+1}} \int_{\mathbb{R}^{2d}} [\delta(\xi) - \delta(\xi - u(z,t))]$$

$$\times \rho(y,z,\xi,t)\rho(y,x,\xi,t)\, m(x,\xi,t)dzdxdyd\xi$$

$$= -\int_{\mathbb{R}^d} m(x,0,t)dx$$

$$+ \int_{\mathbb{R}^{d+1}} \int_{\mathbb{R}^{2d}} \delta(\xi - u(z,t))\rho(y,z,\xi,t)\rho(y,x,\xi,t)\, m(x,\xi,t)dzdxdyd\xi.$$

Using next (3.288) gives (3.289).

3.15.3 Dissipative Solutions

The notion of dissipative solutions, which was studied by Perthame and Souganidis [91], is equivalent to that of entropy solutions. The interest in them is twofold. Firstly, the definition resembles and enjoys the same flexibility as the one for viscosity solutions in, of course, the appropriate function space. Secondly, in defining them, it is not necessary to talk at all about entropies, shocks, etc.

It is said that $u \in L^\infty((0, T), (L^1 \cap L^\infty)(\mathbb{R}^d))$ is a dissipative solution of (3.268), if, for all $\Psi \in C([0, \infty); C_c^\infty(\mathbb{R}^d))$ and all $\psi \in C_c^\infty(\mathbb{R}; [0, \infty))$, where the subscript c means compactly supported, in the sense of distributions,

$$\frac{d}{dt} \int_{\mathbb{R}^d} \int_{\mathbb{R}} \psi(k)(u - k - \Psi)_+ dx dk \leq \int_{\mathbb{R}^d} \int_{\mathbb{R}} \psi(k)\text{sign}_+(u - k - \Psi)$$

$$\times (-\Psi_t - \sum_{i=1}^{d} \partial_{x_i}(A^i(\Psi)) \cdot dB_i) dx dk.$$

To provide an equivalent definition which will allow to go around the difficulties with inequalities mentioned earlier, it is necessary to take a small detour to recall the classical fact that, under our regularity assumptions on the flux and paths, for any $\phi \in C_c^\infty(\mathbb{R}^d)$ and any $t_0 > 0$, there exists $h > 0$, which depends on ϕ, such that the problem

$$d\bar{\Psi} + \sum_{i=1}^{d} \partial_{x_i}(A^i(\bar{\Psi})) \cdot dB_i = 0 \text{ in } \mathbb{R}^d \times (t_0 - h, t_0 + h) \qquad \bar{\Psi}(\cdot t_0) = \phi, \qquad (3.290)$$

has a smooth solution given by the method of characteristics.

It is left up to the reader to check that the definition of the dissipative solution is equivalent to saying that, for $\phi \in C_c^\infty(\mathbb{R}^d)$, $\psi \in C_c^\infty(\mathbb{R}; [0, \infty))$ and any $t_0 > 0$, there exists $h > 0$, which depends on ϕ, such that, if $\bar{\Psi}$ and $h > 0$ are as in (3.290), then in the sense of distributions

$$\frac{d}{dt} \int_{\mathbb{R}^d} \int_{\mathbb{R}} \psi(k)(u - k - \bar{\Psi})_+ dx dk \leq 0 \text{ in } (t_0 - h, t_0 + h).$$

3.15.4 Pathwise Kinetic/Entropy Solutions

The following definition is motivated by the theory of pathwise viscosity solutions.

Definition 15.1 Assume (3.269) and (3.270). Then $u \in (L^1 \cap L^\infty)(Q_T)$ is a pathwise kinetic/entropy solution to (3.268), if there exists a nonnegative bounded measure m on $\mathbb{R}^d \times \mathbb{R} \times (0, \infty)$ such that, for all test functions ρ given by (3.284) with ρ_0 satisfying (3.283), in the sense of distributions in $\mathbb{R} \times (0, \infty)$,

$$\frac{d}{dt} \int_{\mathbb{R}^d} \chi(x, \xi, t) \rho(y, x, \xi, t) dx = \int_{\mathbb{R}^d} \rho(y, x, \xi, t) \partial_\xi m(x, \xi, t) dx. \qquad (3.291)$$

The main result is:

Theorem 15.1 *Assume (3.269), (3.270) and $u_0 \in (L^1 \cap L^\infty)(\mathbb{R}^d)$. For all $T > 0$ there exists a unique pathwise entropy/kinetic solution $u \in C([0, \infty); L^1(\mathbb{R}^d)) \cap L^\infty(Q_T)$ to (3.268) and (3.277), (3.75), (3.280), (3.281) and (3.282) hold. In addition, any pathwise entropy solutions $u_1, u_2 \in C([0, \infty); L^1(\mathbb{R}^d))$ to (3.268) satisfy, for all $t > 0$, the contraction property*

$$\|u_2(\cdot, t) - u_1(\cdot, t)\|_{L^1(\mathbb{R}^d)} \le \|u_2(\cdot, 0) - u_1(\cdot, 0)\|_{L^1(\mathbb{R}^d)}. \qquad (3.292)$$

Moreover, there exists a uniform constant $C > 0$ such that, if, for $i = 1, 2$, u_i is the pathwise entropy/kinetic solution to (3.268) with path $\mathbf{B_i}$ and $u_{i,0} \in BV(\mathbb{R}^d)$, then u_1 and u_2 satisfy, for all $t > 0$, the contraction property

$$\|u_2(\cdot, t) - u_1(\cdot, t)\|_{L^1(\mathbb{R}^d)} \le \|u_{2,0} - u_{1,0}\|_{L^1(\mathbb{R}^d)}$$

$$+ C[\|\mathbf{a}\|(|u_{1,0}|_{BV(\mathbb{R}^d)} + |u_{2,0}|_{BV(\mathbb{R}^d)})|(\mathbf{B_1} - \mathbf{B_2})(t)| \qquad (3.293)$$

$$+ (\sup_{s \in (0,t)} |(\mathbf{B_1} - \mathbf{B_2})(s)| \|\mathbf{a}'\|[\|u_{1,0}\|^2_{L^2(\mathbb{R}^N)} + \|u_{2,0}\|^2_{L^2(\mathbb{R}^N)}])^{1/2}].$$

Looking carefully into the proof of (3.293) for smooth paths, it is possible to establish, after some approximations, an estimate similar to (3.293), for non BV-data, with a rate that depends on the modulus of continuity in L^1 of the initial data. It is also possible to obtain an error estimate for different fluxes. The details for both are left to the interested reader.

3.15.5 Estimates for Regular Paths

Following ideas from the earlier parts of the notes, the solution operator of (3.268) may be thought of as the unique extension of the solution operators with regular paths. It is therefore necessary to study first (3.268) with smooth paths and to obtain estimates that allow to prove that the solutions corresponding to any regularization of the same path converge to the same limit, which is a pathwise entropy/kinetic solution. The intrinsic uniqueness for the latter is proved later.

The key step is a new estimate, which depends only on the sup-norm of $\mathbf{B}$ and yields compactness with respect to time.

Theorem 15.2 *Assume* (3.269) *and, for* $i = 1, 2$, $u_{i,0} \in (L^1 \cap L^\infty \cap BV)(\mathbb{R}^d)$. *Consider two smooth paths* $\mathbf{B_1}$ *and* $\mathbf{B_2}$ *and the corresponding solutions* u_1 *and* u_2 *to* (3.268). *There exists a uniform constant* $C > 0$ *such that, for all* $t > 0$, (3.293) *holds.*

The proof of Theorem 15.2, which is long and technical, can be found in [65]. It combines the uniqueness proof for scalar conservation laws based on the kinetic formulation of [89, 90] and the regularization method along the characteristics introduced for Hamilton-Jacobi equations in [71, 74–77].

3.15.6 The Proof of Theorem 15.1

The existence of a pathwise kinetic/entropy solution follows easily. Indeed, the estimate of Theorem 15.2 implies that, for every $u_0 \in (L^1 \cap L^\infty \cap BV)(\mathbb{R}^d)$ and for every $T > 0$, the mapping $\mathbf{B} \in C([0, T]; \mathbb{R}^d) \mapsto u \in C([0, T]; L^1(\mathbb{R}^d))$ is well defined and uniformly continuous with the respect to the norm of $C([0, T]; \mathbb{R}^d)$. Therefore, by density, it has a unique extension to $C([0, T])$. Passing to the limit gives the contraction properties (3.292) and (3.293) as well as (15.1). Once (3.292) is available for initial data in $BV(\mathbb{R}^d)$, the extension to general data is immediate by density.

The next step is to show that pathwise kinetic/entropy satisfying (15.1) are intrinsically unique in an intrinsic sense. The contraction property only proves uniqueness of the solution built by the above regularization process. It is, however, possible to prove that (3.291) implies uniqueness. Indeed, for BV-data, the estimates in the proof of Theorem 15.2 only use the equality of Definition 15.1. From there the only nonlinear manipulation needed is to check that

$$
\frac{1}{2} \frac{d}{dt} \int_{\mathbb{R}^{d+1}} \left(\int_{\mathbb{R}^d} \chi(x, \xi, t) \rho(y, x, \xi, t) dx \right)^2
$$
$$
= \int_{\mathbb{R}^{d+1}} \left(\int_{\mathbb{R}^d} \chi(x, \xi, t) \rho(y, x, \xi, t) dx \right)
$$
$$
\times \frac{d}{dt} \int_{\mathbb{R}^{d+1}} \left(\int_{\mathbb{R}^d} \chi(x, \xi, t) \rho(y, x, \xi, t) dx \right).
$$

This is justified after time regularization by convolution because it has been assumed that solutions belong to $C([0, T); L^1(\mathbb{R}^d))$ for all $T > 0$. This fact also allows to justify that the right hand side

$$
\int_{\mathbb{R}^{d+1}} \left(\int_{\mathbb{R}^d} \chi(x, \xi, t) \rho(y, x, \xi, t) dx \right) \int_{\mathbb{R}^{d+1}} \int_{\mathbb{R}^d} \chi(z, \xi, t) \rho(y, z, \xi, t) dz
$$
$$
\int_{\mathbb{R}^d} \rho(y, x, \xi, t) \partial_{x_i} \chi(x, \xi, t) \, dx
$$

can be analyzed by a usual integration by parts, because it is possible to incorporate a convolution in ξ before forming the square. All these technicalities are standard and I omit them. The uniqueness for general data requires one more layer of approximation.

3.15.7   The Semilinear Problem

Based on the results of Sect. 3.5, it is natural to expect that the approach developed earlier will also be applicable to the semilinear problem (3.271) to yield a pathwise theory of stochastic entropy solutions. It turns out, however, that this not the case.

To keep things simple, here it is assumed that $d = 1$, $\mathbf{B} = t$ and $\tilde{\mathbf{B}} \in C([0, \infty); \mathbb{R})$ is a single continuous path. Consider, for $\Phi \in C^2(\mathbb{R}; \mathbb{R})$, the problem

$$du + \operatorname{div} A(u)dt = \Phi(u) \cdot dB \text{ in } Q_T \quad u = u_0. \tag{3.294}$$

Following the earlier considerations as well as the analogous problem for Hamilton-Jacobi equations, it is assumed that, for each $v \in \mathbb{R}$ and $T > 0$, the initial value problem

$$d\Psi = \Phi(\Psi) \cdot d\tilde{B} \text{ in } (0, \infty) \quad \Psi(0) = v, \tag{3.295}$$

has a unique solution $\Psi(v; \cdot) \in C([0, T]; \mathbb{R})$ such that, for all $t \in [0, T]$,

$$\Psi(\cdot, t) \in C^1(\mathbb{R}; \mathbb{R}). \tag{3.296}$$

According to [77], to study (3.294) it is natural to consider a change of unknown given by the Doss-Sussman-type transformation

$$u(x, t) = \Psi(v(x, t), t). \tag{3.297}$$

Assuming for a moment that $\tilde{B}$ and, hence, Ψ are smooth with respect to t and (3.294) and (3.295) have classical solutions, it follows, after a straightforward calculation, that

$$v_t + \operatorname{div} \tilde{A}(v, t) = 0 \text{ in } Q_T \quad v = u_0, \tag{3.298}$$

where $\tilde{A} \in C^{0,1}(\mathbb{R} \times [0, T])$ is given by $\tilde{A}'(v, t) = A'(\Psi(v, t))$.

Under the above assumptions on the flux and the forcing term, the theory of entropy solutions of scalar conservation laws applies to (3.298) and yields the existence of a unique entropy solution.

Hence, exactly as in Sect. 3.5, it is tempting to define $u \in (L^1 \cap L^\infty)(Q_T)$, for all $T > 0$, to be a pathwise entropy/kinetic solution of (3.294) if $v \in (L^1 \cap L^\infty)(Q_T)$ defined, for all $T > 0$, by (3.297) is an entropy solution of (3.298).

This does not, however, lead to a well-posed theory. The difficulty is best seen when adding a small viscosity v to (3.294), and, hence, considering the approximate equation

$$u_t + \text{div} A(u) = \Phi(u) \cdot dB + v \Delta u,$$

and, after the transformation (3.297), the problem

$$v_t + \langle \mathbf{a}(\Psi(v(x,t), t)), Dv \rangle = \frac{v}{\Psi_v(v(x,t), t)} \Delta \Psi(v(x,t), t)$$

$$= v \Delta v + v(\frac{\Psi_{vv}}{\Psi_v})(v(x,t), t)|Dv|^2.$$

If the approach based on (3.297) were correct, one would expect to get, after letting $v \to 0$, (3.298). This, however, does not seem to be the case due to the lack of the necessary a priori bounds to pass to the limit.

The problem is, however, not just a technicality but something deeper. Indeed the transformation (3.297) does not, in general, preserve the shocks unless, as an easy calculation shows, the forcing is linear.

Assume that $d = 1$ and $B(t) = t$, let H be the Heaviside step function and consider the semilinear Burgers equation

$$u_t + \frac{1}{2}(u^2)_x = \Phi(u) \text{ in } Q_T, \quad u_0 = H, \tag{3.299}$$

with Φ such that

$$\Phi(0) = 0, \quad \Phi(1) = 0, \text{ and } \Phi(u) > 0 \text{ for } u \in (0, 1). \tag{3.300}$$

It is easily seen that the entropy solution of (3.299) is

$$u(x,t) = \begin{cases} 1 \text{ for } x < t/2, \\ 0 \text{ for } x > t/2. \end{cases}$$

Next consider the transformation $u = \Psi(v, t)$ with $\dot{\Psi}(v; t) = \Phi(\Psi(v; t))$, and $\Psi(v; 0) = v$.

Since, in view of (3.300), $\Psi(0; t) = \Psi(1; t) \equiv 1$, and $\Psi(v; t) > v$ for $v \in (0, 1)$, it follows that the flux for the equation for v is

$$\tilde{A}(v, t) = \int_0^v \Psi(w; t) dw,$$

and the entropy solution with initial data u_0 is $v(x, t) = H(x - \bar{x}(t))$ with the Rankine-Hugoniot condition

$$\dot{\bar{x}}(t) = \int_0^1 \Psi(w; t)dw > \int_0^1 wdw = \frac{1}{2},$$

which shows that the shock waves are not preserved.

The final point is that, when **B** is a Brownian path, it is more natural to consider contractions in $L^1(\mathbb{R}^d \times \Omega)$ instead of $L^1(\mathbb{R}^d)$ a.s. in ω for (3.271). To fix the ideas take $\mathbf{A} = 0$ and B a Brownian motion and consider the stochastic initial value problem

$$du = \Phi(u) \circ dB \text{ in } (0, \infty) \quad u(\cdot, 0) = u_0. \tag{3.301}$$

If u_1, u_2 are solutions to (3.301) with initial data $u_{1,0}, u_{2,0}$ respectively, then, subtracting the two equations, multiplying by sign($u_1 - u_2$), taking expectations and using Itô's calculus, gives, for some $C > 0$ depending on bounds on Φ and its derivatives,

$$E \int |u_1(x, t) - u_2(x, t)|dx \leq \exp(Ct)E \int |u_1^0(x) - u_2^0(x)|dx,$$

while it is not possible, in general, to get an almost sure inequality on $\int |u_1(x, t; \omega) - u_2(x, t, \omega)|dx$.

Acknowledgements I would like to thank Ben Seeger for his help in preparing these notes.

This work was partially supported by the National Science Foundation Grants DMS-1266383 and DMS-1600129, the Office for Naval Research grant N000141712095 and the Air Force Office for Scientific Research grant FA9550-18-1-0494.

Appendix: A Brief Review of the Theory of Viscosity Solutions in the Deterministic Setting

This is a summary of several facts about the theory of viscosity solutions of Hamilton-Jacobi equations that are used in these notes. At several places, an attempt is made to motivate the definitions and the arguments. This review is very limited in scope. Good references are the books by Bardi and Capuzzo-Dolceta [2], Barles [4], Fleming and Soner [29], the CIME notes [3] and the "User's Guide" by Crandall et al. [17].

The Method of Characteristics

Consider the initial value problem

$$u_t = H(Du, x) \text{ in } Q_T \quad u(\cdot, 0) = u_0. \tag{3.302}$$

The classical method of characteristics yields, for smooth H and u_0, short time smooth solutions of (3.302). Indeed, assume that $H, u_0 \in C^2$. The characteristics associated with (3.302) are the solutions of the system of odes

$$\begin{cases} \dot{X} = -D_p H(P, X), \\ \dot{P} = D_x H(P, X), \\ \dot{U} = (H(P, X) - \langle D_p H(P, X), P \rangle) \end{cases} \tag{3.303}$$

with initial conditions

$$X(x, 0) = x, \quad P(x, 0) = Du_0(x) \quad \text{and} \quad U(x, 0) = u_0(x). \tag{3.304}$$

The connection between (3.302) and (3.303) is made through the relationship

$$U(t) = u(X(x, t), t) \quad \text{and} \quad P(t) = Du(X(x, t), t). $$

The issue is then the invertibility, with respect to x, of the map $x \mapsto X(x, t)$. A simple calculation involving the Jacobian of X shows that $x \mapsto X(t, x)$ is a diffeomorphism in $(-T^*, T^*)$ with

$$T^* = (\|D^2 H\| \, \|D^2 u_0\|)^{-1}. $$

Viscosity Solutions and Comparison Principle

Passing next to the issues of the definition and well-posedness of weak solutions, to keep the ideas simple, it is convenient to consider the two simple problems

$$u_t = H(Du)\dot{B} \text{ in } Q_T \quad u(\cdot, 0) = u_0, \tag{3.305}$$

and

$$u + H(Du) = f \text{ in } \mathbb{R}^d, \tag{3.306}$$

and to assume that $H \in C(\mathbb{R}^d)$, $B \in C^1(\mathbb{R})$ and $f \in BUC(\mathbb{R}^d)$.

Nonlinear first-order equations do not have in general smooth solutions. This can be easily seen with explicit examples. On the other hand, it is natural to expect, in view of the many applications, like control theory, front propagation, etc., that global, not necessarily smooth solutions, must exist for all time and must satisfy a comparison principle. For (3.305) this will mean that if $u_0 \leqq v_0$, then $u(\cdot, t) \leqq v(\cdot, t)$ for all $t > 0$ and, for (3.306), if $f \leqq g$, then $u \leqq v$.

To motivate the definition of the viscosity solutions it is useful to proceed, in a formal way, to prove this comparison principle.

Beginning with (3.306), it is assumed that, for $i = 1, 2$, u_i solves (3.306) with right hand side f_i. To avoid further technicalities it is further assumed that the u_i's and f_i's are periodic in the unit cube. The goal is to show that if $f_1 \leqq f_2$, then $u_1 \leqq u_2$.

The "classical" proof consists of looking at $\max(u_1 - u_2)$ which, in view of the assumed periodicity, is attained at some $x_0 \in \mathbb{R}^d$, that is,

$$(u_1 - u_2)(x_0) = \max(u_1 - u_2) .$$

If both u_1 and u_2 are differentiable at x_0, then $Du_1(x_0) = Du_2(x_0)$, and then it follows from the equations that

$$(u_1 - u_2)(x_0) \leqq (f_1 - f_2)(x_0) .$$

Observe that to prove that $u_1 \leqq u_2$, it is enough to have that

$$u_1 + H(Du_1) \leqq f_1 \quad \text{and} \quad u_2 + H(Du_2) \geqq f_2 ,$$

that is, it suffices for u_1 and u_2 to be respectively a subsolution and a supersolution.

Turning now to (3.305), it is again assumed the data is periodic in space. If, for $i = 1, 2$, u_i solves (3.305) and $u_1(\cdot, 0) \leqq u_2(\cdot, 0)$, the aim is to show that, for all $t > 0$, $u_1(\cdot, t) \leqq u_2(\cdot, t)$.

Fix $\delta > 0$ and let (x_0, t_0) be such that

$$(u_1 - u_2)(x_0, t_0) - \delta t_0 = \max_{(x,t) \in \mathbb{R}^d \times [0,T]} (u_1(x, t) - u_2(x, t) - \delta t) .$$

If $t_0 \in (0, T]$ and u_1, u_2 are differentiable at (x_0, t_0), then

$$Du_1(x_0, t_0) = Du_2(x_0, t_0) \quad \text{and} \quad u_{1,t}(x_0, t_0) \geq u_{2,t}(x_0, t_0) + \delta.$$

Since evaluating the equations at (x_0, t_0) yields a contradiction, it must be that $t_0 = 0$, and, hence,

$$\max_{(x,t) \in \mathbb{R}^d \times [0,T]} ((u_1 - u_2)(x, t) - \delta t) \leqq \max_{\mathbb{R}^d}(u_1(\cdot, 0) - u_2(\cdot, 0)) \leqq 0 .$$

Letting $\delta \to 0$ leads to the desired conclusion.

The previous arguments use, of course, strongly the fact that u_1 and u_2 are both differentiable at the maximum of $u_1 - u_2$, which is not the case in general. This is a major difficulty that is overcome using the notion of viscosity solution, which relaxes the need to have differentiable solutions.

The definition of the viscosity solutions for the general problems

$$u_t = F(D^2u, Du, u, x, t) \text{ in } U \times (0, T], \tag{3.307}$$

and

$$F(D^2u, Du, u, x) = 0 \quad \text{in} \quad U, \tag{3.308}$$

where U is an open subset of $\mathbb{R}^d$, is introduced next.

Definition 1

(i) $u \in C(U \times (0, T])$ (resp. $u \in C(U)$) is a viscosity subsolution of (3.307) (resp. (3.308)), if, for all smooth test functions of $u - \phi$ and all maximum points $(x_0, t_0) \in U \times (0, T]$ (resp. resp. $x_0 \in U$) of $u - \phi$

$$\phi_t + F(D^2\phi, D\phi, u, x_0, t_0) \leqq 0 \quad (\text{resp. } F(D^2\phi, D\phi, u, x_0) \leqq 0).$$

(ii) $u \in C(U \times (0, T])$ (resp. $u \in C(U)$) is a viscosity supersolution of (3.307) (resp. (3.308)) if, for all smooth test functions ϕ and all minimum points $(x_0, t_0) \in U \times (0, T]$ (resp. $x_0 \in U$) of $u - \phi$,

$$\phi_t + F(D^2\phi, D\phi, u, x_0, t_0) \geqq 0 \quad (\text{resp. } F(D^2\phi, D\phi, u, x_0) \geqq 0).$$

(iii) $u \in C(U \times (0, T))$ (resp. $u \in C(U)$) is a viscosity solution of (3.307) (resp. (3.308)) if it is both a sub- and super-solution. of (3.307) (resp. (3.308)).

In the definition above, maxima (resp. minima) can be either global or local. Moreover, ϕ may have any regularity, C^1 being the least required for first-order and $C^{2,1}$ for second-order equations.

Using the definition of viscosity solution, it is possible to make the previous heuristic proof rigorous and to show the well-posedness of the solutions.

A general comparison result for (3.305) and (3.306) is stated and proved next.

Theorem 1

(i) *Assume* $H \in C(\mathbb{R}^d)$, $f, g \in BUC(\mathbb{R}^d)$ *and let* $u, v \in BUC(\mathbb{R}^d)$ *be respectively viscosity subsolution and supersolution of* (3.306) *with right hand side* f *and* g *respectively. Then* $\sup_{\mathbb{R}^d}(u - v)_+ \leqq \sup(f - g)_+$.

(ii) *Assume* $H \in \mathbb{R}^d$, $B \in C^1(\mathbb{R})$, $u_0, v_0 \in BUC(\mathbb{R}^d)$ *and let* $u, v \in BUC(\overline{Q_T})$ *be respectively viscosity sub- and super-solutions of* (3.305) *with initial data* u_0 *and* v_0 *respectively. Then, for all* $t \in [0, T]$, $\sup_{\mathbb{R}^d}(u(\cdot, t) - v(\cdot, t))_+ \leqq \sup(u_0 - v_0)_+$.

Proof To simplify the argument it is assumed throughout the proof that f, g, u_0, v_0, u and v are periodic in the unit cube. This assumption guarantees that all suprema in the statement are actually achieved and are therefore maxima. The general result is proved by introducing appropriate penalization at infinity, i.e., considering, in the case of (3.306) for example, $\sup(u(x) - v(x) - \alpha|x|^2)$ and then letting $\alpha \to 0$; see [4] and [17] for all the arguments and variations.

Consider first (3.306). The key technical step is to double the variables by introducing the new function $z(x, y) = u(x) - v(y)$ which solves the doubled equation

$$z + H(D_x z) - H(-D_y z) \leq f(x) - g(y) \text{ in } \mathbb{R}^d \times \mathbb{R}^d. \tag{3.309}$$

Indeed if, for a test function ϕ, $z - \phi$ attains a maximum at (x_0, y_0), then $u(x) - \phi(x, y_0)$ and $v(y) + \phi(x_0, y)$ attain respectively a maximum at x_0, and a minimum at y_0. Therefore

$$u(x_0) + H(D_x\phi(x_0, y_0)) \leq f(x_0) \quad \text{and} \quad v(y_0) + H(-D_y\phi(x_0, y_0)) \geq f(y_0),$$

and the claim follows by subtracting these two inequalities.

To prove the comparison result, z is compared with a smooth function, which is "almost" a solution, that is, in the case at hand, a function of $x - y$.

It turns out that the most convenient choice is, for an appropriate a_ε,

$$\phi_\varepsilon(x, y) = \frac{1}{2\varepsilon}|x - y|^2 + a_\varepsilon.$$

Indeed

$$\phi_\varepsilon + H(D_x\phi_\varepsilon)) - H(-D_y\phi_\varepsilon) - (f(x) - g(y)) = \frac{1}{2\varepsilon}|x-y|^2 + a_\varepsilon - (f(x) - g(y)) \geq 0,$$

if

$$a_\varepsilon = \max(f - g) + v_\varepsilon \quad \text{and} \quad v_\varepsilon = \max\left(g(x) - g(y) - \frac{1}{2\varepsilon}|x - y|^2\right);$$

note that, since g is uniformly continuous, $\lim_{\varepsilon \to 0} v_\varepsilon = 0$.

Let $(x_\varepsilon, y_\varepsilon)$ be such that

$$z(x_\varepsilon, y_\varepsilon) - \phi(x_\varepsilon, y_\varepsilon) = \max_{\mathbb{R}^d \times \mathbb{R}^d} (z - \phi).$$

Then

$$z(x_\varepsilon, y_\varepsilon) + H(D_x\phi(x_\varepsilon, y_\varepsilon) - H(-D_y\phi(x_\varepsilon, y_\varepsilon)) \leq f(x_\varepsilon) - g(y_\varepsilon).$$

On the other hand, it is known that

$$\phi(x_\varepsilon, y_\varepsilon) + H(D_x\phi(x_\varepsilon, y_\varepsilon)) - H(-D_y\phi(x_\varepsilon, y_\varepsilon)) \geq f(x_\varepsilon) - g(y_\varepsilon) .$$

It follows that

$$z(x_\varepsilon, y_\varepsilon) \leq \phi(x_\varepsilon, y_\varepsilon)$$

and, hence,

$$z \leq \phi \ \text{ in } \ \mathbb{R}^d \times \mathbb{R}^d .$$

Letting $x = y$ in the above inequality yields

$$u(x) - v(x) \leq \phi(x, x) = a_\varepsilon = \max(f - g) + v_\varepsilon$$

and, after sending $\varepsilon \to 0$,

$$\max(u - v) \leq \max(f - g) .$$

The comparison for (3.305) is proved similarly. In the course of the proof, however, it is not necessary to double the t-variable, since the equation is linear in the time derivative. This fact plays an important role in the analysis of the pathwise pde when B is merely continuous.

To this end, define the function

$$z(x, y, t, s) = u(x, t) - v(y, s) ,$$

and observe, as before, that

$$z_t - z_s \leq H(D_x z)\dot{B}(t) - H(-D_y z)\dot{B}(s) \ \text{ in } \ \mathbb{R}^d \times \mathbb{R}^d \times (0, \infty) \times (0, \infty) .$$

On the other hand, it is possible to show that

$$z(x, y, t) = u(x, t) - v(y, t)$$

actually satisfies

$$z_t \leq (H(D_x z) - H(-D_y z))\dot{B} \ \text{ in } \ \mathbb{R}^d \times \mathbb{R}^d \times (0, \infty) . \tag{3.310}$$

Indeed, fix a smooth ϕ and let (x_0, y_0, t_0) be a (strict) local maximum of $(x, y, t) \to z(x, y, t) - \phi(x, y, t)$. Since all functions are assumed to be periodic with respect to the spatial variable, the penalized function

$$u(x, t) - v(y, s) - \phi(x, y, t) - \frac{1}{2\theta}(t - s)^2$$

achieves a local maximum at $(x_\theta, y_\theta, t_\theta, s_{t}heta)$. It follows that, as $\theta \to 0$, $(x_\theta, y_\theta, t_\theta, s_\theta) \to (x_0, y_0, t_0, t_0)$.

Applying the definition to the function $u(x, t) - v(y, s)$ gives at $(x_\theta, y_\theta, t_\theta, s_\theta)$

$$\phi_t + \frac{1}{\theta}(t_\theta - s_\theta) - \frac{1}{\theta}(t_\theta - s_\theta) \leqq H(D_x\phi)\dot{B}(t_\theta) - H(-D_y\phi)\dot{B}(s_\theta)$$

and, after letting $\theta \to 0$ and using the assumption that $B \in C^1$, at (x_0, y_0, t_0),

$$\phi_t \leqq (H(D_x\phi) - H(-D_y\phi))\dot{B}.$$

Since

$$\phi_\varepsilon(x, y) = \frac{1}{2\varepsilon}|x - y|^2$$

is a smooth supersolution of (3.310), it follows immediately, after repeating an earlier argument, that

$$u(x, t) - v(y, t) \leqq \frac{1}{2\varepsilon}|x - y|^2 + \max_{x, y \in \mathbb{R}^d} (u(x, 0) - v(y, 0) - \frac{1}{2\varepsilon}|x - y|^2)$$

and, after letting $\varepsilon \to 0$,

$$\max_{\mathbb{R}^d}(u(x, t) - v(x, t)) \leqq \max_{\mathbb{R}^d}(u(x, 0) - v(x, 0))).$$

Formulae for Solutions

The next item in this review is the control interpretation of Hamilton-Jacobi equation. For simplicity here $\dot{B} \equiv 1$.

Consider the controlled system of ode

$$\dot{x}(t) = b(x(t), \alpha(t)) \quad x(0) = x \in \mathbb{R}^d,$$

where $b : \mathbb{R}^d \times A \to \mathbb{R}^d$ is bounded and Lipschitz continuous with respect to x uniformly in α, A is a compact subset of $\mathbb{R}^M$ for some M, the measurable map $t \mapsto \alpha_t \in A$ is the control, and $(x_t)_{t \geq 0}$ is the state variable.

The associated cost function is given by

$$J(x, t, \alpha.) = \int_0^t f(x(s), \alpha(s))dt + u_0(x(t)),$$

where $u_0 \in BUC(\mathbb{R}^d)$ is the terminal cost and $f : \mathbb{R}^d \times A \to \mathbb{R}$ is the running cost, which is also assumed to be bounded and Lipschitz continuous with respect to x uniformly in α.

The goal is to minimize—one can, of course, consider maximization—the cost function J over all possible controls. The value function is

$$u(x, t) = \inf_{\alpha.} J(x, t, \alpha.) . \tag{3.311}$$

The key tool to study u is the dynamic programming principle, which is nothing more than the semigroup property. It states that, for any $\tau \in (0, t)$,

$$u(x, t) = \inf_{\alpha.} \left[\int_0^\tau f(x(s), \alpha(s))ds + u(x(\tau), t - \tau) \right] . \tag{3.312}$$

Its proof, which is straightforward, is based on the elementary observation that when pieced together, optimal controls and paths in $[a, b]$ and $[b, c]$ form an optimal path for $[a, c]$.

The following formal argument, which can be made rigorous using viscosity solutions and test functions shows the connection between the dynamic programming and the Hamilton-Jacobi equation.

Using the dynamic programming identity, with $\tau = h$ small, yields

$$u(x, t) \approx \inf_{\alpha.}(hf(x, \alpha) + D_x u(x, t) \cdot b(x, \alpha)h) + u(x, t) - u_t(x, t)h ,$$

and, hence,

$$u_t + \sup_\alpha[-D_x u \cdot b(x, \alpha) - f(x, \alpha)] = 0 ,$$

that is

$$u_t + H(Du, x) = 0 \quad \text{in} \quad \mathbb{R}^d \times (0, \infty) ,$$

where the (convex) Hamiltonian H is given by the formula

$$H(p, x) = \sup_\alpha[-\langle p, b(x, \alpha)\rangle - f(x, \alpha)] .$$

Recall that, if $H : \mathbb{R}^d \to \mathbb{R}$ is convex, then

$$H(p) = \sup_{q \in \mathbb{R}^d} (\langle p, q \rangle - H^*(p)),$$

where H^* is the Legendre transform of H defined by

$$H^*(q) = \sup_{p \in \mathbb{R}^d} (\langle q, p \rangle - H(p)).$$

The Legendre transform H^* of any continuous, not necessarily convex, $H : \mathbb{R}^d \to \mathbb{R}$ is convex.

When $H : \mathbb{R}^d \times \mathbb{R}^d \to \mathbb{R}$ is convex, the previous discussion provides a formula for the viscosity solution of the Hamilton-Jacobi equation

$$u_t + H(Du, x) = 0 \text{ in } Q_T, \quad u(\cdot, 0) = u_0 \text{ on } \mathbb{R}^d. \tag{3.313}$$

Indeed recall that

$$H(p, x) = \sup_q [\langle p, q \rangle - H^*(q, x)]$$

and consider the controlled system

$$\dot{x}(t) = q(t) \quad x(0) = x,$$

and the pay-off

$$\mathcal{J}(x, t, q.) = u_0(x(t)) + \int_0^t H^*(q(s), x(s))ds.$$

The theory of viscosity solutions (see [4, 58]) yields that

$$u(x, t) = \inf_{q.} \left[u_0(x(t)) + \int_0^t H^*(q(s), x(s)) \, ds \right]. \tag{3.314}$$

When H does not depend on x, then (3.314) can be simplified considerably. Indeed, applying Jensen's inequality to the representation formula (3.314) of the viscosity solution u of

$$u_t + H(Du) = 0 \text{ in } Q_T \quad u(\cdot, 0) = u_0 \text{ on } \mathbb{R}^d, \tag{3.315}$$

yields the Lax-Oleinik formula

$$u(x, t) = \inf_{y \in \mathbb{R}^d} \left[u_0(y) + t H^*(\frac{x - y}{t}) \right]. \tag{3.316}$$

A similar argument, when H is concave, yields

$$u(x, t) = \sup_{y \in \mathbb{R}^d} \left[u_0(y) - t H^*(\frac{x - y}{t}) \right]. \tag{3.317}$$

The existence of viscosity solution follows either directly using Perron's method (see [17]), which yields the solution as the maximal (resp. minimal) subsolution (resp. supersolution) or indirectly by considering regularizations of the equation, the most commonly used consisting of "adding" $-\varepsilon \Delta u^\varepsilon$ to the equation and passing to the limit $\varepsilon \to 0$.

A summary follows of some of the key facts about viscosity solutions of the initial value problem (3.315), which are used in the notes, for $H \in C(\mathbb{R}^d)$ and $u_0 \in BUC(\mathbb{R}^d)$.

The results discussed earlier yield that there exists a unique solution $u \in BUC(Q_T)$. In particular, $u = S_H(t)u_0$, with the solution operator $S_H(t)$: $BUC(\mathbb{R}^d) \to BUC(\mathbb{R}^d)$ a strongly continuous semigroup, that is, for $s, t > 0$,

$$S_H(t + s) = S_H(t)S_H(s).$$

The time homogeneity of the equation also yields, for $t > 0$, the identity

$$S_H(t) = S_{tH}(1).$$

Moreover, S_H commutes with translations, additions of constants and is order-preserving, and, hence, a contraction in the sup-norm, that is,

$$\|(S_H(t)u - S_H(t)v)_\pm\|_\infty \leqq \|(u - v)_\pm\|_\infty.$$

If $u_0 \in C^{0,1}(\mathbb{R}^d)$, the space homogeneity of H and the contraction property yield, that, for all $t > 0$, $S_H(t)u \in C^{0,1}(\mathbb{R}^d)$ and, moreover,

$$\|DS_H(t)u_0\| \leqq \|Du_0\|.$$

It also follows from the order preserving property that, for all $u, v \in BUC(\mathbb{R}^d)$ and $t > 0$,

$$S_H(t) \max(u, v) \geqq \max(S_H(t)u, S_H(t)v)$$

and

$$S_H(t) \min(u, v) \leqq \min(S_H(t)u, S_H(t)v) .$$

Finally, it can be easily seen from the definition of viscosity solutions that, if, for $i \in I$, u_i is a sub-(resp. super-solution), then $\sup_i u_i$ is a subsolution (resp. $\inf_i u_i$ a supersolution).

A natural question is whether there are any other explicit formulae for the solutions of (3.315); recall that for H convex/concave, the solutions satisfy the Lax-Oleinik formula.

It turns out there exists another formula, known as the Hopf formula, which does not require H to have any concavity/convexity property as long as the initial datum is convex/concave.

For definiteness, here it is assumed that u_0 is convex, and denote by u_0^* its Legendre transform.

It is immediate that, for any $p \in \mathbb{R}^d$, the function $u_p(x,t) = \langle p, x \rangle + tH(p)$ is a viscosity solution of (3.315) and, hence, in view of the previous discussion,

$$v(x,t) = \sup_{p \in \mathbb{R}^d} \left[\langle p, x \rangle + tH(p) - u_0^*(p) \right], \tag{3.318}$$

is a subsolution of (3.315).

The claim is that, if u_0 is convex, then v is actually a solution. Since this fact plays an important role in the analysis, it is stated as a separate proposition.

Proposition 1 *Let $H \in C(\mathbb{R}^d)$ and assume that $u_0 \in BUC(\mathbb{R}^d)$ is convex. The unique viscosity solution $u \in BUC(Q_T)$ of (3.315) is given by (3.318).*

Proof If H is either convex or concave, the claim follows using the Lax-Oleinik formula. Assume, for example, that H is convex. Then

$$\sup_{p \in \mathbb{R}^d} \left[\langle p, x \rangle + tH(p) - u_0^*(p) \right] = \sup_{p \in \mathbb{R}^d} \left[\langle p, x \rangle + t \sup_{q \in \mathbb{R}^d} \left((p, q) - H^*(q) \right) - u_0^*(p) \right]$$

$$= \sup_{p \in \mathbb{R}^d} \sup_{q \in \mathbb{R}^d} \left[\langle p, x \rangle + t(p, q) - tH^*(q) - u_0^*(p) \right]$$

$$= \sup_{q \in \mathbb{R}^d} \sup_{p \in \mathbb{R}^d} \left[\langle p, x + tq \rangle - u_0^*(p) - tH^*(q) \right]$$

$$= \sup_{q \in \mathbb{R}^d} \left[u_0(x + tq) - tH^*(q) \right]$$

$$= \sup_{y \in \mathbb{R}^d} \left[u_0(y) - tH^*(\frac{y-x}{t}) \right].$$

If H is concave, the argument is similar, provided the min-max theorem is used to interchange the sup and inf that appear in the formula.

For the general case the first step is that the map $F(t) : BUC(\mathbb{R}^d) \to BUC(\mathbb{R}^d)$ defined by

$$F(t)u_0(x) = \sup_{p \in \mathbb{R}^d} \left[\langle p, x \rangle + tH(p) - u_0^*(p) \right]$$

has the semi-group property, that is,

$$F(t+s) = F(t)F(s).$$

If u_0 is convex, then $F(t)u_0$ is also convex, since it is the sup of linear functions, and, moreover,

$$F(t)u_0 = (u_0^* - tH)^*.$$

In view of this observation and the fact that, if w is convex then $w = w^{**}$, the semigroup identity follows if it shown that

$$(u_0^* - (t+s)H)^{**} = ((u_0^* - sH)^{**} - tH)^{**}.$$

On the other hand, the definition of the Legendre transform, the min-max theorems and the fact that

$$\sup_{x \in \mathbb{R}^d} \langle z, x \rangle = \begin{cases} +\infty & \text{if } z \neq 0, \\ 0 & \text{if } z = 0, \end{cases}$$

yield, for $\tau > 0$, the following sequence of equalities:

$$\begin{aligned}
(u_0^* - \tau H)^{**}(y) &= \sup_{x \in \mathbb{R}^d} [\langle y, x \rangle - (u_0^* - \tau H)^*(x)] \\
&= \sup_{x \in \mathbb{R}^d} [\langle y, x \rangle - \sup_{p \in \mathbb{R}^d} [\langle x, p \rangle + \tau H(p) - u_0^*(p)]] \\
&= \sup_{x \in \mathbb{R}^d} \inf_{p \in \mathbb{R}^d} [\langle y - p, x \rangle - \tau H(p) + u_0^*(p)] \\
&= \inf_{p \in \mathbb{R}^d} \sup_{x \in \mathbb{R}^d} [\langle y - p, x \rangle - \tau H(p) + u_0^*(p)] = u_0^*(y) - \tau H(p).
\end{aligned}$$

It follows that

$$\begin{aligned}
(u_0^* - (t+s)H)^{**} &= u_0^* - (t+s)H \\
&= u_0^* - sH - tH = (u_0^* - sH)^{**} - tH \\
&= ((u_0^* - sH)^{**} - tH)^{**}.
\end{aligned}$$

Next it is shown that actually (3.318) is a viscosity solution. In view of the previous discussion, it is only needed to check the super-solution property.

Assume that, for some smooth ϕ, $v - \phi$ attains a minimum at (x_0, t_0) with $t_0 > 0$. Let $p = D\phi(x_0, t_0)$ and $\lambda = \phi_t(x_0, t_0)$. The convexity of v yields that, for all (x, t) and $h \in (0, t_0)$,

$$v(x, t_0 - h) \geq v(x_0, t_0) + \langle p, x - x_0 \rangle - \lambda h + o(h).$$

Since

$$v(x_0, t_0) = F(h)v(\cdot, t_0 - h)(x_0),$$

it follows that

$$v(x_0, t_0) = F(h)(v(x_0, t_0) + \langle p, \cdot - x_0 \rangle)(x_0) - \lambda h + o(h),$$

and, finally,

$$\lambda h \geq hH(p) + o(h).$$

Dividing by h and letting $h \to 0$ gives $\lambda \geq H(p)$.

The above proof is a typical argument in the theory of viscosity solutions which has been used by Lions [59] to give a characterization of viscosity solutions and Souganidis [102] and Barles and Souganidis [5] to prove convergence of approximations to viscosity solutions. Similar arguments were also used by Lions [60] in image processing and Barles and Souganidis [6] to study front propagation.

It is a natural question to investigate whether the Hopf formula can be used for more general Hamilton-Jacobi equations with possible dependence on (u, x).

A first requirement for such formula to hold is that the equation must preserve convexity, that is, if u_0 is convex, then $u(\cdot, t)$ must be convex for all $t > 0$.

It turns out that the general form of Hamiltonian's satisfying this latter property is

$$H(p, u, x) = \sum_{j=1}^{d} x_j H_j(Du) + u H_0(Du) + G(Du).$$

To establish a Hopf-type formula, it is necessary to look at solutions starting with linear initial data, that is, for some $p \in \mathbb{R}^d$ and $a \in \mathbb{R}$,,

$$u_0(x) = \langle p, x \rangle + a.$$

If there is a Hopf-type formula, the solution u starting with u_0 as above must be of the form

$$u(x, t) = P(t)x + A(t) \quad \text{with} \quad A(0) = a \quad \text{and} \quad P(0) = p.$$

A straightforward computation yields that P and A must satisfy, for $H = (H_1, \ldots, H_N)$, the ode

$$\dot{P} = H(P) + H_0(P)P \quad \text{and} \quad \dot{A} = H_0(P)A + G(P).$$

Whether the function

$$\sup_{p \in \mathbb{R}^d} [\langle P(t), x \rangle + A(t)]$$

with $A(0) = -u_0^*(p)$ is a solution of the Hamilton-Jacobi equation is an open question in general. Some special cases can be analyzed under additional assumptions on the H^i's, H_0, etc.

References

1. M. Alfaro, D. Antonopoulou, G. Karali, H. Matano, Generation of fine transition layers and their dynamics for the stochastic Allen–Cahn equation (2018). e-prints arXiv:1812.03804
2. M. Bardi, I. Capuzzo-Dolcetta, Optimal control and viscosity solutions of Hamilton-Jacobi-Bellman equations, in *Systems & Control: Foundations & Applications* (Birkhäuser Boston, Inc., Boston, 1997). With appendices by M. Falcone and P. Soravia. https://doi.org/10.1007/978-0-8176-4755-1
3. M. Bardi, M.G. Crandall, L.C. Evans, H.M. Soner, P.E. Souganidis, *Viscosity Solutions and Applications*. Lecture Notes in Mathematics, vol. 1660 (Springer/Centro Internazionale Matematico Estivo (C.I.M.E.), Berlin/Florence, 1997). Lectures given at the 2nd C.I.M.E. Session held in Montecatini Terme, June 12–20, 1995, Edited by I. Capuzzo Dolcetta and P. L. Lions, Fondazione C.I.M.E.. [C.I.M.E. Foundation]. https://doi.org/10.1007/BFb0094293
4. G. Barles, *Solutions de viscosité des équations de Hamilton-Jacobi*. Mathématiques & Applications (Berlin) [Mathematics & Applications], vol. 17 (Springer, Paris, 1994)
5. G. Barles, P.E. Souganidis, Convergence of approximation schemes for fully nonlinear second order equations. Asymptot. Anal. **4**(3), 271–283 (1991)
6. G. Barles, P.E. Souganidis, A new approach to front propagation problems: theory and applications. Arch. Ration. Mech. Anal. **141**(3), 237–296 (1998). https://doi.org/10.1007/s002050050077
7. G. Barles, H.M. Soner, P.E. Souganidis, Front propagation and phase field theory. SIAM J. Control Optim. **31**(2), 439–469 (1993). https://doi.org/10.1137/0331021
8. P. Billingsley, *Convergence of Probability Measures*, 2nd edn. Wiley Series in Probability and Statistics: Probability and Statistics (Wiley, New York, 1999). https://doi.org/10.1002/9780470316962
9. R. Buckdahn, J. Ma, Stochastic viscosity solutions for nonlinear stochastic partial differential equations. II. Stoch. Process. Appl. **93**(2), 205–228 (2001). https://doi.org/10.1016/S0304-4149(00)00092-2
10. R. Buckdahn, J. Ma, Pathwise stochastic control problems and stochastic HJB equations. SIAM J. Control Optim. **45**(6), 2224–2256 (2007, electronic). https://doi.org/10.1137/S036301290444335X
11. G.Q. Chen, Q. Ding, K.H. Karlsen, On nonlinear stochastic balance laws. Arch. Ration. Mech. Anal. **204**(3), 707–743 (2012)
12. I.D. Chueshov, P.A. Vuillermot, On the large-time dynamics of a class of parabolic equations subjected to homogeneous white noise: Stratonovitch's case. C. R. Acad. Sci. Paris Sér. I Math. **323**(1), 29–33 (1996)
13. I.D. Chueshov, P.A. Vuillermot, On the large-time dynamics of a class of random parabolic equations. C. R. Acad. Sci. Paris Sér. I Math. **322**(12), 1181–1186 (1996)

14. M.G. Crandall, P.L. Lions, Hamilton-Jacobi equations in infinite dimensions. I. Uniqueness of viscosity solutions. J. Funct. Anal. **62**(3), 379–396 (1985). https://doi.org/10.1016/0022-1236(85)90011-4

15. M.G. Crandall, P.L. Lions, Hamilton-Jacobi equations in infinite dimensions. II. Existence of viscosity solutions. J. Funct. Anal. **65**(3), 368–405 (1986). https://doi.org/10.1016/0022-1236(86)90026-1

16. M.G. Crandall, P.L. Lions, P.E. Souganidis, Maximal solutions and universal bounds for some partial differential equations of evolution. Arch. Ration. Mech. Anal. **105**(2), 163–190 (1989). https://doi.org/10.1007/BF00250835

17. M.G. Crandall, H. Ishii, P.L. Lions, User's guide to viscosity solutions of second order partial differential equations. Bull. Am. Math. Soc. (N.S.) **27**(1), 1–67 (1992). https://doi.org/10.1090/S0273-0979-1992-00266-5

18. G. Da Prato, M. Iannelli, L. Tubaro, Some results on linear stochastic differential equations in Hilbert spaces. Stochastics **6**(2), 105–116 (1981/1982)

19. C.M. Dafermos, *Hyperbolic Conservation Laws in Continuum Physics*. Grundlehren der Mathematischen Wissenschaften (Fundamental Principles of Mathematical Sciences), vol. 325, 4th edn. (Springer, Berlin, 2016). https://doi.org/10.1007/978-3-662-49451-6

20. A. Debussche, J. Vovelle, Long-time behavior in scalar conservation laws. Differ. Integr. Equ. **22**(3-4), 225–238 (2009)

21. A. Debussche, J. Vovelle, Scalar conservation laws with stochastic forcing. J. Funct. Anal. **259**(4), 1014–1042 (2010)

22. A. Debussche, J. Vovelle, Invariant measure of scalar first-order conservation laws with stochastic forcing. Probab. Theory Relat. Fields **163**(3–4), 575–611 (2015)

23. N. Dirr, S. Luckhaus, M. Novaga, A stochastic selection principle in case of fattening for curvature flow. Calc. Var. Partial Differ. Equ. **13**(4), 405–425 (2001). https://doi.org/10.1007/s005260100080

24. L.C. Evans, The perturbed test function method for viscosity solutions of nonlinear PDE. Proc. R. Soc. Edinb. Sect. A **111**(3–4), 359–375 (1989). https://doi.org/10.1017/S0308210500018631

25. L.C. Evans, Periodic homogenisation of certain fully nonlinear partial differential equations. Proc. R. Soc. Edinb. Sect. A **120**(3–4), 245–265 (1992). https://doi.org/10.1017/S0308210500032121

26. L.C. Evans, P.E. Souganidis, Differential games and representation formulas for solutions of Hamilton-Jacobi-Isaacs equations. Indiana Univ. Math. J. **33**(5), 773–797 (1984). https://doi.org/10.1512/iumj.1984.33.33040

27. L.C. Evans, H.M. Soner, P.E. Souganidis, Phase transitions and generalized motion by mean curvature. Commun. Pure Appl. Math. **45**(9), 1097–1123 (1992). https://doi.org/10.1002/cpa.3160450903

28. J. Feng, D. Nualart, Stochastic scalar conservation laws. J. Funct. Anal. **255**(2), 313–373 (2008)

29. W.H. Fleming, H.M. Soner, *Controlled Markov Processes and Viscosity Solutions*. Stochastic Modelling and Applied Probability, vol. 25, 2nd edn. (Springer, New York, 2006)

30. P.K. Friz, P. Gassiat, P.L. Lions, P.E. Souganidis, Eikonal equations and pathwise solutions to fully non-linear SPDEs. Stoch. Partial Differ. Equ. Anal. Comput. **5**(2), 256–277 (2017). https://doi.org/10.1007/s40072-016-0087-9

31. T. Funaki, The scaling limit for a stochastic PDE and the separation of phases. Probab. Theory Relat. Fields **102**(2), 221–288 (1995). https://doi.org/10.1007/BF01213390

32. T. Funaki, Singular limit for stochastic reaction-diffusion equation and generation of random interfaces. Acta Math. Sin. (Engl. Ser.) **15**(3), 407–438 (1999). https://doi.org/10.1007/BF02650735

33. P. Gassiat, A stochastic Hamilton-Jacobi equation with infinite speed of propagation. C. R. Math. Acad. Sci. Paris **355**(3), 296–298 (2017). https://doi.org/10.1016/j.crma.2017.01.021

34. P. Gassiat, B. Gess, Regularization by noise for stochastic Hamilton-Jacobi equations. Probab. Theory Relat. Fields **173**(3–4), 1063–1098 (2019). https://doi.org/10.1007/s00440-018-0848-7

35. P. Gassiat, B. Gess, P.L. Lions, P.E. Souganidis, Speed of propagation for Hamilton-Jacobi equations with multiplicative rough time dependence and convex Hamiltonians (2019). ArXiv:1805.08477 [math.PR]

36. P. Gassiat, P.L. Lions, P.E. Souganidis, in preparation

37. M. Gerencsér, I. Gyöngy, N. Krylov, On the solvability of degenerate stochastic partial differential equations in Sobolev spaces. Stoch. Partial Differ. Equ. Anal. Comput. **3**(1), 52–83 (2015). https://doi.org/10.1007/s40072-014-0042-6

38. B. Gess, P.E. Souganidis, Scalar conservation laws with multiple rough fluxes. Commun. Math. Sci. **13**(6), 1569–1597 (2015). https://doi.org/10.4310/CMS.2015.v13.n6.a10

39. B. Gess, P.E. Souganidis, Long-time behavior, invariant measures, and regularizing effects for stochastic scalar conservation laws. Commun. Pure Appl. Math. **70**(8), 1562–1597 (2017). https://doi.org/10.1002/cpa.21646

40. B. Gess, P.E. Souganidis, Stochastic non-isotropic degenerate parabolic-hyperbolic equations. Stoch. Process. Appl. **127**(9), 2961–3004 (2017). https://doi.org/10.1016/j.spa.2017.01.005

41. B. Gess, B. Perthame, P.E. Souganidis, Semi-discretization for stochastic scalar conservation laws with multiple rough fluxes. SIAM J. Numer. Anal. **54**(4), 2187–2209 (2016). https://doi.org/10.1137/15M1053670

42. H. Hoel, K.H. Karlsen, N.H. Risebro, E.B. Storrø sten, Path-dependent convex conservation laws. J. Differ. Equ. **265**(6), 2708–2744 (2018). https://doi.org/10.1016/j.jde.2018.04.045

43. M. Hofmanová, Strong solutions of semilinear stochastic partial differential equations. Nonlinear Differ. Equ. Appl. **20**(3), 757–778 (2013). https://doi.org/10.1007/s00030-012-0178-x

44. M. Hofmanová, Scalar conservation laws with rough flux and stochastic forcing. Stoch. Partial Differ. Equ. Anal. Comput. **4**(3), 635–690 (2016). https://doi.org/10.1007/s40072-016-0072-3

45. H. Huang, H.J. Kushner, Weak convergence and approximations for partial differential equations with stochastic coefficients. Stochastics **15**(3), 209–245 (1985). https://doi.org/10.1080/17442508508833357

46. H. Ishii, Hamilton-Jacobi equations with discontinuous Hamiltonians on arbitrary open sets. Bull. Fac. Sci. Eng. Chuo Univ. **28**, 33–77 (1985)

47. S.N. Kružkov, First order quasilinear equations with several independent variables. Mat. Sb. (N.S.) **81**(123), 228–255 (1970)

48. N.V. Krylov, On L_p-theory of stochastic partial differential equations in the whole space. SIAM J. Math. Anal. **27**(2), 313–340 (1996). https://doi.org/10.1137/S0036141094263317

49. N.V. Krylov, On the foundation of the L_p-theory of stochastic partial differential equations, in *Stochastic Partial Differential Equations and Applications—VII*. Lecture Notes in Pure and Applied Mathematics, vol. 245 (Chapman & Hall/CRC, Boca Raton, 2006), pp. 179–191. https://doi.org/10.1201/9781420028720.ch16

50. N.V. Krylov, M. Röckner, Strong solutions of stochastic equations with singular time dependent drift. Probab. Theory Relat. Fields **131**(2), 154–196 (2005)

51. H. Kunita, *Stochastic Flows and Stochastic Differential Equations*. Cambridge Studies in Advanced Mathematics, vol. 24 (Cambridge University Press, Cambridge, 1997). Reprint of the 1990 original

52. J.M. Lasry, P.L. Lions, A remark on regularization in Hilbert spaces. Israel J. Math. **55**(3), 257–266 (1986). https://doi.org/10.1007/BF02765025

53. J.M. Lasry, P.L. Lions, Jeux à champ moyen. I. Le cas stationnaire. C. R. Math. Acad. Sci. Paris **343**(9), 619–625 (2006)

54. J.M. Lasry, P.L. Lions, Jeux à champ moyen. II. Horizon fini et contrôle optimal. C. R. Math. Acad. Sci. Paris **343**(10), 679–684 (2006)

55. J.M. Lasry, P.L. Lions, Mean field games. Jpn. J. Math. **2**(1), 229–260 (2007). https://doi.org/10.1007/s11537-007-0657-8

56. A. Lejay, T.J. Lyons, On the importance of the Lévy area for studying the limits of functions of converging stochastic processes. Application to homogenization, in *Current Trends in Potential Theory*. Theta Series in Advanced Mathematics, vol. 4 (Theta, Bucharest, 2005), pp. 63–84

57. P.L. Lions, Mean field games. College de France course

58. P.L. Lions, *Generalized Solutions of Hamilton-Jacobi Equations*. Research Notes in Mathematics, vol. 69 (Pitman (Advanced Publishing Program), Boston, 1982)

59. P.L. Lions, Some properties of the viscosity semigroups for Hamilton-Jacobi equations, in *Nonlinear Differential Equations (Granada, 1984)*. Research Notes in Mathematics, vol. 132 (Pitman, Boston, 1985), pp. 43–63

60. P.L. Lions, Axiomatic derivation of image processing models. Math. Models Methods Appl. Sci. **4**(4), 467–475 (1994). https://doi.org/10.1142/S0218202594000261

61. P.L. Lions, B. Perthame, Remarks on Hamilton-Jacobi equations with measurable time-dependent Hamiltonians. Nonlinear Anal. **11**(5), 613–621 (1987). https://doi.org/10.1016/0362-546X(87)90076-9

62. P.L. Lions, B. Perthame, E. Tadmor, Kinetic formulation of the isentropic gas dynamics and *p*-systems. Commun. Math. Phys. **163**(2), 415–431 (1994). http://projecteuclid.org/euclid.cmp/1104270470

63. P.L. Lions, G. Papanicolaou, S.R.S. Varadhan, Homogenization of Hamilton-Jacobi equations (1996). Preprint

64. P.L. Lions, B. Perthame, P.E. Souganidis, Existence and stability of entropy solutions for the hyperbolic systems of isentropic gas dynamics in Eulerian and Lagrangian coordinates. Commun. Pure Appl. Math. **49**(6), 599–638 (1996). https://doi.org/10.1002/(SICI)1097-0312(199606)49:6<599::AID-CPA2>3.0.CO;2-5

65. P.L. Lions, B. Perthame, P.E. Souganidis, Scalar conservation laws with rough (stochastic) fluxes. Stoch. Partial Differ. Equ. Anal. Comput. **1**(4), 664–686 (2013)

66. P.L. Lions, B. Perthame, P.E. Souganidis, Stochastic averaging lemmas for kinetic equations, in *Séminaire Laurent Schwartz—Équations aux dérivées partielles et applications*. Année 2011–2012, Sémin. Équ. Dériv. Partielles, pp. Exp. No. XXVI, 17. (École Polytech., Palaiseau, 2013)

67. P.L. Lions, B. Perthame, P.E. Souganidis, Scalar conservation laws with rough (stochastic) fluxes: the spatially dependent case. Stoch. Partial Differ. Equ. Anal. Comput. **2**(4), 517–538 (2014). https://doi.org/10.1007/s40072-014-0038-2

68. P.L. Lions, B. Seeger, P.E. Souganidis, in preparation

69. P.L. Lions, P.E. Souganidis, The asymptotics of stochastically perturbed reaction-diffusion equations and front propagation. Preprint

70. P.L. Lions, P.E. Souganidis, Ill-posedness of fronts moving with space-time white noise (in preparation)

71. P.L. Lions, P.E. Souganidis, Pathwise solutions for nonlinear partial differential equations with rough signals (in preparation)

72. P.L. Lions, P.E. Souganidis, Well posedness of pathwise solutions of fully nonlinear pde with multiple rough signals (in preparation)

73. P.L. Lions, P.E. Souganidis, Well posedness of pathwise solutions of Hamilton-Jacobi equations with convex Hamiltonians (in preparation)

74. P.L. Lions, P.E. Souganidis, Fully nonlinear stochastic partial differential equations. C. R. Acad. Sci. Paris Sér. I Math. **326**(9), 1085–1092 (1998). https://doi.org/10.1016/S0764-4442(98)80067-0

75. P.L. Lions, P.E. Souganidis, Fully nonlinear stochastic partial differential equations: non-smooth equations and applications. C. R. Acad. Sci. Paris Sér. I Math. **327**(8), 735–741 (1998). https://doi.org/10.1016/S0764-4442(98)80161-4

76. P.L. Lions, P.E. Souganidis, Équations aux dérivées partielles stochastiques nonlinéaires et solutions de viscosité, in Seminaire: Équations aux Dérivées Partielles, 1998–1999, Sémin. Équ. Dériv. Partielles, pp. Exp. No. I, 15 (École Polytech., Palaiseau, 1999)

77. P.L. Lions, P.E. Souganidis, Fully nonlinear stochastic pde with semilinear stochastic dependence. C. R. Acad. Sci. Paris Sér. I Math. **331**(8), 617–624 (2000). https://doi.org/10.1016/S0764-4442(00)00583-8
78. P.L. Lions, P.E. Souganidis, Uniqueness of weak solutions of fully nonlinear stochastic partial differential equations. C. R. Acad. Sci. Paris Sér. I Math. **331**(10), 783–790 (2000). https://doi.org/10.1016/S0764-4442(00)01597-4
79. P.L. Lions, P.E. Souganidis, Viscosity solutions of fully nonlinear stochastic partial differential equations. Sūrikaisekikenkyūsho Kōkyūroku **1287**, 58–65 (2002). Viscosity solutions of differential equations and related topics (Japanese) (Kyoto, 2001)
80. P.L. Lions, P. Souganidis, New regularity results and long time behavior of pathwise (stochastic) Hamilton-Jacobi equations (2018). Preprint
81. T.J. Lyons, Differential equations driven by rough signals. Rev. Mat. Iberoamericana **14**(2), 215–310 (1998)
82. T.J. Lyons, Z. Qian, Flow equations on spaces of rough paths. J. Funct. Anal. **149**(1), 135–159 (1997). https://doi.org/10.1006/jfan.1996.3088
83. T.J. Lyons, Z. Qian, *System Control and Rough Paths*. Oxford Mathematical Monographs (Oxford University Press, Oxford, 2002)
84. T. Otha, D. Jasnow, K. Kawasaki, Universal scaling in the motion of random interfaces. Phys. Rev. Lett. **49**, 1223–1226 (1982)
85. E. Pardoux, Sur des équations aux dérivées partielles stochastiques monotones. C. R. Acad. Sci. Paris Sér. A-B **275**, A101–A103 (1972)
86. E. Pardoux, Équations aux dérivées partielles stochastiques de type monotone, in *Séminaire sur les Équations aux Dérivées Partielles (1974–1975)*, III, Exp. No. 2 (Collège de France, Paris, 1975), p. 10
87. E. Pardoux, Stochastic partial differential equations and filtering of diffusion processes. Stochastics **3**(2), 127–167 (1979). https://doi.org/10.1080/17442507908833142
88. Y. Peres, Points of increase for random walks. Israel J. Math. **95**, 341–347 (1996). https://doi.org/10.1007/BF02761045
89. B. Perthame, Uniqueness and error estimates in first order quasilinear conservation laws via the kinetic entropy defect measure. J. Math. Pures Appl. (9) **77**(10), 1055–1064 (1998). https://doi.org/10.1016/S0021-7824(99)80003-8
90. B. Perthame, *Kinetic Formulation of Conservation Laws*. Oxford Lecture Series in Mathematics and its Applications, vol. 21 (Oxford University Press, Oxford, 2002)
91. B. Perthame, P.E. Souganidis, Dissipative and entropy solutions to non-isotropic degenerate parabolic balance laws. Arch. Ration. Mech. Anal. **170**(4), 359–370 (2003). https://doi.org/10.1007/s00205-003-0282-5
92. B. Perthame, E. Tadmor, A kinetic equation with kinetic entropy functions for scalar conservation laws. Commun. Math. Phys. **136**(3), 501–517 (1991). http://projecteuclid.org/euclid.cmp/1104202434
93. F. Rezakhanlou, J.E. Tarver, Homogenization for stochastic Hamilton-Jacobi equations. Arch. Ration. Mech. Anal. **151**(4), 277–309 (2000). https://doi.org/10.1007/s002050050198
94. B.L. Rozovskiĭ, Stochastic partial differential equations that arise in nonlinear filtering problems. Usp. Mat. Nauk **27**(3(165)), 213–214 (1972)
95. B.L. Rozovskiĭ, Stochastic partial differential equations. Mat. Sb. (N.S.) **96**(138), 314–341, 344 (1975)
96. B. Seeger, Approximation schemes for viscosity solutions of fully nonlinear stochastic partial differential equations (2018). ArXiv:1802.04740 [math.AP]
97. B. Seeger, Scaling limits and homogenization of stochastic Hamilton-Jacobi equations (in preparation)
98. B. Seeger, Homogenization of pathwise Hamilton-Jacobi equations. J. Math. Pures Appl. (9) **110**, 1–31 (2018). https://doi.org/10.1016/j.matpur.2017.07.012
99. B. Seeger, Perron's method for pathwise viscosity solutions. Commun. Partial Differ. Equ. **43**(6), 998–1018 (2018). https://doi.org/10.1080/03605302.2018.1488262

100. B. Seeger, Fully nonlinear stochastic partial differential equations, Thesis (Ph.D.), The University of Chicago, 2019
101. D. Serre, *Systems of Conservation Laws 1* (Cambridge University Press, Cambridge, 1999). Hyperbolicity, entropies, shock waves. Translated from the 1996 French original by I. N. Sneddon. https://doi.org/10.1017/CBO9780511612374
102. P.E. Souganidis, Approximation schemes for viscosity solutions of Hamilton-Jacobi equations. J. Differ. Equ. **59**(1), 1–43 (1985). https://doi.org/10.1016/0022-0396(85)90136-6
103. P.E. Souganidis, Stochastic homogenization of Hamilton-Jacobi equations and some applications. Asymptot. Anal. **20**(1), 1–11 (1999)
104. P.E. Souganidis, N.K. Yip, Uniqueness of motion by mean curvature perturbed by stochastic noise. Ann. Inst. H. Poincaré Anal. Non Linéaire **21**(1), 1–23 (2004). https://doi.org/10.1016/S0294-1449(03)00029-5
105. H. Watanabe, On the convergence of partial differential equations of parabolic type with rapidly oscillating coefficients to stochastic partial differential equations. Appl. Math. Optim. **20**(1), 81–96 (1989). https://doi.org/10.1007/BF01447648
106. N.K. Yip, Stochastic motion by mean curvature. Arch. Ration. Mech. Anal. **144**(4), 313–355 (1998). https://doi.org/10.1007/s002050050120

Chapter 4
Random Data Wave Equations

Nikolay Tzvetkov

Abstract Nowadays we have many methods allowing to exploit the regularising properties of the linear part of a nonlinear dispersive equation (such as the KdV equation, the nonlinear wave or the nonlinear Schrödinger equations) in order to prove well-posedness in low regularity Sobolev spaces. By well-posedness in low regularity Sobolev spaces we mean that less regularity than the one imposed by the energy methods is required (the energy methods do not exploit the dispersive properties of the linear part of the equation). In many cases these methods to prove well-posedness in low regularity Sobolev spaces lead to optimal results in terms of the regularity of the initial data. By optimal we mean that if one requires slightly less regularity then the corresponding Cauchy problem becomes ill-posed in the Hadamard sense. We call the Sobolev spaces in which these ill-posedness results hold *spaces of supercritical regularity*. More recently, methods to prove *probabilistic well-posedness* in Sobolev spaces of supercritical regularity were developed. More precisely, by probabilistic well-posedness we mean that one endows the corresponding Sobolev space of supercritical regularity with a non degenerate probability measure and then one shows that almost surely with respect to this measure one can define a (unique) global flow. However, in most of the cases when the methods to prove probabilistic well-posedness apply, there is no information about the measure transported by the flow. Very recently, a method to prove that the transported measure is absolutely continuous with respect to the initial measure was developed. In such a situation, we have a measure which is quasi-invariant under the corresponding flow.

The aim of these lectures is to present all of the above described developments in the context of the nonlinear wave equation.

N. Tzvetkov (✉)
Department of Mathematics, Universite de Cergy-Pontoise, Cergy-Pontoise Cedex, France
e-mail: nikolay.tzvetkov@u-cergy.fr

© Springer Nature Switzerland AG 2019   221
F. Flandoli et al. (eds.), *Singular Random Dynamics*, Lecture Notes
in Mathematics 2253, https://doi.org/10.1007/978-3-030-29545-5_4

4.1 Deterministic Cauchy Theory for the 3*d* Cubic Wave Equation

4.1.1 Introduction

In this section, we consider the cubic defocusing wave equation

$$(\partial_t^2 - \Delta)u + u^3 = 0, \tag{4.1.1}$$

where $u = u(t, x)$ is real valued, $t \in \mathbb{R}$, $x \in \mathbb{T}^3 = (\mathbb{R}/(2\pi\mathbb{Z}))^3$ (the 3*d* torus). In (4.1.1), Δ denotes the Laplace operator, namely

$$\Delta = \partial_{x_1}^2 + \partial_{x_2}^2 + \partial_{x_3}^2.$$

Since (4.1.1) is of second order in time, it is natural to complement it with two initial conditions

$$u(0, x) = u_0(x), \quad \partial_t u(0, x) = u_1(x). \tag{4.1.2}$$

In this section, we will be studying the local and global well-posedness of the initial value problem (4.1.1)–(4.1.2) in Sobolev spaces via deterministic methods.

The Sobolev spaces $H^s(\mathbb{T}^3)$ are defined as follows. For a function f on $\mathbb{T}^3$ given by its Fourier series

$$f(x) = \sum_{n \in \mathbb{Z}^3} \hat{f}(n) \, e^{in \cdot x},$$

we define the Sobolev norm $H^s(\mathbb{T}^3)$ of f as

$$\|f\|_{H^s}^2 = \sum_{n \in \mathbb{Z}^3} \langle n \rangle^{2s} \, |\hat{f}(n)|^2,$$

where $\langle n \rangle = (1 + |n|^2)^{1/2}$. On has that

$$\|f\|_{H^s} \approx \|D^s f\|_{L^2}, \quad D \equiv (1 - \Delta)^{1/2}.$$

For integer values of s one can also give an equivalent norm in the physical space as follows

$$\|f\|_{H^s(\mathbb{T}^3)} \approx \sum_{|\alpha| \leq s} \|\partial_{x_1}^{\alpha_1} \partial_{x_2}^{\alpha_2} \partial_{x_3}^{\alpha_3} f\|_{L^2(\mathbb{T}^3)},$$

where the summation is taken over all multi-indexes $\alpha = (\alpha_1, \alpha_2, \alpha_3) \in \mathbb{N}^3$.

As we shall see, it will be of importance to understand the interplay between the linear and the nonlinear part of (4.1.1). Indeed, let us first consider the Cauchy problem

$$\partial_t^2 u + u^3 = 0, \quad u(0, x) = u_0(x), \quad \partial_t u(0, x) = u_1(x)$$

which is obtained from (4.1.1) by neglecting the Laplacian. If we set

$$U = (U_1, U_2) \equiv (u, \partial_t u)^t$$

then the last problem can be written as

$$\partial_t U = F(U), \quad F(U) = (U_2, -U_1^3)^t.$$

On may wish to solve, at least locally, the last problem via the Cauchy-Lipschitz argument in the spaces $H^{s_1}(\mathbb{T}^3) \times H^{s_2}(\mathbb{T}^3)$. For such a purpose one should check that the vector field $F(U)$ is locally Lipschitz on these spaces. Thanks to the Sobolev embedding $H^s(\mathbb{T}^3) \subset L^\infty(\mathbb{T}^3)$, $s > 3/2$ we can see that the map $U_1 \mapsto U_1^3$ is locally Lipschitz on $H^s(\mathbb{T}^3)$, $s > 3/2$. It is also easy to check that the map $U_1 \mapsto U_1^3$ is not continuous on $H^s(\mathbb{T}^3)$, $s < 3/2$. A more delicate argument shows that it is not continuous on $H^{3/2}(\mathbb{T}^3)$ either. Therefore, if we impose that $F(u)$ is locally Lipschitz on $H^{s_1}(\mathbb{T}^3) \times H^{s_2}(\mathbb{T}^3)$ than we necessarily need to impose a regularity assumption $s_1 > 3/2$. As we shall see below the term containing the Laplacian in (4.1.1) will allow as to significantly relax this regularity assumption.

On the other hand if we neglect the nonlinear term u^3 in (4.1.1), we get the linear wave equation which is well-posed in $H^s(\mathbb{T}^3) \times H^{s-1}(\mathbb{T}^3)$ for *any* $s \in \mathbb{R}$, as it can be easily seen by the Fourier series description of the solutions of the linear wave equation (see the next section). In other words the absence of a nonlinearity allows us to solve the problem in arbitrary singular Sobolev spaces.

In summary, we expect that the Laplacian term in (4.1.1) will help us to prove the well-posedness of the problem (4.1.1) in singular Sobolev spaces while the nonlinear term u^3 will be responsible for the lack of well-posedness in singular spaces.

4.1.2   Local and Global Well-Posedness in $H^1 \times L^2$

4.1.2.1   The Free Evolution

We first define the free evolution, i.e. the map defining the solutions of the linear wave equation

$$(\partial_t^2 - \Delta)u = 0, \quad u(0, x) = u_0(x), \quad \partial_t u(0, x) = u_1(x). \tag{4.1.3}$$

Using the Fourier transform and solving the corresponding second order linear ODE's, we obtain that the solutions of (4.1.3) are generated by the map $S(t)$, defined as follows

$$S(t)(u_0, u_1) \equiv \cos(t\sqrt{-\Delta})(u_0) + \frac{\sin(t\sqrt{-\Delta})}{\sqrt{-\Delta}}(u_1),$$

where

$$\cos(t\sqrt{-\Delta})(u_0) \equiv \sum_{n \in \mathbb{Z}^3} \cos(t|n|)\widehat{u_0}(n)\, e^{in \cdot x}$$

and

$$\frac{\sin(t\sqrt{-\Delta})}{\sqrt{-\Delta}}(u_1) \equiv t\widehat{u_1}(0) + \sum_{n \in \mathbb{Z}^3_*} \frac{\sin(t|n|)}{|n|}\widehat{u_1}(n)\, e^{in \cdot x}, \quad \mathbb{Z}^3_* = \mathbb{Z}^3 \backslash \{0\}.$$

We have that $S(t)(u_0, u_1)$ solves (4.1.3) and if $(u_0, u_1) \in H^s \times H^{s-1}$, $s \in \mathbb{R}$ then $S(t)(u_0, u_1)$ is the unique solution of (4.1.3) in $C(\mathbb{R}; H^s(\mathbb{T}^3))$ such that its time derivative is in $C(\mathbb{R}; H^{s-1}(\mathbb{T}^3))$. It follows directly from the definition that the operator $\bar{S}(t) \equiv (S(t), \partial_t S(t))$ is bounded on $H^s \times H^{s-1}$, $\bar{S}(0) = \text{Id}$ and $\bar{S}(t+\tau) = \bar{S}(t) \circ \bar{S}(\tau)$, for every real numbers t and τ. In the proof of the boundedness on $H^s \times H^{s-1}$, we only use the boundedness of $\cos(t|n|)$ and $\sin(t|n|)$. As we shall see below one may use the oscillations of $\cos(t|n|)$ and $\sin(t|n|)$ for $|n| \gg 1$ in order to get more involved L^p, $p > 2$ properties of the map $S(t)$.

Let us next consider the non homogeneous problem

$$(\partial_t^2 - \Delta)u = F(t, x), \quad u(0, x) = 0, \quad \partial_t u(0, x) = 0. \tag{4.1.4}$$

Using the variation of the constants method, we obtain that the solutions of (4.1.4) are given by

$$u(t) = \int_0^t \frac{\sin((t-\tau)\sqrt{-\Delta})}{\sqrt{-\Delta}}((F(\tau))d\tau.$$

As a consequence, we obtain that the solution of the non homogeneous problem (4.1.4) is one derivative smoother than the source term F. More precisely, for every $s \in \mathbb{R}$, the solution of (4.1.4) satisfies the bound

$$\|u\|_{L^\infty([0,1]; H^{s+1}(\mathbb{T}^3))} \leq C\|F\|_{L^1([0,1]; H^s(\mathbb{T}^3))}. \tag{4.1.5}$$

4.1.2.2   The Local Well-Posedness

We state the local well-posedness result.

Proposition 4.1.1 (Local Well-Posedness) *Consider the cubic defocusing wave equation*

$$(\partial_t^2 - \Delta)u + u^3 = 0, \tag{4.1.6}$$

posed on $\mathbb{T}^3$. *There exist constants c and C such that for every $a \in \mathbb{R}$, every $\Lambda \geq 1$, every*

$$(u_0, u_1) \in H^1(\mathbb{T}^3) \times L^2(\mathbb{T}^3)$$

satisfying

$$\|u_0\|_{H^1} + \|u_1\|_{L^2} \leq \Lambda \tag{4.1.7}$$

there exists a unique solution of (4.1.6) on the time interval $[a, a + c\Lambda^{-2}]$ of (4.1.6) with initial data

$$u(a, x) = u_0(x), \quad \partial_t u(a, x) = u_1(x).$$

Moreover the solution satisfies

$$\|(u, \partial_t u)\|_{L^\infty([a, a+c\Lambda^{-2}], H^1(\mathbb{T}^3) \times L^2(\mathbb{T}^3))} \leq C\Lambda,$$

$(u, \partial_t u)$ is unique in the class $L^\infty([a, a + c\Lambda^{-2}], H^1(\mathbb{T}^3) \times L^2(\mathbb{T}^3))$ and the dependence with respect to the initial data and with respect to the time is continuous. Finally, if

$$(u_0, u_1) \in H^s(\mathbb{T}^3) \times H^{s-1}(\mathbb{T}^3)$$

for some $s \geq 1$ then there exists $c_s > 0$ such that

$$(u, \partial_t u) \in C([a, a + c_s \Lambda^{-2}]; H^s(\mathbb{T}^3) \times H^{s-1}(\mathbb{T}^3)).$$

Proof If $u(t, x)$ is a solution of (4.1.6) then so is $u(t + a, x)$. Therefore, it suffices to consider the case $a = 0$.

Thanks to the analysis of the previous section, we obtain that we should solve the integral equation

$$u(t) = S(t)(u_0, u_1) - \int_0^t \frac{\sin((t - \tau)\sqrt{-\Delta})}{\sqrt{-\Delta}} ((u^3(\tau)) d\tau. \tag{4.1.8}$$

Set

$$\Phi_{u_0,u_1}(u) \equiv S(t)(u_0, u_1) - \int_0^t \frac{\sin((t-\tau)\sqrt{-\Delta})}{\sqrt{-\Delta}}((u^3(\tau))d\tau.$$

Then for $T \in (0, 1]$, we define X_T as

$$X_T \equiv C([0, T]; H^1(\mathbb{T}^3)),$$

endowed with the natural norm

$$\|u\|_{X_T} = \sup_{0 \le t \le T} \|u(t)\|_{H^1(\mathbb{T}^3)}.$$

Using the boundedness properties of $\bar{S}$ on $H^s \times H^{s-1}$ explained in the previous section and using the Sobolev embedding $H^1(\mathbb{T}^3) \subset L^6(\mathbb{T}^3)$, we get

$$\|\Phi_{u_0,u_1}(u)\|_{X_T} \le C\big(\|u_0\|_{H^1} + \|u_1\|_{L^2} + T \sup_{\tau \in [0,T]} \|u(\tau)\|_{L^6}^3\big)$$

$$\le C\big(\|u_0\|_{H^1} + \|u_1\|_{L^2} + CT\|u\|_{X_T}^3\big).$$

It is now clear that for $T = c\Lambda^{-2}$, $c \ll 1$ the map Φ_{u_0,u_1} sends the ball

$$B \equiv (u : \|u\|_{X_T}) \le 2C\Lambda)$$

into itself. Moreover, by a similar arguments involving the Sobolev embedding $H^1(\mathbb{T}^3) \subset L^6(\mathbb{T}^3)$ and the Hölder inequality, we obtain the estimate

$$\|\Phi_{u_0,u_1}(u) - \Phi_{u_0,u_1}(\tilde{u})\|_{X_T} \le CT\|u - \tilde{u}\|_{X_T}\big(\|u\|_{X_T}^2 + \|\tilde{u}\|_{X_T}^2\big). \qquad (4.1.9)$$

Therefore, with our choice of T, we get that

$$\|\Phi_{u_0,u_1}(u) - \Phi_{u_0,u_1}(\tilde{u})\|_{X_T} \le \frac{1}{2}\|u - \tilde{u}\|_{X_T}, \quad u, \tilde{u} \in B.$$

Consequently the map Φ_{u_0,u_1} is a contraction on B. The fixed point of this contraction defines the solution u on $[0, T]$ we are looking for. The estimate of $\|\partial_t u\|_{L^2}$ follows by differentiating in t the Duhamel formula (4.1.8). Let us now turn to the uniqueness. Let $u, \tilde{u}$ be two solutions of (4.1.6) with the same initial data in the space X_T for some $T > 0$. Then for $\tau \le T$, we can write similarly to (4.1.9)

$$\|\Phi_{u_0,u_1}(u) - \Phi_{u_0,u_1}(\tilde{u})\|_{X_\tau} \le C\tau\|u - \tilde{u}\|_{X_\tau}\big(\|u\|_{X_T}^2 + \|\tilde{u}\|_{X_T}^2\big). \qquad (4.1.10)$$

Let us take τ such that

$$C\tau\left(\|u\|_{X_T}^2 + \|\tilde{u}\|_{X_T}^2\right) < \frac{1}{2}.$$

This fixes the value of τ. Thanks to (4.1.10), we obtain that u and $\tilde{u}$ are the same on $[0, \tau]$. Next, we cover the interval $[0, T]$ by intervals of size τ and we inductively obtain that u and $\tilde{u}$ are the same on each interval of size τ. This yields the uniqueness statement.

The continuous dependence with respect to time follows from the Duhamel formula representation of the solution of (4.1.8). The continuity with respect to the initial data follows from the estimates on the difference of two solutions we have just performed. Notice that we also obtain uniform continuity of the map data-solution on bounded subspaces of $H^1 \times L^2$.

Let us finally turn to the propagation of higher regularity. Let $(u_0, u_1) \in H^1 \times L^2$ such that (4.1.7) holds satisfy the additional regularity property $(u_0, u_1) \in H^s \times H^{s-1}$ for some $s > 1$. We will show that the corresponding solution remains in $H^s \times H^{s-1}$ in the (essentially) whole time of existence. For $s \geq 1$, we define X_T^s as

$$X_T^s \equiv C([0, T]; H^s(\mathbb{T}^3)),$$

endowed with the norm

$$\|u\|_{X_T^s} = \sup_{0 \leq t \leq T} \|u(t)\|_{H^s(\mathbb{T}^3)}.$$

We have that the solution with data $(u_0, u_1) \in H^s \times H^{s-1}$ remains in this space for time intervals of order $(1 + \|u_0\|_{H^s} + \|u_1\|_{H^{s-1}})^{-2}$ by a fixed point argument, similar to the one we performed for data in $H^1 \times L^2$. We now show that the regularity is preserved for (the longer) time intervals of order $(1 + \|u_0\|_{H^1} + \|u_1\|_{L^2})^{-2}$. Coming back to (4.1.8), we can write

$$\|\Phi_{u_0, u_1}(u)\|_{X_T^s} \leq C\left(\|u_0\|_{H^s} + \|u_1\|_{H^{s-1}} + T \sup_{\tau \in [0,T]} \|u^3(\tau)\|_{H^{s-1}}\right).$$

Now using the Kato-Ponce product inequality, we can obtain that for $\sigma \geq 0$, one has the bound

$$\|v^3\|_{H^\sigma(\mathbb{T}^3)} \leq C\|D^\sigma v\|_{L^6(\mathbb{T}^3)} \|v\|_{L^6(\mathbb{T}^3)}^2. \tag{4.1.11}$$

Using (4.1.11) and applying the Sobolev embedding $H^1(\mathbb{T}^3) \subset L^6(\mathbb{T}^3)$, we infer that

$$\|u^3(\tau)\|_{H^{s-1}} \lesssim \|D^{s-1}u(\tau)\|_{L^6} \|u(\tau)\|_{L^6}^2 \lesssim \|D^s u(\tau)\|_{L^2} \|u(\tau)\|_{H^1}^2.$$

Therefore, we arrive at the bound

$$\|\Phi_{u_0,u_1}(u)\|_{X_T^s} \leq C\big(\|u_0\|_{H^s} + \|u_1\|_{H^{s-1}} + C_s T \sup_{\tau \in [0,T]} \|D^s u(\tau)\|_{L^2} \|u(\tau)\|_{H^1}^2\big).$$

By construction of the solution we infer that if $T \leq c_s \Lambda^{-2}$ with c_s small enough, we have that

$$\|u\|_{X_T^s} = \|\Phi_{u_0,u_1}(u)\|_{X_T^s} \leq C\big(\|u_0\|_{H^s} + \|u_1\|_{H^{s-1}}\big) + \frac{1}{2}\|u\|_{X_T^s}$$

which implies the propagation of the regularity statement for u. Strictly speaking, one should apply a bootstrap argument starting from the propagation of the regularity on times of order $(1 + \|u_0\|_{H^s} + \|u_1\|_{H^{s-1}})^{-2}$ and then extend the regularity propagation to the longer interval $[0, c_s \Lambda^{-2}]$. One estimates similarly $\partial_t u$ in H^{s-1} by differentiating the Duhamel formula with respect to t. The continuous dependence with respect to time in $H^s \times H^{s-1}$ follows once again from the Duhamel formula (4.1.8). This completes the proof of Proposition 4.1.1. $\qquad\square$

Theorem 4.1.2 (Global Well-Posedness) *For every* $(u_0, u_1) \in H^1(\mathbb{T}^3) \times L^2(\mathbb{T}^3)$ *the local solution of the cubic defocusing wave equation*

$$(\partial_t^2 - \Delta)u + u^3 = 0, \quad u(0, x) = u_0(x), \quad \partial_t u(0, x) = u_1(x)$$

can be extended globally in time. It is unique in the class $C(\mathbb{R}; H^1(\mathbb{T}^3) \times L^2(\mathbb{T}^3))$ *and there exists a constant C depending only on* $\|u_0\|_{H^1}$ *and* $\|u_1\|_{L^2}$ *such that for every* $t \in \mathbb{R}$,

$$\|u(t)\|_{H^1(\mathbb{R})} \leq C.$$

If in addition $(u_0, u_1) \in H^s(\mathbb{T}^3) \times H^{s-1}(\mathbb{T}^3)$ *for some* $s \geq 1$ *then*

$$(u, \partial_t u) \in C(\mathbb{R}; H^s(\mathbb{T}^3) \times H^{s-1}(\mathbb{T}^3)).$$

Remark 4.1.3 One may obtain global weak solutions of the cubic defocusing wave equation for data in $H^1 \times L^2$ via compactness arguments. The uniqueness and the propagation of regularity statements of Theorem 4.1.2 are the major differences with respect to the weak solutions.

Proof of Theorem 4.1.2 The key point is the conservation of the energy displayed in the following lemma.

Lemma 4.1.4 *There exist* $c > 0$ *and* $C > 0$ *such that for every* $(u_0, u_1) \in H^1(\mathbb{T}^3) \times L^2(\mathbb{T}^3)$ *the local solution of the cubic defocusing wave equation, with data* (u_0, u_1), *constructed in Proposition 4.1.1 is defined on* $[0, T]$ *with*

$$T = c(1 + \|u_0\|_{H^1(\mathbb{T}^3)} + \|u_1\|_{L^2(\mathbb{T}^3)})^{-2}$$

and

$$\int_{\mathbb{T}^3} \left((\partial_t u(t, x))^2 + |\nabla_x u(t, x)|^2 + \frac{1}{2} u^4(t, x) \right) dx$$

$$= \int_{\mathbb{T}^3} \left((u_1(x))^2 + |\nabla_x u_0(x)|^2 + \frac{1}{2} u_0^4(x) \right) dx, \quad t \in [0, T]. \quad (4.1.12)$$

As a consequence, for $t \in [0, T]$,

$$\|u(t)\|_{H^1(\mathbb{T}^3)} + \|\partial_t u(t)\|_{L^2(\mathbb{T}^3)} \leq C \left(1 + \|u_0\|_{H^1(\mathbb{T}^3)}^2 + \|u_1\|_{L^2(\mathbb{T}^3)} \right).$$

Remark 4.1.5 Using the invariance with respect to translations in time, we can state Lemma 4.1.4 with initial data at an arbitrary initial time.

Proof of Lemma 4.1.4 We apply Proposition 4.1.1 with $\Lambda = \|u_0\|_{H^1} + \|u_1\|_{L^2}$ and we take $T = c_{10}\Lambda^{-2}$, where c_{10} is the small constant involved in the propagation of the $H^{10} \times H^9$ regularity. Let $(u_{0,n}, u_{1,n})$ be a sequence in $H^{10} \times H^9$ which converges to (u_0, u_1) in $H^1 \times L^2$ and such that

$$\|u_{0,n}\|_{H^1} + \|u_{1,n}\|_{L^2} \leq \|u_0\|_{H^1} + \|u_1\|_{L^2}.$$

Let $u_n(t)$ be the solution of the cubic defocusing wave equation, with data $(u_{0,n}, u_{1,n})$. By Proposition 4.1.1 these solutions are defined on $[0, T]$ and they keep their $H^{10} \times H^9$ regularity *on the same time interval*. We multiply the equation

$$(\partial_t^2 - \Delta) u_n + u_n^3 = 0$$

by $\partial_t u_n$. Using the regularity properties of $u_n(t)$, after integrations by parts, we arrive at

$$\frac{d}{dt} \left[\int_{\mathbb{T}^3} \left((\partial_t u_n(t, x))^2 + |\nabla_x u_n(t, x)|^2 + \frac{1}{2} u_n^4(t, x) \right) dx \right] = 0$$

which implies the identity

$$\int_{\mathbb{T}^3} \left((\partial_t u_n(t, x))^2 + |\nabla_x u_n(t, x)|^2 + \frac{1}{2} u_n^4(t, x) \right) dx$$

$$= \int_{\mathbb{T}^3} \left((u_{1,n}(x))^2 + |\nabla_x u_{0,n}(x)|^2 + \frac{1}{2} u_{0,n}^4(x) \right) dx, \quad t \in [0, T]. \quad (4.1.13)$$

We now pass to the limit $n \longrightarrow \infty$ in (4.1.13). The right hand-side converges to

$$\int_{\mathbb{T}^3} \left((u_1(x))^2 + |\nabla_x u_0(x)|^2 + \frac{1}{2} u_0^4(x) \right) dx$$

by the definition of $(u_{0,n}, u_{1,n})$ (we invoke the Sobolev embedding for the convergence of the L^4 norms) . The right hand-side of (4.1.13) converges to

$$\int_{\mathbb{T}^3} \left((\partial_t u(t,x))^2 + |\nabla_x u(t,x)|^2 + \frac{1}{2} u^4(t,x)\right) dx$$

by the continuity of the flow map established in Proposition 4.1.1. Using the compactness of $\mathbb{T}^3$ and the Hölder inequality, we have that

$$\|u\|_{L^2(\mathbb{T}^3)} \leq C \|u\|_{L^4(\mathbb{T}^3)} \leq C(1 + \|u\|_{L^4(\mathbb{T}^3)}^2)$$

and therefore

$$\|u(t)\|_{H^1(\mathbb{T}^3)}^2 + \|\partial_t u(t)\|_{L^2(\mathbb{T}^3)}^2$$

is bounded by

$$C \int_{\mathbb{T}^3} \left(1 + (\partial_t u(t,x))^2 + |\nabla_x u(t,x)|^2 + \frac{1}{2} u^4(t,x)\right) dx \, .$$

Now, using (4.1.12) and the Sobolev inequality

$$\|u\|_{L^4(\mathbb{T}^3)} \leq C \|u\|_{H^1(\mathbb{T}^3)} \, ,$$

we obtain that for $t \in [0, T]$,

$$\|u(t)\|_{H^1(\mathbb{T}^3)}^2 + \|\partial_t u(t)\|_{L^2(\mathbb{T}^3)}^2 \leq C\left(1 + \|u_0\|_{H^1(\mathbb{T}^3)}^4 + \|u_1\|_{L^2(\mathbb{T}^3)}^2\right).$$

This completes the proof of Lemma 4.1.4.   □

Let us now complete the proof of Theorem 4.1.2. Let $(u_0, u_1) \in H^1(\mathbb{T}^3) \times L^2(\mathbb{T}^3)$. Set

$$T = c\left(C\left(1 + \|u_0\|_{H^1(\mathbb{T}^3)}^2 + \|u_1\|_{L^2(\mathbb{T}^3)}\right)\right)^{-2},$$

where the constants c and C are defined in Lemma 4.1.4. We now observe that we can use Proposition 4.1.1 and Lemma 4.1.4 on the intervals $[0, T]$, $[T, 2T]$, $[2T, 3T]$, and so on and therefore we extend the solution with data (u_0, u_1) on $[0, \infty)$. By the time reversibility of the wave equation we similarly can construct the solution for negative times. More precisely, the free evolution $S(t)(u_0, u_1)$ well-defined for all $t \in \mathbb{R}$ and one can prove in the same way the natural counterparts of Proposition 4.1.1 and Lemma 4.1.4 for negative times. The propagation of higher Sobolev regularity globally in time follows from Proposition 4.1.1 while the H^1 a priori bound on the solutions follows from Lemma 4.1.4. This completes the proof of Theorem 4.1.2.   □

Remark 4.1.6 One may proceed slightly differently in the proof of Theorem 4.1.2 by observing that as a consequence of Proposition 4.1.1, if a local solution with $H^1 \times L^2$ data blows-up at time $T^* < \infty$ then

$$\lim_{t \to T^*} \|(u(t), \partial_t u(t))\|_{H^1(\mathbb{T}^3) \times L^2(\mathbb{T}^3)} = \infty. \tag{4.1.14}$$

The statement (4.1.14) is in contradiction with the energy conservation law.

Remark 4.1.7 Observe that the nonlinear problem

$$(\partial_t^2 - \Delta)u + u^3 = 0 \tag{4.1.15}$$

behaves better than the linear problem

$$(\partial_t^2 - \Delta)u = 0 \tag{4.1.16}$$

with respect to the H^1 global in time bounds. Indeed, Theorem 4.1.2 establishes that the solutions of (4.1.15) are bounded in H^1 as far as the initial data is in $H^1 \times L^2$. On the other hand one can consider $u(t, x) = t$ which is a solution of the linear wave equation (4.1.16) on $\mathbb{T}^3$ with data in $H^1 \times L^2$ and its H^1 norm is clearly growing in time.

Remark 4.1.8 The sign in front of the nonlinearity is not of importance for Proposition 4.1.1. One can therefore obtain the local well-posedness of the cubic focusing wave equation

$$(\partial_t^2 - \Delta)u - u^3 = 0, \tag{4.1.17}$$

posed on $\mathbb{T}^3$, with data in $H^1(\mathbb{T}^3) \times L^2(\mathbb{T}^3)$. However, the sign in front of the nonlinearity is of crucial importance in the proof of Theorem 4.1.2. Indeed, one has that

$$u(t, x) = \frac{\sqrt{2}}{1 - t}$$

is a solution of (4.1.17), posed on $\mathbb{T}^3$ with data $(\sqrt{2}, -\sqrt{2})$ which is not defined globally in time (it blows-up in $H^1 \times L^2$ at $t = 1$).

4.1.3   The Strichartz Estimates

In the previous section, we solved globally in time the cubic defocusing wave equation in $H^1 \times L^2$. One may naturally ask whether it is possible to extend these results to the more singular Sobolev spaces $H^s \times H^{s-1}$ for some $s < 1$. It turns out that this is possible by invoking more refined properties of the map $S(t)$ defining the free evolution. The proof of these properties uses in an essential way the time oscillations in $S(t)$ and can be quantified as the L^p, $p > 2$ mapping properties of $S(t)$ (cf. [19, 30]).

Theorem 4.1.9 (Strichartz Inequality for the Wave Equation) *Let $(p, q) \in \mathbb{R}^2$ be such that $2 < p \leq \infty$ and $\frac{1}{p} + \frac{1}{q} = \frac{1}{2}$. Then we have the estimate*

$$\|S(t)(u_0, u_1)\|_{L^p([0,1]; L^q(\mathbb{T}^3))} \leq C\left(\|u_0\|_{H^{\frac{2}{p}}(\mathbb{T}^3)} + \|u_1\|_{H^{\frac{2}{p}-1}(\mathbb{T}^3)}\right).$$

We shall use that the solutions of the wave equation satisfy a finite propagation speed property which will allow us to deduce the result of Theorem 4.1.9 from the corresponding Strichartz estimate for the wave equation on the Euclidean space. Consider therefore the wave equation

$$(\partial_t^2 - \Delta)u = 0, \quad u(0, x) = u_0(x), \quad \partial_t u(0, x) = u_1(x), \tag{4.1.18}$$

where now the spatial variable x belongs to $\mathbb{R}^3$ and the initial data (u_0, u_1) belong to $H^s(\mathbb{R}^3) \times H^{s-1}(\mathbb{R}^3)$. Using the Fourier transform on $\mathbb{R}^3$, we can solve (4.1.18) and obtain that the solutions are generated by the map $S_e(t)$, defined as

$$S_e(t)(u_0, u_1) \equiv \cos(t\sqrt{-\Delta_{\mathbb{R}^3}})(u_0) + \frac{\sin(t\sqrt{-\Delta_{\mathbb{R}^3}})}{\sqrt{-\Delta_{\mathbb{R}^3}}}(u_1),$$

where for u_0 and u_1 in the Schwartz class,

$$\cos(t\sqrt{-\Delta_{\mathbb{R}^3}})(u_0) \equiv \int_{\mathbb{R}^3} \cos(t|\xi|)\widehat{u_0}(\xi)\, e^{i\xi \cdot x} d\xi$$

and

$$\frac{\sin(t\sqrt{-\Delta_{\mathbb{R}^3}})}{\sqrt{-\Delta_{\mathbb{R}^3}}}(u_1) \equiv \int_{\mathbb{R}^3} \frac{\sin(t|\xi|)}{|\xi|}\widehat{u_1}(\xi)\, e^{i\xi \cdot x} d\xi,$$

where $\widehat{u_0}$ and $\widehat{u_1}$ are the Fourier transforms of u_0 and u_1 respectively. By density, one then extends $S_e(t)(u_0, u_1)$ to a bounded map from $H^s(\mathbb{R}^3) \times H^{s-1}(\mathbb{R}^3)$ to $H^s(\mathbb{R}^3)$ for any $s \in \mathbb{R}$. The next lemma displays the finite propagation speed property of $S_e(t)$.

Proposition 4.1.10 (Finite Propagation Speed) *Let $(u_0, u_1) \in H^s(\mathbb{R}^3) \times H^{s-1}(\mathbb{R}^3)$ for some $s \geq 0$ be such that*

$$\mathrm{supp}(u_0) \cup \mathrm{supp}(u_1) \subset \{x \in \mathbb{R}^3 : |x - x_0| \leq R\},$$

for some $R > 0$ and $x_0 \in \mathbb{R}^3$. Then for $t \geq 0$,

$$\text{supp}(S_e(t)(u_0, u_1)) \subset \{x \in \mathbb{R}^3 : |x - x_0| \leq t + R\}.$$

Proof The statement of Proposition 4.1.10 (and even more precise localisation property) follows from the Kirchoff formula representation of the solutions of the three dimensional wave equation. Here we will present another proof which has the advantage to extend to an arbitrary dimension and to variable coefficient settings. By the invariance of the wave equation with respect to spatial translations, we can assume that $x_0 = 0$. We need to prove Proposition 4.1.10 only for (say) $s \geq 100$ which ensures by the Sobolev embedding that the solutions we study are of class $C^2(\mathbb{R}^4)$. We than can treat the case of an arbitrary $(u_0, u_1) \in H^s(\mathbb{R}^3) \times H^{s-1}(\mathbb{R}^3)$, $s \geq 0$ by observing that

$$\rho_\varepsilon \star S_e(t)(u_0, u_1) = S_e(t)(\rho_\varepsilon \star u_0, \rho_\varepsilon \star u_1), \tag{4.1.19}$$

where $\rho_\varepsilon(x) = \varepsilon^{-3} \rho(x/\varepsilon)$, $\rho \in C_0^\infty(\mathbb{R}^3)$, $0 \leq \rho \leq 1$, $\int \rho = 1$. It suffices then to pass to the limit $\varepsilon \to 0$ in (4.1.19). Indeed, for $\varphi \in C_0^\infty(|x| > t + R)$, $S_e(t)(\rho_\varepsilon \star u_0, \rho_\varepsilon \star u_1)(\varphi)$ is zero for ε small enough while $\rho_\varepsilon \star S_e(t)(u_0, u_1)(\varphi)$ converges to $S_e(t)(u_0, u_1)(\varphi)$.

Therefore, in the remaining of the proof of Proposition 4.1.10, we shall assume that $S_e(t)(u_0, u_1)$ is a C^2 solution of the $3d$ wave equation. The main point in the proof is the following lemma.

Lemma 4.1.11 *Let $x_0 \in \mathbb{R}^3$, $r > 0$ and let $S_e(t)(u_0, u_1)$ be a C^2 solution of the 3d linear wave equation. Suppose that $u_0(x) = u_1(x) = 0$ for $|x - x_0| \leq r$. Then $S_e(t)(u_0, u_1) = 0$ in the cone C defined by*

$$C = \{(t, x) \in \mathbb{R}^4 : 0 \leq t \leq r, \ |x - x_0| \leq r - t\}.$$

Proof Let $u(t, x) = S_e(t)(u_0, u_1)$. For $t \in [0, r]$, we set

$$E(t) \equiv \frac{1}{2} \int_{B(x_0, r-t)} \left((\partial_t u)^2(t, x) + |\nabla_x u(t, x)|^2\right) dx,$$

where $B(x_0, r - t) = \{x \in \mathbb{R}^3 : |x| \leq r - t\}$. Then using the Gauss-Green theorem and the equation solved by u, we obtain that

$$\dot{E}(t) = -\frac{1}{2} \int_{\partial B} \left((\partial_t u)^2(t, y) + |\nabla_x u(t, y)|^2 - 2\partial_t u(t, y)\nabla_x u(t, y) \cdot v(y)\right) dS(y),$$

where $\partial B \equiv \{x \in \mathbb{R}^3 : |x| = r - t\}$, $dS(y)$ is the volume element associated with ∂B and $v(y)$ is the outer unit normal to ∂B. We clearly have

$$2\partial_t u(t, y)\nabla_x u(t, y) \cdot v(y) \leq (\partial_t u)^2(t, y) + |\nabla_x u(t, y)|^2,$$

which implies that $\dot{E}(t) \leq 0$. Since $E(0) = 0$ we obtain that $E(t) = 0$ for every $t \in [0, r]$. This in turn implies that $u(t, x)$ is a constant in C. We also know that $u(0, x) = 0$ for $|x - x_0| \leq r$. Therefore $u(t, x) = 0$ in C. This completes the proof of Lemma 4.1.11. $\square$

Let us now complete the proof of Proposition 4.1.10. Let $t_0 \in \mathbb{R}$ and $y \in \mathbb{R}^3$ such that $|y| > R + t_0$. We need to show that $u(t_0, y) = 0$. Consider the cone C defined by

$$C = \{(t, x) \in \mathbb{R}^4 : 0 \leq t \leq t_0, \ |x - y| \leq t_0 - t\}.$$

Set $B \equiv C \cap \{(t, x) \in \mathbb{R}^4 : t = 0\}$. We have that

$$B = \{(t, x) \in \mathbb{R}^4 : t = 0, \ |x - y| \leq t_0\}$$

and therefore by the definition of t_0 and y we have that

$$B \cap \{(t, x) \in \mathbb{R}^4 : t = 0, \ |x| \leq R\} = \emptyset. \tag{4.1.20}$$

Therefore $u(0, x) = \partial_t u(0, x)$ for $|x - y| \leq t_0$. Using Lemma 4.1.11, we obtain that $u(t, x) = 0$ in C. In particular $u(t_0, y) = 0$. This completes the proof of Proposition 4.1.10. $\square$

Using Proposition 4.1.10 and a decomposition of the initial data associated with a partition of unity corresponding to a covering of $\mathbb{T}^3$ by sufficiently small balls, we obtain that the result of Theorem 4.1.9 is a consequence of the following statement.

Proposition 4.1.12 (Local in Time Strichartz Inequality for the Wave Equation on $\mathbb{R}^3$) *Let $(p, q) \in \mathbb{R}^2$ be such that $2 < p \leq \infty$ and $\frac{1}{p} + \frac{1}{q} = \frac{1}{2}$. Then we have the estimate*

$$\|S_e(t)(u_0, u_1)\|_{L^p([0,1]; L^q(\mathbb{R}^3))} \leq C\left(\|u_0\|_{H^{\frac{2}{p}}(\mathbb{R}^3)} + \|u_1\|_{H^{\frac{2}{p}-1}(\mathbb{R}^3)}\right).$$

Proof Let $\chi \in C_0^\infty(\mathbb{R}^3)$ be such that $\chi(x) = 1$ for $|x| < 1$. We then define the Fourier multiplier $\chi(D_x)$ by

$$\chi(D_x)(f) = \int_{\mathbb{R}^3} \chi(\xi) \widehat{f}(\xi) \, e^{i\xi \cdot x} d\xi. \tag{4.1.21}$$

Using a suitable Sobolev embedding in $\mathbb{R}^3$, we obtain that for every $\sigma \in \mathbb{R}$,

$$\left\|\frac{\sin(t\sqrt{-\Delta_{\mathbb{R}^3}})}{\sqrt{-\Delta_{\mathbb{R}^3}}}(\chi(D_x)u_1)\right\|_{L^p([0,1]; L^q(\mathbb{R}^3))} \leq C\|u_1\|_{H^\sigma(\mathbb{R}^3)}.$$

Therefore, by splitting u_1 as

$$u_1 = \chi(D_x)(u_1) + (1 - \chi(D_x))(u_1)$$

and by expressing the sin and cos functions as combinations of exponentials, we observe that Proposition 4.1.12 follows from the following statement.

Proposition 4.1.13 *Let $(p,q) \in \mathbb{R}^2$ be such that $2 < p \leq \infty$ and $\frac{1}{p} + \frac{1}{q} = \frac{1}{2}$. Then we have the estimate*

$$\left\| e^{\pm it\sqrt{-\Delta_{\mathbb{R}^3}}}(f) \right\|_{L^p([0,1];L^q(\mathbb{R}^3))} \leq C\|f\|_{H^{\frac{2}{p}}(\mathbb{R}^3)}.$$

Remark 4.1.14 Let us make an important remark. As a consequence of Proposition 4.1.13 and a suitable Sobolev embedding, we obtain the estimate

$$\left\| e^{\pm it\sqrt{-\Delta_{\mathbb{R}^3}}}(f) \right\|_{L^2([0,1];L^\infty(\mathbb{R}^3))} \leq C\|f\|_{H^s(\mathbb{R}^3)}, \quad s > 1. \tag{4.1.22}$$

Therefore, we obtain that for $f \in H^s(\mathbb{R}^3)$, $s > 1$, the function $e^{it\sqrt{-\Delta_{\mathbb{R}^3}}}(f)$ which is a priori defined as an element of $C([0,1]; H^s(\mathbb{R}^3))$ has the remarkable property that

$$e^{it\sqrt{-\Delta_{\mathbb{R}^3}}}(f) \in L^\infty(\mathbb{R}^3)$$

for almost every $t \in [0,1]$. Recall that the Sobolev embedding requires the condition $s > 3/2$ in order to ensure that an $H^s(\mathbb{R}^3)$ function is in $L^\infty(\mathbb{R}^3)$. Therefore, one may wish to see (4.1.22) as an almost sure in t improvement (with $1/2$ derivative) of the Sobolev embedding $H^{\frac{3}{2}+}(\mathbb{R}^3) \subset L^\infty(\mathbb{R}^3)$, under the evolution of the linear wave equation.

Proof of Proposition 4.1.13 Consider a Littlewood-Paley decomposition of the unity

$$\text{Id} = P_0 + \sum_N P_N, \tag{4.1.23}$$

where the summation is taken over the dyadic values of N, i.e. $N = 2^j$, $j = 0, 1, 2, \ldots$ and P_0, P_N are Littlewood-Paley projectors. More precisely they are defined as Fourier multipliers by $\Delta_0 = \psi_0(D_x)$ and for $N \geq 1$, $P_N = \psi(D_x/N)$, where $\psi_0 \in C_0^\infty(\mathbb{R}^3)$ and $\psi \in C_0^\infty(\mathbb{R}^3\backslash\{0\})$ are suitable functions such that (4.1.23) holds. The maps $\psi(D_x/N)$ are defined similarly to (4.1.21) by

$$\psi(D_x/N)(f) = \int_{\mathbb{R}^3} \psi(\xi/N)\widehat{f}(\xi)\, e^{i\xi \cdot x}d\xi.$$

Set

$$u(t, x) \equiv e^{\pm it \sqrt{-\Delta_{\mathbb{R}^3}}}(f).$$

Our goal is to evaluate $\|u\|_{L^p([0,1]L^q(\mathbb{R}^3))}$. Thanks to the Littlewood-Paley square function theorem, we have that

$$\|u\|_{L^q(\mathbb{R}^3)} \approx \left\| \left(|P_0 u|^2 + \sum_N |P_N u|^2 \right)^{\frac{1}{2}} \right\|_{L^q(\mathbb{R}^3)}. \tag{4.1.24}$$

The proof of (4.1.24) can be obtained as a combination of the Mikhlin-Hörmander multiplier theorem and the Khinchin inequality for Bernouli variables.[1] Using the Minkowski inequality, since $p \geq 2$ and $q \geq 2$, we can write

$$\|u\|_{L_t^p L_x^q} \lesssim \|P_0 u\|_{L_t^p L_x^q} + \|P_N u\|_{L_t^p L_x^q l_N^2} \leq \|P_0 u\|_{L_t^p L_x^q} + \|P_N u\|_{l_N^2 L_t^p L_x^q} \tag{4.1.25}$$

Therefore, it suffices to prove that for every $\psi \in C_0^\infty(\mathbb{R}^3 \backslash \{0\})$ there exists $C > 0$ such that for every N and every $f \in L^2(\mathbb{R}^3)$,

$$\|\psi(D_x/N) e^{\pm it \sqrt{-\Delta_{\mathbb{R}^3}}}(f)\|_{L^p([0,1];L^q(\mathbb{R}^3))} \leq C N^{\frac{2}{p}} \|f\|_{L^2(\mathbb{R}^3)}. \tag{4.1.26}$$

Indeed, suppose that (4.1.26) holds true. Then, we define $\tilde{P}_N$ as $\tilde{P}_N = \tilde{\psi}(D_x/N)$, where $\tilde{\psi} \in C_0^\infty(\mathbb{R}^3 \backslash \{0\})$ is such that $\tilde{\psi} \equiv 1$ on the support of ψ. Then $P_N = \tilde{P}_N P_N$. Now, coming back to (4.1.25), using the Sobolev inequality to evaluate $\|P_0 u\|_{L_t^p L_x^q}$ and (4.1.26) to evaluate $\|P_N u\|_{l_N^2 L_t^p L_x^q}$, we arrive at the bound

$$\|u\|_{L_t^p L_x^q} \lesssim \|f\|_{L^2} + \|N^{\frac{2}{p}} \|P_N f\|_{L_x^2} \|_{l_N^2} \lesssim \|f\|_{H^{\frac{2}{p}}}.$$

Therefore, it remains to prove (4.1.26). Set

$$T \equiv \psi(D_x/N) e^{\pm it \sqrt{-\Delta_{\mathbb{R}^3}}}.$$

Our goal is to study the mapping properties of T from L_x^2 to $L_t^p L_x^q$. We can write

$$\|Tf\|_{L_t^p L_x^q} = \sup_{\|G\|_{L_t^{p'} L_x^{q'}} \leq 1} \left| \int_{t,x} Tf \overline{G} \right|, \tag{4.1.27}$$

---

[1] Interestingly, variants of the Khinchin inequality will be essentially used in our probabilistic approach to the cubic defocusing wave equation with data of super-critical regularity.

where $\frac{1}{p} + \frac{1}{p'} = \frac{1}{q} + \frac{1}{q'} = 1$. Note that in order to write (4.1.27) the values 1 and ∞ of p and q are allowed. Next, we can write

$$\int_{t,x} Tf\overline{G} = \int_x f\overline{T^\star G}, \tag{4.1.28}$$

where $T^\star$ is defined by

$$T^\star G \equiv \int_0^1 \psi(D_x/N)e^{\mp i\tau\sqrt{-\Delta_{\mathbb{R}^3}}}G(\tau)d\tau.$$

Indeed, we have

$$\int_{t,x} Tf\overline{G} = \int_0^1 \int_{\mathbb{R}^3} \psi(D_x/N)e^{\pm it\sqrt{-\Delta_{\mathbb{R}^3}}}f\,\overline{G(t)}dxdt$$

$$= \int_0^1 \int_{\mathbb{R}^3} f\,\overline{\psi(D_x/N)e^{\mp it\sqrt{-\Delta_{\mathbb{R}^3}}}G(t)}dxdt$$

$$= \int_{\mathbb{R}^3} f\int_0^1 \overline{\psi(D_x/N)e^{\mp it\sqrt{-\Delta_{\mathbb{R}^3}}}G(t)}dt\,dx.$$

Therefore (4.1.28) follows. But thanks to the Cauchy-Schwarz inequality we can write

$$\left|\int_x f\overline{T^\star G}\right| \leq \|f\|_{L_x^2}\|T^\star G\|_{L_x^2}.$$

Therefore, in order to prove (4.1.26), it suffices to prove the bound

$$\|T^\star G\|_{L_x^2} \lesssim N^{\frac{2}{p}}\|G\|_{L_t^{p'}L_x^{q'}}.$$

Next, we can write

$$\|T^\star G\|_{L_x^2}^2 = \int_x T^\star G\,\overline{T^\star G}$$

$$= \int_{t,x} T(T^\star(G))\overline{G}$$

$$\leq \|T(T^\star(G))\|_{L_t^p L_x^q}\|G\|_{L_t^{p'}L_x^{q'}}.$$

Therefore, estimate (4.1.26) would follow from the estimate

$$\|T(T^\star(G))\|_{L_t^p L_x^q} \lesssim N^{\frac{4}{p}}\|G\|_{L_t^{p'}L_x^{q'}}. \tag{4.1.29}$$

An advantage of (4.1.29) with respect to (4.1.26) is that we have the same number of variables in both sides of the estimates. Coming back to the definition of T and $T^\star$, we can write

$$T(T^\star(G)) = \int_0^1 \psi^2(D_x/N)e^{\pm i(t-\tau)\sqrt{-\Delta_{\mathbb{R}^3}}}G(\tau)d\tau .$$

Now by using the triangle inequality, for a fixed $t \in [0, 1]$, we can write

$$\|T(T^\star(G))\|_{L_x^q} \leq \int_0^1 \left\|\psi^2(D_x/N)e^{\pm i(t-\tau)\sqrt{-\Delta_{\mathbb{R}^3}}}G(\tau)\right\|_{L_x^q}d\tau. \qquad (4.1.30)$$

On the other hand, using the Fourier transform, we can write

$$\psi^2(D_x/N)e^{\pm it\sqrt{-\Delta_{\mathbb{R}^3}}}(f) = \int_{\mathbb{R}^3} \psi^2(\xi/N)e^{\pm it|\xi|}e^{ix\cdot\xi}\hat{f}(\xi)d\xi .$$

Therefore,

$$\psi^2(D_x/N)e^{\pm it\sqrt{-\Delta_{\mathbb{R}^3}}}(f) = \int_{\mathbb{R}^3} K(t, x - x')f(x')dx,$$

where

$$K(t, x - x') = \int_{\mathbb{R}^3} \psi^2(\xi/N)e^{\pm it|\xi|}e^{i(x-x')\cdot\xi}d\xi .$$

A simple change of variable leads to

$$K(t, x - x') = N^3 \int_{\mathbb{R}^3} \psi^2(\xi)e^{\pm itN|\xi|}e^{iN(x-x')\cdot\xi}d\xi .$$

In order to estimate $K(t, x - x')$, we invoke the following proposition.

Proposition 4.1.15 (Soft Stationary Phase Estimate) *Let $d \geq 1$. For every $\Lambda > 0$, $N \geq 1$ there exists $C > 0$ such that for every $\lambda \geq 1$, every $a \in C_0^\infty(\mathbb{R}^d)$, satisfying*

$$\sup_{|\alpha|\leq 2N} \sup_{x\in\mathbb{R}^d} |\partial^\alpha a(x)| \leq \Lambda,$$

every $\varphi \in C^\infty(\text{supp}(a))$ satisfying

$$\sup_{2\leq|\alpha|\leq 2N+2} \sup_{x\in\text{supp}(a)} |\partial^\alpha \varphi(x)| \leq \Lambda \qquad (4.1.31)$$

one has the bound

$$\left| \int_{\mathbb{R}^d} e^{i\lambda\varphi(x)} a(x)\, dx \right| \le C \int_{\mathrm{supp}(a)} \frac{dx}{(1 + \lambda |\nabla\varphi(x)|^2)^N}. \qquad (4.1.32)$$

Remark 4.1.16 Observe that in (4.1.31), we do not require upper bounds for the first derivatives of φ.

We will give the proof of Proposition 4.1.15 later. Let us first show how to use it in order to complete the proof of (4.1.26). We claim that

$$|K(t, x - x')| \lesssim N^3 (tN)^{-1} = N^2 t^{-1}. \qquad (4.1.33)$$

Estimate (4.1.33) trivially follows from the expression defining $K(t, x - x')$ for $|tN| \le 1$ (one simply ignores the oscillation term). For $|Nt| \ge 1$, using Proposition 4.1.15 (with $a = \psi^2$, $N = 2$ and $d = 3$), we get the bound

$$|K(t, x - x')| \lesssim N^3 \int_{\mathrm{supp}(\psi)} \frac{d\xi}{(1 + |tN||\nabla\varphi(\xi)|^2)^2},$$

where

$$\varphi(\xi) = \pm|\xi| + \frac{N(x - x') \cdot \xi}{t}.$$

Observe that φ is C^∞ on the support of ψ and moreover it satisfies the assumptions of Proposition 4.1.15. We next observe that

$$\int_{\mathrm{supp}(\psi)} \frac{d\xi}{(1 + |tN||\nabla\varphi(\xi)|^2)^2} \lesssim (tN)^{-1}. \qquad (4.1.34)$$

Indeed, since $\nabla\varphi(\xi) = \pm\frac{\xi}{|\xi|} + t^{-1}N(x - x')$ we obtain that one can split the support of integration in regions such that there are two different $j_1, j_2 \in \{1, 2, 3\}$ such that one can perform the change of variable

$$\eta_{j_1} = \partial_{\xi_{j_1}} \varphi(\xi), \quad \eta_{j_2} = \partial_{\xi_{j_2}} \varphi(\xi),$$

with a non-degenerate Hessian. More precisely, we have

$$\det \begin{pmatrix} \partial^2_{\xi_1} \varphi(\xi) & \partial^2_{\xi_1, \xi_2} \varphi(\xi) \\ \partial^2_{\xi_1, \xi_2} \varphi(\xi) & \partial^2_{\xi_2} \varphi(\xi) \end{pmatrix} = \frac{\xi_3^2}{|\xi|^4}$$

which is not degenerate for $\xi_3 \ne 0$. Therefore for $\xi_3 \ne 0$, we can choose $j_1 = 1$ and $j_2 = 2$. Similarly, $\xi_1 \ne 0$, we can choose $j_1 = 2$ and $j_2 = 3$ and for $\xi_2 \ne 0$, we can choose $j_1 = 1$ and $j_2 = 3$. Therefore, using that the support of ψ does not meet zero, after splitting the support of the integration in three regions, by choosing the two "good" variables and by neglecting the integration with respect to the remaining variable, we obtain that

$$\int_{\text{supp}(\psi)} \frac{d\xi}{(1+|tN||\nabla\varphi(\xi)|^2)^2} \lesssim \int_{\mathbb{R}^2} \frac{d\eta_{j_1} d\eta_{j_2}}{(1+|tN|(|\eta_{j_1}|^2+|\eta_{j_2}|^2)^2} \lesssim (tN)^{-1}.$$

Thus, we have (4.1.34) which in turn implies (4.1.33).

Thanks to (4.1.33), we arrive at the estimate

$$\left\| \psi^2(D_x/N)e^{\pm i(t-\tau)\sqrt{-\Delta_{\mathbb{R}^3}}} G(\tau) \right\|_{L_x^\infty} \lesssim N^2 |t-\tau|^{-1} \|G(\tau)\|_{L_x^1}.$$

On the other hand, we also have the trivial bound

$$\left\| \psi^2(D_x/N)e^{\pm i(t-\tau)\sqrt{-\Delta_{\mathbb{R}^3}}} G(\tau) \right\|_{L_x^2} \lesssim \|G(\tau)\|_{L_x^2}.$$

Therefore using the basic Riesz-Torin interpolation theorem, we arrive at the bound

$$\left\| \psi^2(D_x/N)e^{\pm i(t-\tau)\sqrt{-\Delta_{\mathbb{R}^3}}} G(\tau) \right\|_{L_x^q} \lesssim \frac{N^{\frac{4}{p}}}{|t-\tau|^{\frac{2}{p}}} \|G(\tau)\|_{L_x^{q'}}.$$

Therefore coming back to (4.1.30), we get

$$\|T(T^\star(G))\|_{L_x^q} \lesssim \int_0^1 \frac{N^{\frac{4}{p}}}{|t-\tau|^{\frac{2}{p}}} \|G(\tau)\|_{L_x^{q'}} d\tau.$$

Therefore, the estimate (4.1.29) would follow from the one dimensional estimate

$$\left\| \int_{\mathbb{R}} \frac{f(\tau)}{|t-\tau|^{\frac{2}{p}}} d\tau \right\|_{L^p(\mathbb{R})} \lesssim \|f\|_{L^{p'}(\mathbb{R})}. \tag{4.1.35}$$

Thanks to our assumption, one has $\frac{2}{p} < 1$ and also

$$1 + \frac{1}{p} = \frac{1}{p'} + \frac{2}{p}.$$

Therefore estimate (4.1.35) is precisely the Hardy-Littlewood-Sobolev inequality (cf. [29]). This completes the proof of (4.1.26), once we provide the proof of Proposition 4.1.15.

Proof of Proposition 4.1.15 We follow [17]. Consider the first order differential operator defined by

$$L \equiv \frac{1}{i(1+\lambda|\nabla\varphi|^2)} \sum_{j=1}^{d} \partial_j\varphi\partial_j + \frac{1}{1+\lambda|\nabla\varphi|^2}.$$

which satisfies $L(e^{i\lambda\varphi}) = e^{i\lambda\varphi}$. We have that

$$\int_{\mathbb{R}^d} e^{i\lambda\varphi(x)}a(x)dx = \int_{\mathbb{R}^d} L(e^{i\lambda\varphi(x)})a(x)dx = \int_{\mathbb{R}^d} e^{i\lambda\varphi(x)}\tilde{L}(a(x))dx,$$

where $\tilde{L}$ is defined by

$$\tilde{L}(u) = -\sum_{j=1}^{d} \frac{\partial_j\varphi}{i(1+\lambda|\nabla\varphi|^2)}\partial_j u$$

$$+ \left(-\sum_{j=1}^{d} \frac{\partial_j^2\varphi}{i(1+\lambda|\nabla\varphi|^2)} + \sum_{j=1}^{d} \frac{2\lambda\partial_j\varphi\,(\nabla\varphi\cdot\nabla\partial_j\varphi)}{i(1+\lambda|\nabla\varphi|^2)^2}\right)u + \frac{1}{1+\lambda|\nabla\varphi|^2}u.$$

As a consequence, we get the bound

$$\left|\int_{\mathbb{R}^d} e^{i\lambda\varphi(x)}a(x)dx\right| \leq \int_{\mathbb{R}^d} |\tilde{L}^N a|, \qquad (4.1.36)$$

where $N \in \mathbb{N}$. To conclude, we need to estimate the coefficients of $\tilde{L}$. We shall use the notation $\langle u\rangle = (1+|u|^2)^{\frac{1}{2}}$ and we set $\lambda = \mu^2$. At first, we consider

$$F(x) = Q(\mu^2|\nabla\varphi(x)|^2), \qquad Q(u) = \frac{1}{1+u}, \ u \geq 0.$$

We clearly have

$$F \lesssim \langle\mu\nabla\varphi\rangle^{-2} \qquad (4.1.37)$$

and we shall estimate the derivatives of F. Set

$$\Lambda^k(x) = \sup_{2\leq|\alpha|\leq k} |\partial^\alpha\varphi(x)|.$$

We have the following statement.

Lemma 4.1.17 *For $|\alpha| = k \geq 1$, we have the bound*

$$|\partial^\alpha F(x)| \lesssim C(\Lambda^{k+1}(x))\left(\frac{1}{\langle\mu\nabla\varphi(x)\rangle^2} + \frac{\mu^k}{\langle\mu\nabla\varphi(x)\rangle^{k+2}}\right), \qquad (4.1.38)$$

where $C : \mathbb{R}^+ \to \mathbb{R}^+$ is a suitable continuous increasing function (which can change from line to line and can always be taken of the form $C(t) = (1+t)^M$ for a sufficiently large M).

Proof Using an induction on k, we get that $\partial^\alpha F$ for $|\alpha| = k \geq 1$ is a linear combination of terms under the form

$$T_q = Q^{(m)}(\mu^2|\nabla\varphi|^2)\left(\partial^{\gamma_1}(\mu^2|\nabla\varphi|^2)\right)^{q_1} \cdots \left(\partial^{\gamma_k}(\mu^2|\nabla\varphi|^2)\right)^{q_k}$$

where

$$q_1 + \cdots + q_k = m \quad \text{and} \quad \sum |\gamma_i|q_i = k, \quad q_i \geq 0. \tag{4.1.39}$$

Since $|Q^{(m)}(u)| \lesssim \langle u \rangle^{-m-1}$, we get

$$|T_q| \lesssim \frac{1}{\langle \mu\nabla\varphi \rangle^2} \left(\frac{\mu}{\langle \mu\nabla\varphi \rangle}\right)^{2m} \left|\left(\partial^{\gamma_1}(|\nabla\varphi|^2)\right)^{q_1} \cdots \left(\partial^{\gamma_k}(|\nabla\varphi|^2)\right)^{q_k}\right|.$$

Moreover, by the Leibnitz formula

$$\partial^{\gamma_i}(|\nabla\varphi|^2) \leq \begin{cases} C(\Lambda^2)|\nabla\varphi|, & \text{if } |\gamma_i| = 1, \\ C(\Lambda^{|\gamma_i|+1})(|\nabla\varphi| + 1), & \text{if } |\gamma_i| > 1. \end{cases}$$

We therefore have the following bound for T_q

$$|T_q| \lesssim C(\Lambda^{k+1}) \frac{1}{\langle \mu\nabla\varphi \rangle^2} \left(\frac{\mu}{\langle \mu\nabla\varphi \rangle}\right)^{2m} \left(|\nabla\varphi|^m + |\nabla\varphi|^{\sum_{|\gamma_i|=1} q_i}\right)$$

$$\lesssim C(\Lambda^{k+1}) \frac{1}{\langle \mu\nabla\varphi \rangle^2} \left[\left(\frac{\mu}{\langle \mu\nabla\varphi \rangle}\right)^m + \left(\frac{\mu}{\langle \mu\nabla\varphi \rangle}\right)^{m+\sum_{|\gamma_i|>1} q_i}\right].$$

Next, by using (4.1.39), we note that

$$m + \sum_{|\gamma_i|>1} q_i = \sum_{|\gamma_i|>1} 2q_i + \sum_{|\gamma_i|=1} q_i \leq \sum |\gamma_i|q_i = k.$$

Therefore, we get

$$|T_q| \lesssim C(\Lambda^{k+1})\left(\frac{1}{\langle \mu\nabla\varphi \rangle^2} + \frac{\mu^k}{\langle \mu\nabla\varphi \rangle^{k+2}}\right).$$

This completes the proof of Lemma 4.1.17. $\qquad\qquad\qquad\qquad\qquad\qquad\square$

We are now in position to prove the following statement.

Lemma 4.1.18 *For $N \in \mathbb{N}$, we can write $\tilde{L}^N$ under the form*

$$\tilde{L}^N u = \sum_{|\alpha| \leq N} a_\alpha^{(N)} \partial^\alpha u \tag{4.1.40}$$

with the estimates

$$|a_\alpha^{(N)}(x)| \lesssim C(\Lambda^{N+2}(x)) \frac{1}{\langle \mu \nabla \varphi(x) \rangle^N} \tag{4.1.41}$$

and more generally for $|\beta| = k$,

$$|\partial^\beta a_\alpha^{(N)}(x)| \lesssim C(\Lambda^{N+k+2}(x)) \left(\frac{1}{\langle \mu \nabla \varphi(x) \rangle^N} + \frac{\mu^k}{\langle \mu \nabla \varphi(x) \rangle^{N+k}} \right). \tag{4.1.42}$$

Proof We reason by induction on N. First, we notice that $\tilde{L}$ is under the form

$$\tilde{L} = \sum_{j=1}^d a_j \partial_j + b,$$

where

$$a_j = i \partial_j \varphi F, \quad b = F + i \sum_{j=1}^d \partial_j (\partial_j \varphi F) = F + \sum_{j=1}^d \partial_j a_j.$$

Consequently, by using (4.1.37), we get that

$$|a_j| \lesssim \frac{1}{\mu} \frac{1}{\langle \mu \nabla \varphi \rangle} \tag{4.1.43}$$

and by the Leibnitz formula, since $\partial^\alpha a_j$ for $|\alpha| \geq 1$ is a linear combination of terms under the form

$$(\partial^\beta \partial_j \varphi) \partial^\gamma F, \quad |\beta| + |\gamma| = |\alpha|,$$

we get by using (4.1.38) that for $|\alpha| = k \geq 1$,

$$|\partial^\alpha a_j| \lesssim C(\Lambda^{k+1}) \left(\frac{1}{\langle \mu \nabla \varphi \rangle} + \frac{\mu^{k-1}}{\langle \mu \nabla \varphi \rangle^{k+1}} \right). \tag{4.1.44}$$

Consequently, we also find thanks to (4.1.44), (4.1.38) that for $|\alpha| = k \geq 0$,

$$|\partial^\alpha b| \lesssim C(\Lambda^{k+2}) \left(\frac{1}{\langle \mu \nabla \varphi \rangle} + \frac{\mu^k}{\langle \mu \nabla \varphi \rangle^{k+2}} \right). \tag{4.1.45}$$

Using (4.1.44) (4.1.45), we obtain that the assertion of the lemma holds true for $N = 1$. Next, let us assume that it is true at the order N. We have

$$(\tilde{L})^{N+1} u = \sum_{j=1}^{d} \sum_{|\alpha| \le N} \left(a_j a_\alpha^{(N)} \partial_j \partial^\alpha u + a_j \partial_j a_\alpha^{(N)} \partial^\alpha u \right) + \sum_{|\alpha| \le N} b a_\alpha^{(N)} \partial^\alpha u.$$

Consequently, we get that the coefficients are under the form

$$a_\alpha^{(N+1)} = a_j a_\beta^{(N)}, \quad |\alpha| = N + 1, \ |\beta| = N,$$

$$a_\alpha^{(N+1)} = a_j \partial_j a_\beta^{(N)} + a_j a_\gamma^{(N)} + b a_\delta^{(N)}, \quad |\beta| = |\delta| = |\alpha|, \ |\gamma| = |\alpha| - 1.$$

Therefore, by using (4.1.43) and (4.1.42), we get that (4.1.41) is true for $N + 1$. In order to prove (4.1.42) for $N + 1$, we need to evaluate $\partial^\gamma a_\alpha^{(N+1)}$. The estimate of the contribution of all terms except $\partial^\gamma (a_j \partial_j a_\beta^{(N)})$ follows directly from the induction hypothesis. In order to estimate $\partial^\gamma (a_j \partial_j a_\beta^{(N)})$, we need to invoke (4.1.43) and (4.1.44) and the induction hypothesis. This completes the proof of Lemma 4.1.18.   □

Finally, thanks to (4.1.36) and Lemma 4.1.18, we get

$$\left| \int_{\mathbb{R}^d} e^{i\lambda\varphi(x)} a(x)\, dx \right| \lesssim K \int_{\text{supp}(a)} \frac{dx}{(1 + \lambda |\nabla\varphi|^2)^{\frac{N}{2}}}\, dx,$$

where

$$K \equiv \left(\sup_{x \in \text{supp}(a)} \Lambda^{N+2}(x) \right) \left(\sup_{x \in \mathbb{R}^d} \sup_{|\alpha| \le N} |\partial^\alpha a(x)| \right).$$

This completes the proof of Proposition 4.1.15.   □
    This completes the proof of Proposition 4.1.13.   □
    This completes the proof of Proposition 4.1.12.   □

Remark 4.1.19 If in the proof of the Strichartz estimates, we use the triangle inequality instead of the square function theorem and the Young inequality instead of the Hardy-Littlewood-Sobolev inequality, we would obtain slightly less precise estimates. These estimates are sufficient to get all sub-critical well-posedness results. However in the case of initial data with critical Sobolev regularity the finer arguments using the square function and the Hardy-Littlewood-Sobolev inequality are essentially needed.

4.1.4 Local Well-Posedness in $H^s \times H^{s-1}$, $s \geq 1/2$

In this section, we shall use the Strichartz estimates in order to improve the well-posedness result of Proposition 4.1.1. We shall be able to consider initial data in the more singular Sobolev spaces $H^s \times H^{s-1}$, $s \geq 1/2$. We start by a definition.

Definition 4.1.20 For $0 \leq s < 1$, a couple of real numbers (p, q), $\frac{2}{s} \leq p \leq +\infty$ is s-admissible if

$$\frac{1}{p} + \frac{3}{q} = \frac{3}{2} - s.$$

For $T > 0, 0 \leq s < 1$, we define the spaces

$$X_T^s = C([0, T]; H^s(\mathbb{T}^3)) \bigcap_{(p,q) \ s\text{-admissible}} L^p((0, T); L^q(\mathbb{T}^3)) \tag{4.1.46}$$

and its "dual space"

$$Y_T^s = \bigcup_{(p,q) \ s\text{-admissible}} L^{p'}((0, T); L^{q'}(\mathbb{T}^3)) \tag{4.1.47}$$

(p', q') being the conjugate couple of (p, q), equipped with their natural norms (notice that to define these spaces, we can keep only the extremal couples corresponding to $p = 2/s$ and $p = +\infty$ respectively).

We can now state the non homogeneous Strichartz estimates for the three dimensional wave equation on the torus $\mathbb{T}^3$.

Theorem 4.1.21 For every $0 < s < 1$, every s-admissible couple (p, q), there exists $C > 0$ such that for every $T \in]0, 1]$, every $F \in Y_T^{1-s}$, every $(u_0, u_1) \in H^s(\mathbb{T}^3) \times H^{s-1}(\mathbb{T}^3)$ one has

$$\|S(t)(u_0, u_1)\|_{X_T^s} \leq C(\|u_0\|_{H^s(\mathbb{T}^3)} + \|u_1\|_{H^{s-1}(\mathbb{T}^3)}) \tag{4.1.48}$$

and

$$\left\| \int_0^t \frac{\sin((t - \tau)\sqrt{-\Delta})}{\sqrt{-\Delta}} (F(\tau))d\tau \right\|_{X_T^s} \leq C\|F\|_{Y_T^{1-s}} \tag{4.1.49}$$

Proof Thanks to the Hölder inequality, in order to prove (4.1.48), it suffices the consider the two end point cases for p, i.e. $p = 2/s$ and $p = \infty$ (the estimate in $C([0, T]; H^s(\mathbb{T}^3))$ is straightforward). The case $p = 2/s$ follows from Theorem 4.1.9. The case $p = \infty$ results from the Sobolev embedding. This ends the proof of (4.1.48).

Let us next turn to (4.1.49). We first observe that

$$\left\| \int_0^t \frac{\sin((t-\tau)\sqrt{-\Delta})}{\sqrt{-\Delta}}(F(\tau))d\tau \right\|_{C([0,T];H^s(\mathbb{T}^3))} \le C\|F\|_{Y_T^{1-s}} \qquad (4.1.50)$$

follows by duality from (4.1.48). Thanks to (4.1.50), we obtain that it suffices to show

$$\left\| \int_0^t \frac{\sin((t-\tau)\sqrt{-\Delta})}{\sqrt{-\Delta}}(F(\tau))d\tau \right\|_{L_T^{p_1}L^{q_1}} \le C\|F\|_{L_T^{p_2'}L^{q_2'}}, \qquad (4.1.51)$$

where (p_1, q_1) is s-admissible and (p_2', q_2') are such that (p_2, q_2) are $(1-s)$-admissible and where for shortness we set

$$L_T^p L^q \equiv L^p((0,T); L^q(\mathbb{T}^3)).$$

Denote by Π_0 the projector on the zero Fourier mode on $\mathbb{T}^3$, i.e.

$$\Pi_0(f) = (2\pi)^{-3} \int_{\mathbb{T}^3} f(x)dx.$$

We have the bound

$$\left\| \int_0^t \frac{\sin((t-\tau)\sqrt{-\Delta})}{\sqrt{-\Delta}}(\Pi_0 F(\tau))d\tau \right\|_{L_T^p L^q} \le C\|F\|_{L^1((0,T);L^1(\mathbb{T}^3))}.$$

By the Hölder inequality

$$\|F\|_{L^1((0,T);L^1(\mathbb{T}^3))} \le C\|F\|_{L_T^{p_2'}L^{q_2'}}$$

and therefore, it suffices to show the bound

$$\left\| \int_0^t \frac{\sin((t-\tau)\sqrt{-\Delta})}{\sqrt{-\Delta}}(\Pi_0^\perp F(\tau))d\tau \right\|_{L_T^{p_1}L^{q_1}} \le C\|F\|_{L_T^{p_2'}L^{q_2'}}, \qquad (4.1.52)$$

where

$$\Pi_0^\perp \equiv 1 - \Pi_0.$$

By writing the sin function as a sum of exponentials, we obtain that (4.1.52) follows from

$$\left\| \int_0^t e^{\pm i(t-\tau)\sqrt{-\Delta}}((-\Delta)^{-\frac{1}{2}}\Pi_0^\perp F(\tau))d\tau \right\|_{L_T^{p_1}L^{q_1}} \le C\|F\|_{L_T^{p_2'}L^{q_2'}}. \qquad (4.1.53)$$

Observe that $(-\Delta)^{-\frac{1}{2}}\Pi_0^{\perp}$ is well defined as a bounded operator from $H^s(\mathbb{T}^3)$ to $H^{s+1}(\mathbb{T}^3)$. Set

$$K \equiv e^{\pm it\sqrt{-\Delta}}\Pi_0^{\perp}.$$

Thanks to (4.1.48), by writing

$$e^{\pm it\sqrt{-\Delta}}\Pi_0^{\perp} = \cos(t\sqrt{-\Delta})\Pi_0^{\perp} \pm i\sin(t\sqrt{-\Delta})(-\Delta)^{-\frac{1}{2}}\Pi_0^{\perp}(-\Delta)^{\frac{1}{2}},$$

we see that the map K is bounded from $H^s(\mathbb{T}^3)$ to X_T^s. Consequently, the dual map K^*, defined by

$$K^*(F) = \int_0^T e^{\mp it\sqrt{-\Delta}}\Pi_0^{\perp}(F(\tau))d\tau$$

is bounded from Y^s to $H^{-s}(\mathbb{T}^3)$. Using the last property with s replaced by $1-s$ (which remains in $]0, 1[$ if $s \in]0, 1[$), we obtain the following sequence of continuous mappings

$$L_T^{p_2'}L^{q_2'} \xrightarrow{K^*} H^{s-1}(\mathbb{T}^3) \xrightarrow{(-\Delta)^{-\frac{1}{2}}\Pi_0^{\perp}} H^s(\mathbb{T}^3) \xrightarrow{K} L_T^{p_1}L^{q_1}. \tag{4.1.54}$$

On the other hand, we have

$$\left(K \circ ((-\Delta)^{-\frac{1}{2}}\Pi_0^{\perp}) \circ K^*\right)(F) = \int_0^T e^{\pm i(t-\tau)\sqrt{-\Delta}}((-\Delta)^{-\frac{1}{2}}\Pi_0^{\perp} F(\tau))d\tau$$

Therefore, we obtain the bound

$$\left\| \int_0^T e^{\pm i(t-\tau)\sqrt{-\Delta}}((-\Delta)^{-\frac{1}{2}}\Pi_0^{\perp} F(\tau))d\tau \right\|_{L_T^{p_1}L^{q_1}} \leq C\|F\|_{L_T^{p_2'}L^{q_2'}}. \tag{4.1.55}$$

The passage from (4.1.55) to (4.1.53) can be done by using the Christ-Kiselev [11] argument, as we explain below. By a density argument it suffices to prove (4.1.53) for $F \in C^{\infty}([0, T] \times \mathbb{T}^3)$. We can of course also assume that

$$\|F\|_{L_T^{p_2'}L^{q_2'}} = 1.$$

For $n \geq 1$ an integer and $m = 0, 1, \cdots, 2^n$, we define $t_{n,m}$ as

$$\int_0^{t_{n,m}} \|F(\tau)\|_{L^{q_2'}(\mathbb{T}^3)}^{p_2'} d\tau = m2^{-n}.$$

Of course $0 = t_{n,0} \le t_{n,1} \le \cdots \le t_{n,2^n} = T$. Next, we observe that for $0 \le \alpha < \beta \le 1$ there is a unique n such that $\alpha \in [2m2^{-n}, (2m+1)2^{-n})$ and $\beta \in [(2m+1)2^{-n}, (2m+2)2^{-n})$ for some $m \in \{0, 1, \cdots, 2^{n-1} - 1\}$. Indeed, this can be checked by writing the representations of α and β in base 2 (the number n corresponds to the first different digit of α and β). Therefore, if we denote by $\chi_{\tau < t}(\tau, t)$ the characteristic function of the set $\{(\tau, t) : 0 \le \tau < t \le T\}$ then we can write

$$\chi_{\tau < t}(\tau, t) = \sum_{n=1}^{\infty} \sum_{m=0}^{2^{n-1}-1} \chi_{n,2m}(\tau)\chi_{n,2m+1}(t), \qquad (4.1.56)$$

where $\chi_{n,m}$ $(m = 0, 1, \cdots, 2^n)$ denotes the characteristic function of the interval $[t_{n,m}, t_{n,(m+1)})$. Indeed, in order to achieve (4.1.56), it suffices to apply the previous observation for every : $0 \le \tau < t \le T$ with α and β defined as

$$\alpha = \int_0^\tau \|F(s)\|_{L_2^{q'_2}(\mathbb{T}^3)}^{p'_2} \, ds, \qquad \beta = \int_0^t \|F(s)\|_{L_2^{q'_2}(\mathbb{T}^3)}^{p'_2} \, ds \, .$$

Therefore, thanks to (4.1.56), we can write

$$\int_0^t e^{\pm i(t-\tau)\sqrt{-\Delta}}((-\Delta)^{-\frac{1}{2}}\Pi_0^{\perp} F(\tau))d\tau$$

as

$$\sum_{n=1}^{\infty} \sum_{m=0}^{2^{n-1}-1} \chi_{n,2m+1}(t) \int_0^T e^{\pm i(t-\tau)\sqrt{-\Delta}}((-\Delta)^{-\frac{1}{2}}\Pi_0^{\perp} \chi_{n,2m}(\tau)F(\tau))d\tau \, .$$

The goal is to evaluate the $L_T^{p_1}L^{q_1}$ norm of the last expression. Using that for a fixed n, $\chi_{n,2m+1}(t)$ have disjoint supports, we obtain that the $L_T^{p_1}L^{q_1}$ norm of the last expression can be estimated by

$$\sum_{n=1}^{\infty} \left(\sum_{m=0}^{2^{n-1}-1} \left\| \int_0^T e^{\pm i(t-\tau)\sqrt{-\Delta}}((-\Delta)^{-\frac{1}{2}}\Pi_0^{\perp} \chi_{n,2m}(\tau)F(\tau))d\tau \right\|_{L_T^{p_1}L^{q_1}}^{p_1} \right)^{\frac{1}{p_1}} \, .$$

Now, using (4.1.55), we obtain that the last expression is bounded by

$$c \sum_{n=1}^{\infty} \left(\sum_{m=0}^{2^{n-1}-1} \left\| \chi_{n,2m}(\tau)F(\tau) \right\|_{L_T^{p'_2}L^{q'_2}}^{p_1} \right)^{\frac{1}{p_1}} \, . \qquad (4.1.57)$$

By definition

$$\left\| \chi_{n,2m}(\tau) F(\tau) \right\|_{L_T^{p_2'} L^{q_2'}}^{p_2'} = 2^{-n}$$

and therefore (4.1.57) equals to

$$c \sum_{n=1}^{\infty} \Big(\sum_{m=0}^{2^{n-1}-1} 2^{-\frac{np_1}{p_2'}} \Big)^{\frac{1}{p_1}} \leq c \sum_{n=1}^{\infty} 2^{n(\frac{1}{p_1} - \frac{1}{p_2'})}.$$

The last series is convergent since by the definition of admissible pairs it follows that $p_2' < 2 < p_1$. Therefore we proved that (4.1.55) indeed implies (4.1.53). This completes the proof of Theorem 4.1.21.   □

We can now use Theorem 4.1.21 in order to get the following improvement of Proposition 4.1.1.

Theorem 4.1.22 (Low Regularity Local Well-Posedness) *Let $s > 1/2$. Consider the cubic defocusing wave equation*

$$(\partial_t^2 - \Delta)u + u^3 = 0, \tag{4.1.58}$$

posed on $\mathbb{T}^3$. There exist positive constants γ, c and C such that for every $\Lambda \geq 1$, every

$$(u_0, u_1) \in H^s(\mathbb{T}^3) \times H^{s-1}(\mathbb{T}^3)$$

satisfying

$$\|u_0\|_{H^s} + \|u_1\|_{H^{s-1}} \leq \Lambda \tag{4.1.59}$$

there exists a unique solution of (4.1.58) on the time interval $[0, T]$, $T \equiv c\Lambda^{-\gamma}$ with initial data

$$u(0, x) = u_0(x), \quad \partial_t u(0, x) = u_1(x).$$

Moreover the solution satisfies

$$\|(u, \partial_t u)\|_{L^\infty([0,T], H^s(\mathbb{T}^3) \times H^{s-1}(\mathbb{T}^3))} \leq C\Lambda,$$

u is unique in the class X_T^s described in Definition 4.1.20 and the dependence with respect to the initial data and with respect to the time is continuous. More precisely, *if u and $\tilde{u}$ are two solutions of* (4.1.58) *with initial data satisfying* (4.1.59) *then*

$$\|(u - \tilde{u}, \partial_t u - \partial_t \tilde{u})\|_{L^\infty([0,T], H^s(\mathbb{T}^3) \times H^{s-1}(\mathbb{T}^3))}$$

$$\leq C\big(\|u(0) - \tilde{u}(0)\|_{H^s(\mathbb{T}^3)} + \|\partial_t u(0) - \partial_t \tilde{u}(0)\|_{H^{s-1}(\mathbb{T}^3)}\big). \qquad (4.1.60)$$

Finally, if

$$(u_0, u_1) \in H^\sigma(\mathbb{T}^3) \times H^{\sigma-1}(\mathbb{T}^3)$$

for some $\sigma \geq s$ then there exists $c_\sigma > 0$ such that

$$(u, \partial_t u) \in C([0, c_\sigma \Lambda^{-\gamma}]; H^\sigma(\mathbb{T}^3) \times H^{\sigma-1}(\mathbb{T}^3)).$$

Proof We shall suppose that $s \in (1/2, 1)$, the case $s \geq 1$ being already treated in Proposition 4.1.1. As in the proof of Proposition 4.1.1, we solve the integral equation

$$u(t) = S(t)(u_0, u_1) - \int_0^t \frac{\sin((t - \tau)\sqrt{-\Delta})}{\sqrt{-\Delta}}((u^3(\tau))d\tau$$

by a fixed point argument. Recall that

$$\Phi_{u_0, u_1}(u) = S(t)(u_0, u_1) - \int_0^t \frac{\sin((t - \tau)\sqrt{-\Delta})}{\sqrt{-\Delta}}((u^3(\tau))d\tau.$$

We shall estimate $\Phi_{u_0, u_1}(u)$ in the spaces X_T^s introduced in Definition 4.1.20. Thanks to Theorem 4.1.21

$$\|S(t)(u_0, u_1)\|_{X_T^s} \leq C(\|u_0\|_{H^s(\mathbb{T}^3)} + \|u_1\|_{H^{s-1}(\mathbb{T}^3)}).$$

Another use of Theorem 4.1.21 gives

$$\left\| \int_0^t \frac{\sin((t - \tau)\sqrt{-\Delta})}{\sqrt{-\Delta}}((u^3(\tau))d\tau \right\|_{X_T^s} \leq C\|u^3\|_{L_T^{\frac{2}{1+s}} L^{\frac{2}{2-s}}} = C\|u\|^3_{L_T^{\frac{6}{1+s}} L^{\frac{6}{2-s}}}.$$

Observe that the couple $(\frac{2}{1+s}, \frac{2}{2-s})$ is the dual of $(\frac{2}{1-s}, \frac{2}{s})$ which is the end point $(1 - s)$-admissible couple. We also observe that if (p, q) is an s-admissible couple then $\frac{1}{q}$ ranges in the interval $[\frac{1}{2} - \frac{s}{2}, \frac{1}{2} - \frac{s}{3}]$. The assumption $s \in (1/2, 1)$ implies

$$\frac{1}{2} - \frac{s}{2} < \frac{2 - s}{6} < \frac{1}{2} - \frac{s}{3}.$$

Therefore $q^\star \equiv \frac{6}{2-s}$ is such that there exists $p^\star$ such that $(p^\star, q^\star)$ is an s-admissible couple. By definition $p^\star$ is such that

$$\frac{1}{p^\star} + \frac{3}{q^\star} = \frac{3}{2} - s.$$

The last relation implies that

$$\frac{1}{p^\star} = \frac{1}{2} - \frac{s}{2}.$$

Now, using the Hölder inequality in time, we obtain

$$\|u\|_{L_T^{\frac{6}{1+s}} L^{\frac{6}{2-s}}} \leq T^{\frac{2s-1}{3}} \|u\|_{L_T^{p^\star} L^{q^\star}}$$

which in turn implies

$$\left\| \int_0^t \frac{\sin((t-\tau)\sqrt{-\Delta})}{\sqrt{-\Delta}} ((u^3(\tau))d\tau \right\|_{X_T^s} \leq C T^{2s-1} \|u\|_{X_T^s}^3.$$

Consequently

$$\|\Phi_{u_0,u_1}(u)\|_{X_T^s} \leq C(\|u_0\|_{H^s(\mathbb{T}^3)} + \|u_1\|_{H^{s-1}(\mathbb{T}^3)}) + C T^{2s-1} \|u\|_{X_T^s}^3.$$

A similar argument yields

$$\|\Phi_{u_0,u_1}(u) - \Phi_{u_0,u_1}(v)\|_{X_T^s} \leq C T^{2s-1} (\|u\|_{X_T^s}^2 + \|v\|_{X_T^s}^2) \|u - v\|_{X_T^s}. \qquad (4.1.61)$$

Now, one obtains the existence and the uniqueness statements as in the proof of Proposition 4.1.1. Estimate (4.1.60) follows from (4.1.61) and a similar estimate obtained after differentiation of the Duhamel formula with respect to t. The propagation of regularity statement can be obtained as in the proof of Proposition 4.1.1. This completes the proof of Theorem 4.1.22. □

Concerning the uniqueness statement, we also have the following corollary which results from the proof of Theorem 4.1.22.

Corollary 4.1.23 *Let $s > 1/2$. Let $(p^\star, q^\star)$ be the s-admissible couple defined by*

$$p^\star = \frac{2}{1-s}, \qquad q^\star \equiv \frac{6}{2-s}.$$

Then the solutions constructed in Theorem 4.1.22 is unique in the class

$$L^{p^\star}([0, T]; L^{q^\star}(\mathbb{T}^3)).$$

Remark 4.1.24 As a consequence of Theorem 4.1.22, we have that for each $(u_0, u_1) \in H^s(\mathbb{T}^3) \times H^{s-1}(\mathbb{T}^3)$ there is a solution with a maximum time existence T^* and if $T^* < \infty$ than necessarily

$$\lim_{t \to T^*} \|(u(t), \partial_t u(t))\|_{H^s(\mathbb{T}^3) \times H^{s-1}(\mathbb{T}^3)} = \infty. \tag{4.1.62}$$

One can also prove a suitable local well-posedness in the case $s = 1/2$ but in this case the dependence of the existence time on the initial data is more involved. Here is a precise statement.

Theorem 4.1.25 *Consider the cubic defocusing wave equation*

$$(\partial_t^2 - \Delta)u + u^3 = 0, \tag{4.1.63}$$

posed on $\mathbb{T}^3$. *For every*

$$(u_0, u_1) \in H^{\frac{1}{2}}(\mathbb{T}^3) \times H^{-\frac{1}{2}}(\mathbb{T}^3)$$

there exists a time $T > 0$ *and a unique solution of* (4.1.63) *in*

$$L^4([0, T] \times \mathbb{T}^3) \times C([0, T]; H^{\frac{1}{2}}(\mathbb{T}^3)),$$

with initial data

$$u(0, x) = u_0(x), \quad \partial_t u(0, x) = u_1(x).$$

Proof For $T > 0$, using the Strichartz estimates of Theorem 4.1.21, we get

$$\|\Phi_{u_0, u_1}(u)\|_{L^4([0,T] \times \mathbb{T}^3)} \le \|S(t)(u_0, u_1)\|_{L^4([0,T] \times \mathbb{T}^3)} + C\|u^3\|_{L^{4/3}([0,T] \times \mathbb{T}^3)}$$

$$= \|S(t)(u_0, u_1)\|_{L^4([0,T] \times \mathbb{T}^3)} + C\|u\|_{L^4([0,T] \times \mathbb{T}^3)}^3.$$

Similarly, we get

$$\|\Phi_{u_0, u_1}(u) - \Phi_{u_0, u_1}(v)\|_{L^4([0,T] \times \mathbb{T}^3)}$$

$$\le C\big(\|u\|_{L^4([0,T] \times \mathbb{T}^3)}^2 + \|v\|_{L^4([0,T] \times \mathbb{T}^3)}^2\big)\|u - v\|_{L^4([0,T] \times \mathbb{T}^3)}.$$

Therefore if T is small enough then we can construct the solution by a fixed point argument in $L^4([0, T] \times \mathbb{T}^3)$. In addition, the Strichartz estimates of Theorem 4.1.21 yield that the obtained solution belongs to $C([0, T]; H^{\frac{1}{2}}(\mathbb{T}^3))$. This completes the proof of Theorem 4.1.25. $\square$

Remark 4.1.26 Observe that for data in $H^{\frac{1}{2}}(\mathbb{T}^3) \times H^{-\frac{1}{2}}(\mathbb{T}^3)$ we no longer have the small factor T^κ, $\kappa > 0$ in the estimates for Φ_{u_0, u_1}. This makes that the dependence of the existence time T on the data (u_0, u_1) is much less explicit. In particular, we can no longer conclude that the existence time is the same for a fixed ball in $H^{\frac{1}{2}}(\mathbb{T}^3) \times H^{-\frac{1}{2}}(\mathbb{T}^3)$ and therefore we do not have the blow-up criterium (4.1.62) (with $s = 1/2$).

4.1.5 A Constructive Way of Seeing the Solutions

In the proof of Theorem 4.1.22, we used the contraction mapping principle in order to construct the solutions. Therefore, one can define the solutions in a constructive way via the Picard iteration scheme. More precisely, for $(u_0, u_1) \in H^s(\mathbb{T}^3) \times H^{s-1}(\mathbb{T}^3)$, we define the sequence $(u^{(n)})_{n \geq 0}$ as $u^{(0)} = 0$ and for a given $u^{(n)}, n \geq 0$, we define $u^{(n+1)}$ as the solutions of the linear wave equation

$$(\partial_t^2 - \Delta)u^{(n+1)} + (u^{(n)})^3 = 0, \quad u(0) = u_0, \ \partial_t u(0) = u_1.$$

Thanks to (the proof of) Theorem 4.1.22 the sequence $(u^{(n)})_{n \geq 0}$ is converging in X_T^s, and in particular in $C([0, T]; H^s(\mathbb{T}^3))$ for

$$T \approx (\|u(0)\|_{H^s(\mathbb{T}^3)} + \|\partial_t u(0)\|_{H^{s-1}(\mathbb{T}^3)})^{-\gamma}, \quad \gamma > 0.$$

One has that

$$u^{(1)} = S(t)(u_0, u_1)$$

and for $n \geq 1$,

$$u^{(n+1)} = u^{(1)} + \mathcal{T}(u^{(n)}, u^{(n)}, u^{(n)}),$$

where the trilinear map $\mathcal{T}$ is defined as

$$\mathcal{T}(u, v, w) = -\int_0^t \frac{\sin((t - \tau)\sqrt{-\Delta})}{\sqrt{-\Delta}}((u(\tau)v(\tau)w(\tau))d\tau.$$

One then may compute

$$u^{(2)} = u^{(1)} + \mathcal{T}(u^{(1)}, u^{(1)}, u^{(1)}).$$

The expression for $u^{(3)}$ is then

$$u^{(3)} = u^{(1)} + \mathcal{T}(u^{(1)}, u^{(1)}, u^{(1)}) + 3\mathcal{T}(u^{(1)}, u^{(1)}, \mathcal{T}(u^{(1)}, u^{(1)}, u^{(1)}))$$
$$+ 3\mathcal{T}(u^{(1)}, \mathcal{T}(u^{(1)}, u^{(1)}, u^{(1)}), \mathcal{T}(u^{(1)}, u^{(1)}, u^{(1)}))$$
$$+ \mathcal{T}(\mathcal{T}(u^{(1)}, u^{(1)}, u^{(1)}), \mathcal{T}(u^{(1)}, u^{(1)}, u^{(1)}), \mathcal{T}(u^{(1)}, u^{(1)}, u^{(1)})).$$

We now observe that for $n \geq 2$, the nth Picard iteration $u^{(n)}$ is a sum from $j = 1$ to $j = 3^{n-1}$ of j-linear expressions of $u^{(1)}$. Moreover the first 3^{n-2} terms of this sum contain the $(n-1)$th iteration. Therefore the solution can be seen as an infinite sum of multi-linear expressions of $u^{(1)}$. The Strichartz inequalities we proved can be used to show that for $s \geq 1/2$,

$$\|\mathcal{T}(u, v, w)\|_{H^s(\mathbb{T}^3)} \lesssim \|u\|_{H^s(\mathbb{T}^3)} \|v\|_{H^s(\mathbb{T}^3)} \|w\|_{H^s(\mathbb{T}^3)}.$$

The last estimate can be used to analyse the multi-linear expressions in the expansion and to show its convergence. Observe that, we do not exploit any regularising effect in the terms of the expansion. The ill-posedness result of the next section, will basically show that such an effect is in fact not possible. In our probabilistic approach in the next section, we will exploit that the trilinear term in the expression defining the solution is more regular in the scale of the Sobolev spaces than the linear one, almost surely with respect to a probability measure on H^s, $s < 1/2$.

4.1.6 Global Well-Posedness in $H^s \times H^{s-1}$, for Some $s < 1$

One may naturally ask whether the solutions obtained in Theorem 4.1.22 can be extended globally in time. Observe that one cannot use the argument of Theorem 4.1.2 because there is no a priori bound available at the H^s, $s \neq 1$ regularity. One however has the following partial answer.

Theorem 4.1.27 (Low Regularity Global Well-Posedness) *Let $s > 13/18$. Then the local solution obtained in Theorem 4.1.22 can be extended globally in time.*

For the proof of Theorem 4.1.27, we refer to [18, 27, 40]. Here, we only present the main idea (introduced in [14]). Let $(u_0, u_1) \in H^s(\mathbb{T}^3) \times H^{s-1}(\mathbb{T}^3)$ for some $s \in (1/2, 1)$. Let $T \gg 1$. For $N \geq 1$, we define a smooth Fourier multiplier acting as 1 for frequencies $n \in \mathbb{Z}^3$ such that $|n| \leq N$ and acting as $N^{1-s}|n|^{s-1}$ for frequencies $|n| \geq 2N$. A concrete choice of I_N is $I_N(D) = I((-\Delta)^{1/2}/N)$, where $I(x)$ is a smooth function which equals 1 for $x \leq 1$ and which equals x^{s-1} for $x \geq 2$. In other words $I(x)$ is one for x close to zero and decays like x^{s-1} for $x \gg 1$. We choose $N = N(T)$ such that for the times of the local existence the modified energy (which is well-defined in $H^s \times H^{s-1}$)

$$\int_{\mathbb{T}^3} \left((\partial_t I_N u)^2 + (\nabla I_N u)^2 + \frac{1}{2}(I_N u)^4 \right)$$

does not vary much. This allows to extend the local solutions up to time $T \gg 1$. The analysis contains two steps, a local existence argument for $I_N u$ under the assumption that the modified energy remains below a fixed size and an energy increase estimate which is the substitute of the energy conservation used in the proof of Theorem 4.1.2. More precisely, we choose N as $N = T^\gamma$ for some $\gamma = \gamma(s) \geq 1$. With this choice of N the initial size of the modified energy is $T^{\gamma(1-s)}$. The local well-posedness argument assures that $I_N u$ (and thus u as well) exists on time of size $T^{-\beta}$ for some $\beta > 0$ as far as the modified energy remains $\lesssim T^{\gamma(1-s)}$. The main part of the analysis is to get an energy increase estimate showing that on the local existence time the modified energy does not increase more then $T^{-\alpha}$ for some $\alpha > 0$. In order to arrive at time T we need to iterate $\approx T^{1+\beta}$ times the local existence argument. In order to ensure that at each step of the iteration the modified energy remains $\lesssim T^{\gamma(1-s)}$, we need to impose the condition

$$T^{1+\beta} T^{-\alpha} \lesssim T^{\gamma(1-s)}, \quad T \gg 1. \tag{4.1.64}$$

As far as (4.1.64) is satisfied, we can extend the local solutions globally in time. The condition (4.1.64) imposes the lower bound on s involved in the statement of Theorem 4.1.27. One may conjecture that the global well-posedness in Theorem 4.1.27 holds for any $s > 1/2$.

4.1.7 Local Ill-Posedness in $H^s \times H^{s-1}$, $s \in (0, 1/2)$

It turns out that the restriction $s > 1/2$ in Theorem 4.1.22 is optimal. Recall that the classical notion of well-posedness in the Hadamard sense requires the existence, the uniqueness and the continuous dependence with respect to the initial data. A very classical example of contradicting the continuous dependence with respect to the initial data for a PDE is the initial value problem for the Laplace equation with data in Sobolev spaces. Indeed, consider

$$(\partial_t^2 + \partial_x^2)v = 0, \quad v : \mathbb{R}_t \times \mathbb{T}_x \longrightarrow \mathbb{R}. \tag{4.1.65}$$

Equation (4.1.65) has the explicit solution

$$v_n(t, x) = e^{-\sqrt{n}}\text{sh}(nt) \cos(nx).$$

Then for every $(s_1, s_2) \in \mathbb{R}^2$, v_n satisfies

$$\|(v_n(0), \partial_t v_n(0))\|_{H^{s_1}(\mathbb{T}) \times H^{s_2}(\mathbb{T})} \lesssim e^{-\sqrt{n}} n^{\max(s_1, s_2+1)} \longrightarrow 0,$$

as n tends to $+\infty$ but for $t \neq 0$,

$$\|(v_n(t), \partial_t v_n(t))\|_{H^{s_1}(\mathbb{T}) \times H^{s_2}(\mathbb{T})} \gtrsim e^{n|t|} e^{-\sqrt{n}} n^{\min(s_1, s_2+1)} \longrightarrow +\infty,$$

as n tends to $+\infty$. Consequently (4.1.65) in not well-posed in $H^{s_1}(\mathbb{T}) \times H^{s_2}(\mathbb{T})$ for every $(s_1, s_2) \in \mathbb{R}^2$ because of the lack of continuous dependence with respect to the initial data $(0, 0)$.

It turns out that a similar phenomenon happens in the context of the cubic defocusing wave equation with low regularity initial data. As we shall see below the mechanism giving the lack of continuous dependence is however quite different compared to (4.1.65). Here is the precise statement.

Theorem 4.1.28 *Let us fix $s \in (0, 1/2)$ and $(u_0, u_1) \in C^\infty(\mathbb{T}^3) \times C^\infty(\mathbb{T}^3)$. Then there exist $\delta > 0$, a sequence $(t_n)_{n=1}^\infty$ of positive numbers tending to zero and a sequence $(u_n(t, x))_{n=1}^\infty$ of $C(\mathbb{R}; C^\infty(\mathbb{T}^3))$ functions such that*

$$(\partial_t^2 - \Delta)u_n + u_n^3 = 0$$

with

$$\|(u_n(0) - u_0, \partial_t u_n(0) - u_1)\|_{H^s(\mathbb{T}^3) \times H^{s-1}(\mathbb{T}^3)} \leq C[\log(n)]^{-\delta} \to_{n \to +\infty} 0$$

but

$$\|(u_n(t_n), \partial_t u_n(t_n))\|_{H^s(\mathbb{T}^3) \times H^{s-1}(\mathbb{T}^3)} \geq C[\log(n)]^\delta \to_{n \to +\infty} +\infty.$$

In particular, for every $T > 0$,

$$\lim_{n \to +\infty} \|(u_n(t), \partial_t u_n(t))\|_{L^\infty([0,T]; H^s(\mathbb{T}^3) \times H^{s-1}(\mathbb{T}^3))} = +\infty.$$

Proof of Theorem 4.1.28 We follow [6, 12, 48]. Consider

$$(\partial_t^2 - \Delta)u + u^3 = 0 \tag{4.1.66}$$

subject to initial conditions

$$(u_0(x) + \kappa_n n^{\frac{3}{2}-s} \varphi(nx), u_1(x)), \qquad n \gg 1, \tag{4.1.67}$$

where φ is a nontrivial bump function on $\mathbb{R}^3$ and

$$\kappa_n \equiv [\log(n)]^{-\delta_1},$$

with $\delta_1 > 0$ to be fixed later. Observe that for $n \gg 1$, we can see $\varphi(nx)$ as a C^∞ function on $\mathbb{T}^3$.

Thanks to Theorem 4.1.2, we obtain that (4.1.66) with data given by (4.1.67) has a unique global smooth solution which we denote by u_n. Moreover $u_n \in C(\mathbb{R}; C^\infty(\mathbb{T}^3))$ thanks the propagation of the higher Sobolev regularity and the Sobolev embeddings.

Next, we consider the ODE

$$V'' + V^3 = 0, \quad V(0) = 1, \ V'(0) = 0. \tag{4.1.68}$$

Lemma 4.1.29 *The Cauchy problem* (4.1.68) *has a global smooth (non constant) solution* $V(t)$ *which is periodic.*

Proof One defines locally in time the solution of (4.1.68) by an application of the Cauchy-Lipschitz theorem. In order to extend the solutions globally in time, we multiply (4.1.68) by V'. This gives that the solutions of (4.1.68) satisfy

$$\frac{d}{dt}\left((V'(t))^2 + \frac{1}{2}(V(t))^4\right) = 0$$

and therefore taking into account the initial conditions, we get

$$(V'(t))^2 + \frac{1}{2}(V(t))^4 = \frac{1}{2}. \tag{4.1.69}$$

The relation (4.1.69) implies that $(V(t), V'(t))$ cannot go to infinity in finite time. Therefore the local solution of (4.1.68) is defined globally in time. Let us finally show that $V(t)$ is periodic in time. We first observe that thanks to (4.1.69), $|V(t)| \leq 1$ for all times t. Therefore $t = 0$ is a local maximum of $V(t)$. We claim that there is $t_0 > 0$ such that $V'(t_0) = 0$. Indeed, otherwise $V(t)$ is decreasing on $[0, +\infty)$ which implies that $V'(t) \leq 0$ and from (4.1.69), we deduce

$$V'(t) = -\sqrt{\frac{(1 - (V(t)))^4}{2}}.$$

Integrating the last relation between zero and a positive t_0 gives

$$t_0 = \sqrt{2} \int_{V(t_0)}^{1} \frac{dv}{\sqrt{1 - v^4}}.$$

Therefore

$$t_0 \leq \sqrt{2} \int_{-1}^{1} \frac{dv}{\sqrt{1 - v^4}}.$$

and we get a contradiction for $t_0 \gg 1$. Hence, we indeed have that there is $t_0 > 0$ such that $V'(t_0) = 0$. Coming back to (4.1.69) and using that $V(t_0) < 1$, we deduce that $V(t_0) = -1$. Therefore $t = t_0$ is a local minimum of $V(t)$. We now can show exactly as before that there exists $t_1 > t_0$ such that $V'(t_1) = 0$ and $V(t_1) > -1$. Once again using (4.1.69), we infer that $V(t_1) = 1$, i.e. $V(0) = V(t_1)$ and $V'(0) = V'(t_1)$. By the uniqueness part of the Cauchy-Lipschitz theorem, we obtain that V is periodic with period t_1. This completes the proof of Lemma 4.1.29.   □

We next denote by v_n the solution of

$$\partial_t^2 v_n + v_n^3 = 0, \quad (v_n(0), \partial_t v_n(0)) = (\kappa_n n^{\frac{3}{2}-s} \varphi(nx), 0). \tag{4.1.70}$$

It is now clear that

$$v_n(t, x) = \kappa_n n^{\frac{3}{2}-s} \varphi(nx) V\left(t \kappa_n n^{\frac{3}{2}-s} \varphi(nx)\right).$$

In the next lemma, we collect the needed bounds on v_n.

Lemma 4.1.30 *Let*

$$t_n \equiv [\log(n)]^{\delta_2} n^{-(\frac{3}{2}-s)}$$

for some $\delta_2 > \delta_1$. Then, we have the following bounds for $t \in [0, t_n]$,

$$\|\Delta(v_n)(t, \cdot)\|_{H^1(\mathbb{T}^3)} \leq C[\log(n)]^{3\delta_2} n^{3-s}, \tag{4.1.71}$$

$$\|\Delta(v_n)(t, \cdot)\|_{L^2(\mathbb{T}^3)} \leq C[\log(n)]^{2\delta_2} n^{2-s}, \tag{4.1.72}$$

$$\|\nabla^k v_n(t, \cdot)\|_{L^\infty(\mathbb{T}^3)} \leq C[\log(n)]^{k\delta_2} n^{\frac{3}{2}-s+k}, \ k = 0, 1, \cdots. \tag{4.1.73}$$

Finally, there exists $n_0 \gg 1$ such that for $n \geq n_0$,

$$\|v_n(t_n, \cdot)\|_{H^s(\mathbb{T}^3)} \geq C\kappa_n (t_n \kappa_n n^{\frac{3}{2}-s})^s = C[\log(n)]^{-(s+1)\delta_1 + s\delta_2}. \tag{4.1.74}$$

Proof Estimates (4.1.71) and (4.1.72) follow from the general bound

$$\|v_n(t, \cdot)\|_{H^\sigma(\mathbb{T}^3)} \leq C\kappa_n (t_n \kappa_n n^{\frac{3}{2}-s})^\sigma n^{\sigma-s}, \tag{4.1.75}$$

where $t \in [0, t_n]$ and $\sigma \geq 0$. For integer values of σ, the bound (4.1.75) is a direct consequence of the definition of v_n. For fractional values of σ one needs to invoke an elementary interpolation inequality in the Sobolev spaces. Estimate (4.1.73) follows directly from the definition of v_n. The proof of (4.1.74) is slightly more delicate. We first observe that for $n \gg 1$, we have the lower bound

$$\|v_n(t_n, \cdot)\|_{H^1(\mathbb{T}^3)} \geq c\kappa_n (t_n \kappa_n n^{\frac{3}{2}-s}) n^{1-s}. \tag{4.1.76}$$

Now, we can obtain (4.1.74) by invoking (4.1.75) (with $\sigma = 2$), the lower bound (4.1.76) and the interpolation inequality

$$\|v_n(t_n, \cdot)\|_{H^1(\mathbb{T}^3)} \leq \|v_n(t_n, \cdot)\|_{H^s(\mathbb{T}^3)}^{\theta} \|v_n(t_n, \cdot)\|_{H^2(\mathbb{T}^3)}^{1-\theta}$$

for some $\theta > 0$. It remains therefore to show (4.1.76). After differentiating once the expression defining v_n, we see that (4.1.76) follows from the following statement.

Lemma 4.1.31 *Consider a smooth not identically zero periodic function V and a non trivial bump function $\phi \in C_0^\infty(\mathbb{R}^d)$. Then there exist $c > 0$ and $\lambda_0 \geq 1$ such that for every $\lambda > \lambda_0$*

$$\|\phi(x)V(\lambda\phi(x))\|_{L^2(\mathbb{R}^d)} \geq c.$$

Proof We can suppose that the period of V is $2\pi L$ for some $L > 0$. Consider the Fourier expansion of V,

$$V(t) = \sum_{n \in \mathbb{Z}} v_n e^{i\frac{n}{L}t}, \quad |v_n| \leq C_N(1 + |n|)^{-N}.$$

We can assume that there is an open ball B of $\mathbb{R}^d$ such that for some $c_0 > 0$, $|\partial_{x_1}\phi(x)| \geq c_0$ on B. Let $0 \leq \psi \leq 1$ be a non trivial $C_0^\infty(B)$ function. We can write

$$\|\phi(x)V(\lambda\phi(x))\|_{L^2(\mathbb{R}^d)}^2 \geq \|\psi(x)\phi(x)V(\lambda\phi(x))\|_{L^2(B)}^2 = I_1 + I_2,$$

where

$$I_1 = \sum_{n \in \mathbb{Z}} |v_n|^2 \int_B (\psi(x)\phi(x))^2 dx,$$

and

$$I_2 = \sum_{n_1 \neq n_2} v_{n_1} \overline{v_{n_2}} \int_B e^{i\lambda\frac{n_1-n_2}{L}\phi(x)} (\psi(x)\phi(x))^2 dx.$$

Clearly $I_1 > 0$ is independent of λ. On the other hand

$$e^{i\lambda\frac{n_1-n_2}{L}\phi(x)} = \frac{L}{i\lambda(n_1 - n_2)\partial_{x_1}\phi(x)} \partial_{x_1}\left(e^{i\lambda\frac{n_1-n_2}{L}\phi(x)}\right).$$

Therefore, after an integration by parts, we obtain that $|I_2| \lesssim \lambda^{-1}$. This completes the proof of Lemma 4.1.31. $\qquad\square$

This completes the proof of Lemma 4.1.30. $\qquad\square$

We next consider the semi-classical energy

$$E_n(u) \equiv n^{-(1-s)}\big(\|\partial_t u\|_{L^2(\mathbb{T}^3)}^2 + \|\nabla u\|_{L^2(\mathbb{T}^3)}^2\big)^{\frac{1}{2}}$$
$$+ n^{-(2-s)}\big(\|\partial_t u\|_{H^1(\mathbb{T}^3)}^2 + \|\nabla u\|_{H^1(\mathbb{T}^3)}^2\big)^{\frac{1}{2}}.$$

We are going to show that for very small times u_n and $v_n + S(t)(u_0, u_1)$ are close with respect to E_n but these small times are long enough to get the needed amplification of the H^s norm. We emphasise that this amplification is a phenomenon only related to the solution of (4.1.70). Here is the precise statement.

Lemma 4.1.32 *There exist $\varepsilon > 0$, $\delta_2 > 0$ and $C > 0$ such that for $\delta_1 < \delta_2$, if we set*

$$t_n \equiv [\log(n)]^{\delta_2} n^{-(\frac{3}{2}-s)}$$

then for every $n \gg 1$, every $t \in [0, t_n]$,

$$E_n\big(u_n(t) - v_n(t) - S(t)(u_0, u_1)\big) \le Cn^{-\varepsilon}.$$

Moreover,

$$\|u_n(t) - v_n(t) - S(t)(u_0, u_1)\|_{H^s(\mathbb{T}^3)} \le Cn^{-\varepsilon}. \tag{4.1.77}$$

Proof Set $u_L = S(t)(u_0, u_1)$ and $w_n = u_n - u_L - v_n$. Then w_n solves the equation

$$(\partial_t^2 - \Delta)w_n = \Delta v_n - 3v_n^2(u_L + w_n) - 3v_n(u_L + w_n)^2 - (u_L + w_n)^3, \tag{4.1.78}$$

with initial data

$$(w_n(0, \cdot), \partial_t w_n(0, \cdot)) = (0, 0).$$

Set

$$F \equiv \Delta v_n - 3v_n^2(u_L + w_n) - 3v_n(u_L + w_n)^2 - (u_L + w_n)^3.$$

Multiplying Eq. (4.1.78) with $\partial_t w_n$ and integrating over $\mathbb{T}^3$ gives

$$\left| \frac{d}{dt}\big(\|\partial_t w_n(t)\|_{L^2(\mathbb{T}^3)}^2 + \|\nabla w_n(t)\|_{L^2(\mathbb{T}^3)}^2\big)\right| \lesssim \|\partial_t w_n(t)\|_{L^2(\mathbb{T}^3)}\|F(t)\|_{L^2(\mathbb{T}^3)}$$

which in turn implies

$$\left| \frac{d}{dt}\big(\|\partial_t w_n(t)\|_{L^2(\mathbb{T}^3)}^2 + \|\nabla w_n(t)\|_{L^2(\mathbb{T}^3)}^2\big)^{\frac{1}{2}}\right| \lesssim \|F(t)\|_{L^2(\mathbb{T}^3)}. \tag{4.1.79}$$

Similarly, by first differentiating (4.1.78) with respect to the spatial variables, we get the bound

$$\left| \frac{d}{dt} \left(\|\partial_t w_n(t)\|^2_{H^1(\mathbb{T}^3)} + \|\nabla w_n(t)\|^2_{H^1(\mathbb{T}^3)} \right)^{\frac{1}{2}} \right| \lesssim \|F(t)\|_{H^1(\mathbb{T}^3)} . \qquad (4.1.80)$$

Now, using (4.1.79) and (4.1.80), we obtain the estimate

$$\left| \frac{d}{dt} \left(E_n(w_n(t)) \right) \right| \le C n^{-(2-s)} \|F(t)\|_{H^1(\mathbb{T}^3)} + C n^{-(1-s)} \|F(t)\|_{L^2(\mathbb{T}^3)} .$$

Therefore using (4.1.71), (4.1.72), we get

$$\left| \frac{d}{dt} \left(E_n(w_n(t)) \right) \right| \le C \Big([\log(n)]^{3\delta_2} n$$
$$+ n^{-(2-s)} \|G(t, \cdot)\|_{H^1(\mathbb{T}^3)} + n^{-(1-s)} \|G(t, \cdot)\|_{L^2(\mathbb{T}^3)} \Big),$$
$$(4.1.81)$$

where $G \equiv G_1 + G_2$ with

$$G_1 = -3 v_n^2 u_L - 3 v_n u_L^2 - u_L^3$$

and

$$G_2 = -3(u_L + v_n)^2 w_n - 3(u_L + v_n) w_n^2 - w_n^3.$$

Since $u_L \in C^\infty(\mathbb{R} \times \mathbb{T}^3)$ is independent of n, using (4.1.73) and (4.1.75) we can estimate G_1 as follows

$$n^{-(l-s)} \|G_1(t, \cdot)\|_{H^{l-1}(\mathbb{T}^3)} \lesssim [\log n]^{\delta_2} n^{\frac{1}{2} - s} \lesssim [\log(n)]^{3\delta_2} n, \quad l = 1, 2.$$

Writing for $t \in [0, t_n]$,

$$w_n(t, x) = \int_0^t \partial_t w_n(\tau, x) d\tau,$$

we obtain

$$\|w_n(t, \cdot)\|_{H^k(\mathbb{T}^3)} \le C [\log(n)]^{\delta_2} n^{-(\frac{3}{2} - s)} \sup_{0 \le \tau \le t} \|\partial_t w_n(\tau, \cdot)\|_{H^k(\mathbb{T}^3)} . \qquad (4.1.82)$$

Set

$$e_n(w_n(t)) \equiv \sup_{0 \le \tau \le t} \tilde{E}_n(w_n(\tau)) .$$

Observe that $e_n(w_n(t))$ is increasing. Using (4.1.82) (with $k = 0, 1$), (4.1.73) and the Leibniz rule, we get that for $t \in [0, t_n]$ and for $l = 1, 2$,

$$n^{-(l-s)} \|(u_L(t) + v_n(t))^2 w_n(t)\|_{H^{l-1}(\mathbb{T}^3)} \leq C[\log(n)]^{l\delta_2} n^{\frac{3}{2}-s} e_n(w_n(t)).$$

Thanks to the Gagliardo-Nirenberg inequality, and (4.1.82) with $k = 0$, we get for $t \in [0, t_n]$,

$$\|w_n(t, \cdot)\|_{L^\infty(\mathbb{T}^3)} \leq C\|w_n(t, \cdot)\|_{H^2(\mathbb{T}^3)}^{\frac{3}{4}} \|w_n(t, \cdot)\|_{L^2(\mathbb{T}^3)}^{\frac{1}{4}} \tag{4.1.83}$$

$$\leq C n^{\frac{3}{2}-s} e_n(w_n(t)).$$

Hence, we can use (4.1.83) to treat the quadratic and cubic terms in w_n and to get the bound

$$n^{-(l-s)} \|G_2(t, \cdot)\|_{H^{l-1}(\mathbb{T}^3)} \leq C[\log(n)]^{l\delta_2} n^{\frac{3}{2}-s} \left(e_n(w_n(t)) + [e_n(w_n(t))]^3\right).$$

Therefore, coming back to (4.1.81), we get for $t \in [0, t_n]$,

$$\left|\frac{d}{dt}\left(E_n(w_n(t))\right)\right| \leq C[\log(n)]^{3\delta_2} n$$

$$+ C[\log(n)]^{2\delta_2} n^{\frac{3}{2}-s} \left(e_n(w_n(t)) + [e_n(w_n(t))]^3\right).$$

We now observe that

$$\frac{d}{dt}\left(e_n(w_n(t))\right) \leq \left|\frac{d}{dt}\left(E_n(w_n(t))\right)\right|$$

is resulting directly from the definition. Therefore, we have the bound

$$\frac{d}{dt}\left(e_n(w_n(t))\right) \leq C[\log(n)]^{3\delta_2} n$$

$$+ C[\log(n)]^{2\delta_2} n^{\frac{3}{2}-s} \left(e_n(w_n(t)) + [e_n(w_n(t))]^3\right). \tag{4.1.84}$$

We first suppose that $e_n(w_n(t)) \leq 1$. This property holds for small values of t since $E_n(w_n(0)) = e_n(w_n(0)) = 0$. In addition, the estimate for $e_n(w_n(t))$ we are looking for is much stronger than $e_n(w_n(t)) \leq 1$. Therefore, once we prove the desired estimate for $e_n(w_n(t))$ under the assumption $e_n(w_n(t)) \leq 1$, we can use a bootstrap argument to get the estimate without the assumption $e_n(w_n(t)) \leq 1$.

Estimate (4.1.84) yields that for $t \in [0, t_n]$,

$$\frac{d}{dt}(e_n(w_n(t))) \leq C[\log(n)]^{3\delta_2} n + C[\log(n)]^{2\delta_2} n^{\frac{3}{2}-s} e_n(w_n(t))$$

and consequently

$$\frac{d}{dt}\left(e^{-Ct[\log(n)]^{2\delta_2}n^{\frac{3}{2}-s}}e_n(w_n(t))\right) \le C[\log(n)]^{3\delta_2}\,n\,e^{-Ct[\log(n)]^{2\delta_2}n^{\frac{3}{2}-s}}.$$

An integration of the last estimate gives that for $t \in [0, t_n]$,

$$e_n(w_n(t)) \le C\left([\log(n)]^{\delta_2}n^{s-\frac{1}{2}}\right)e^{Ct[\log(n)]^{2\delta_2}n^{\frac{3}{2}-s}}$$

$$\le C\left([\log(n)]^{\delta_2}n^{s-\frac{1}{2}}\right)e^{C[\log(n)]^{3\delta_2}}.$$

(one should see δ_2 as $3\delta_2 - 2\delta_2$ and $s - 1/2$ as $1 - (3/2 - s)$). Since $s < 1/2$, by taking $\delta_2 > 0$ small enough, we obtain that there exists $\varepsilon > 0$ such that for $t \in [0, t_n]$,

$$E_n(w_n(t)) \le Cn^{-\varepsilon}$$

and in particular one has for $t \in [0, t_n]$,

$$\|\partial_t w_n(t, \cdot)\|_{L^2(\mathbb{T}^3)} + \|\nabla w_n(t, \cdot)\|_{L^2(\mathbb{T}^3)} \le Cn^{1-s-\varepsilon}. \tag{4.1.85}$$

We next estimate $\|w_n(t, \cdot)\|_{L^2}$. We may write for $t \in [0, t_n]$,

$$\|w_n(t, \cdot)\|_{L^2(\mathbb{T}^3)} = \|\int_0^t \partial_t w_n(\tau, \cdot)d\tau\|_{L^2(\mathbb{T}^3)} \le ct_n \sup_{0 \le \tau \le t} \|\partial_t w_n(\tau, \cdot)\|_{L^2(\mathbb{T}^3)}.$$

Thanks to (4.1.85) and the definition of t_n, we get

$$\|w_n(t, \cdot)\|_{L^2(\mathbb{T}^3))} \le C[\log(n)]^{\delta_2}n^{-(\frac{3}{2}-s)}n^{1-s}n^{-\varepsilon}.$$

Therefore, since $s < 1/2$,

$$\|w_n(t, \cdot)\|_{L^2(\mathbb{T}^3)} \le Cn^{-s-\varepsilon}. \tag{4.1.86}$$

An interpolation between (4.1.85) and (4.1.86) yields (4.1.77). This completes the proof of Lemma 4.1.32.   □

Using Lemma 4.1.32, we may write

$$\|u_n(t_n, \cdot)\|_{H^s(\mathbb{T}^3)} \ge \|v_n(t_n, \cdot)\|_{H^s(\mathbb{T}^3)} - C - Cn^{-\varepsilon}.$$

Recall that (4.1.74) yields

$$\|v_n(t_n, \cdot)\|_{H^s(\mathbb{T}^3)} \ge C[\log(n)]^{-(s+1)\delta_1+s\delta_2},$$

provided $n \gg 1$. Therefore, by choosing δ_1 small enough (depending on δ_2 fixed in Lemma 4.1.32), we obtain that the exists $\delta > 0$ such that

$$\|v_n(t_n, \cdot)\|_{H^s(\mathbb{T}^3)} \geq C[\log(n)]^\delta, \quad n \gg 1$$

which in turn implies that

$$\|u_n(t_n, \cdot)\|_{H^s(\mathbb{T}^3)} \geq C[\log(n)]^\delta, \quad n \gg 1.$$

This completes the proof of Theorem 4.1.28.   □

Theorem 4.1.28 implies that the Cauchy problem associated with the cubic focusing wave equation,

$$(\partial_t^2 - \Delta)u + u^3 = 0$$

is ill-posed in $H^s(\mathbb{T}^3) \times H^{s-1}(\mathbb{T}^3)$ for $s < 1/2$ because of the lack of continuous dependence for any $C^\infty(\mathbb{T}^3) \times C^\infty(\mathbb{T}^3)$ initial data.

For future references, we also state the following consequence of Theorem 4.1.28.

Theorem 4.1.33 *Let us fix* $s \in (0, 1/2)$, $T > 0$ *and*

$$(u_0, u_1) \in H^s(\mathbb{T}^3) \times H^{s-1}(\mathbb{T}^3).$$

Then there exists a sequence $(u_n(t, x))_{n=1}^\infty$ *of* $C(\mathbb{R}; C^\infty(\mathbb{T}^3))$ *functions such that*

$$(\partial_t^2 - \Delta)u_n + u_n^3 = 0$$

with

$$\lim_{n \to +\infty} \|(u_n(0) - u_0, \partial_t u_n(0) - u_1)\|_{H^s(\mathbb{T}^3) \times H^{s-1}(\mathbb{T}^3)} = 0$$

but

$$\lim_{n \to +\infty} \|(u_n(t), \partial_t u_n(t))\|_{L^\infty([0,T]; H^s(\mathbb{T}^3) \times H^{s-1}(\mathbb{T}^3))} = +\infty.$$

Proof Let $(u_{0,m}, u_{1,m})_{m=1}^\infty$ be a sequence of $C^\infty(\mathbb{T}^3) \times C^\infty(\mathbb{T}^3)$ functions such that

$$\lim_{m \to +\infty} \|(u_0 - u_{0,m}, u_1 - u_{1,m})\|_{H^s(\mathbb{T}^3) \times H^{s-1}(\mathbb{T}^3)} = 0.$$

For a fixed m, we apply Theorem 4.1.28 in order to find a sequence $(u_{m,n}(t, x))_{n=1}^\infty$ of $C(\mathbb{R}; C^\infty(\mathbb{T}^3))$ functions such that

$$(\partial_t^2 - \Delta)u_{m,n} + u_{m,n}^3 = 0$$

with

$$\lim_{n \to +\infty} \|(u_{m,n}(0) - u_{0,m}, \partial_t u_{m,n}(0) - u_{1,m})\|_{H^s(\mathbb{T}^3) \times H^{s-1}(\mathbb{T}^3)} = 0$$

and for every $m \geq 1$,

$$\lim_{n \to +\infty} \|(u_{m,n}(t), \partial_t u_{m,n}(t))\|_{L^\infty([0,T]; H^s(\mathbb{T}^3) \times H^{s-1}(\mathbb{T}^3))} = +\infty. \tag{4.1.87}$$

Now, using the triangle inequality, we obtain that for every $l \geq 1$ there is $M_0(l)$ such that for every $m \geq M_0(l)$ there is $N_0(m)$ such that for every $n \geq N_0(m)$,

$$\|(u_{m,n}(0) - u_0, \partial_t u_{m,n}(0) - u_1)\|_{H^s(\mathbb{T}^3) \times H^{s-1}(\mathbb{T}^3)} < \frac{1}{l}.$$

Thanks to (4.1.87), we obtain that for every $m \geq 1$ there exists $N_1(m) \geq N_0(m)$ such that for every $n \geq N_1(m)$,

$$\|(u_{m,n}(t), \partial_t u_{m,n}(t))\|_{L^\infty([0,T]; H^s(\mathbb{T}^3) \times H^{s-1}(\mathbb{T}^3))} > l.$$

We now observe that

$$u_l(t, x) \equiv u_{M_0(l), N_1(M_0(l))}(t, x), \quad l = 1, 2, 3, \cdots$$

is a sequence of solutions of the cubic defocusing wave equation satisfying the conclusions of Theorem 4.1.33. $\qquad \square$

Remark 4.1.34 It is worth mentioning that we arrive without too much complicated technicalities to a sharp local well-posedness result in the context of the cubic wave equation because we do not need a smoothing effect to recover derivative losses neither in the nonlinearity nor in the non homogeneous Strichartz estimates. The $X^{s,b}$ spaces of Bourgain are an efficient tool to deal with these two difficulties. These developments go beyond the scope of these lectures.

4.1.8 Extensions to More General Nonlinearities

One may consider the wave equation with a more general nonlinearity than the cubic one. Namely, let us consider the nonlinear wave equation

$$(\partial_t^2 - \Delta)u + |u|^\alpha u = 0, \tag{4.1.88}$$

posed on $\mathbb{T}^3$ where $\alpha > 0$ measures the "degree" of the nonlinearity. If $u(t, x)$ is a
solution of (4.1.88) posed on $\mathbb{R}^3$, than so is $u_\lambda(t, x) = \lambda^{\frac{2}{\alpha}} u(\lambda t, \lambda x)$. Moreover

$$\|u_\lambda(t, \cdot)\|_{H^s} \approx \lambda^{\frac{2}{\alpha}} \lambda^s \lambda^{-\frac{3}{2}} \|u(\lambda t, \cdot)\|_{H^s}$$

which implies that H^s with $s = \frac{3}{2} - \frac{2}{\alpha}$ is the critical Sobolev regularity for (4.1.88).
Based on this scaling argument one may wish to expect that for $s > \frac{3}{2} - \frac{2}{\alpha}$ the
Cauchy problem associated with (4.1.88) is well-posed in $H^s \times H^{s-1}$ and that for
$s < \frac{3}{2} - \frac{2}{\alpha}$ it is ill-posed in $H^s \times H^{s-1}$. In this section, we verified that this is
indeed the case for $\alpha = 2$. For $2 < \alpha < 4$, a small modification of the proof
of Theorem 4.1.22 shows that (4.1.88) is locally well-posed in $H^s \times H^{s-1}$ for
$s \in (\frac{3}{2} - \frac{2}{\alpha}, \alpha)$. Then, as in the proof of Theorem 4.1.2, we can show that (4.1.88)
is globally well-posed in $H^1 \times L^2$. Moreover a small modification of the proof of
Theorem 4.1.28 shows that for $s \in (0, \frac{3}{2} - \frac{2}{\alpha})$ the Cauchy problem for (4.1.88) is
locally ill-posed in $H^s \times H^{s-1}$. For $\alpha = 4$, we can prove a local well-posedness
statement for (4.1.88) as in Theorem 4.1.25. The global well-posedness in $H^1 \times L^2$
for $\alpha = 4$ is much more delicate than the globalisation argument of Theorem 4.1.2.
It is however possible to show that (4.1.88) is globally well-posed in $H^1 \times L^2$
(see [20, 21, 41, 42]). The new global information for $\alpha = 4$, in addition to the
conservation of the energy, is the Morawetz estimate which is a quantitative way to
contradict the blow-up criterium in the case $\alpha = 4$. For $\alpha > 4$ the Cauchy problem
associated with (4.1.88) is still locally well-posed in $H^s \times H^{s-1}$ for some $s > \frac{3}{2} - \frac{2}{\alpha}$.
The global well-posedness (i.e. global existence, uniqueness and propagation of
regularity) of (4.1.88) for $\alpha > 4$ is an outstanding open problem. For $\alpha > 4$, the
argument used in Theorem 4.1.28 may allow to construct weak solutions in $H^1 \times L^2$
with initial data in H^σ for $1 < \sigma < \frac{3}{2} - \frac{2}{\alpha}$ which are losing their H^σ regularity. See
[28] for such a result for (4.1.88), posed on $\mathbb{R}^3$.

4.2   Probabilistic Global Well-Posedness for the 3d Cubic Wave Equation in H^s, $s \in [0, 1]$

4.2.1   Introduction

Consider again the Cauchy problem for the cubic defocusing wave equation

$$(\partial_t^2 - \Delta)u + u^3 = 0, \quad u : \mathbb{R} \times \mathbb{T}^3 \to \mathbb{R},$$

$$u|_{t=0} = u_0, \ \partial_t u|_{t=0} = u_1, \qquad (u_0, u_1) \in \mathcal{H}^s(\mathbb{T}^3),$$

(4.2.1)

where

$$\mathcal{H}^s(\mathbb{T}^3) \equiv H^s(\mathbb{T}^3) \times H^{s-1}(\mathbb{T}^3).$$

In the previous section, we have shown that (4.2.1) is (at least locally in time) well-posed in $\mathcal{H}^s(\mathbb{T}^3)$, $s \geq 1/2$. The main ingredient in the proof for $s \in [1/2, 1)$ was the Strichartz estimates for the linear wave equation. We have also shown that for $s \in (0, 1/2)$ the Cauchy problem (4.2.1) is ill-posed in $\mathcal{H}^s(\mathbb{T}^3)$.

One may however ask whether some sort of well-posedness for (4.2.1) survives for $s < 1/2$. We will show below that this is indeed possible, if we accept to "randomise" the initial data. This means that we will endow $\mathcal{H}^s(\mathbb{T}^3)$, $s \in (0, 1/2)$ with suitable probability measures and we will show that the Cauchy problem (4.2.1) is well-posed in a suitable sense for initial data (u_0, u_1) on a set of full measure.

Let us now describe these measures. Starting from $(u_0, u_1) \in \mathcal{H}^s$ given by their Fourier series

$$u_j(x) = a_j + \sum_{n \in \mathbb{Z}_*^3} \Big(b_{n,j} \cos(n \cdot x) + c_{n,j} \sin(n \cdot x) \Big), \quad j = 0, 1,$$

we define u_j^ω by

$$u_j^\omega(x) = \alpha_j(\omega) a_j + \sum_{n \in \mathbb{Z}_*^3} \Big(\beta_{n,j}(\omega) b_{n,j} \cos(n \cdot x) + \gamma_{n,j}(\omega) c_{n,j} \sin(n \cdot x) \Big),$$

$$(4.2.2)$$

where $(\alpha_j(\omega), \beta_{n,j}(\omega), \gamma_{n,j}(\omega))$, $n \in \mathbb{Z}_*^3$, $j = 0, 1$ is a sequence of real random variables on a probability space $(\Omega, p, \mathcal{F})$. We assume that the random variables $(\alpha_j, \beta_{n,j}, \gamma_{n,j})_{n \in \mathbb{Z}_*^3, j=0,1}$ are independent identically distributed real random variables with a distribution θ satisfying

$$\exists c > 0, \quad \forall \gamma \in \mathbb{R}, \quad \int_{-\infty}^{\infty} e^{\gamma x} d\theta(x) \leq e^{c\gamma^2} \qquad (4.2.3)$$

(notice that under the assumption (4.2.3) the random variables are necessarily of mean zero). Typical examples (see Remark 4.2.13 below) of random variables satisfying (4.2.3) are the standard Gaussians, i.e.

$$d\theta(x) = (2\pi)^{-\frac{1}{2}} e^{-\frac{x^2}{2}} dx$$

(with an identity in (4.2.3)) or the Bernoulli variables

$$d\theta(x) = \frac{1}{2}(\delta_{-1} + \delta_1).$$

An advantage of the Bernoulli randomisation is that it keeps the $\mathcal{H}^s$ norm of the original function. The Gaussian randomisation has the advantage to "generate" a dense set in $\mathcal{H}^s$ via the map

$$\omega \in \Omega \longmapsto (u_0^\omega, u_1^\omega) \in \mathcal{H}^s \qquad (4.2.4)$$

for most of $(u_0, u_1) \in \mathcal{H}^s$ (see Proposition 4.2.2 below).

Definition 4.2.1 For fixed $(u_0, u_1) \in \mathcal{H}^s$, the map (4.2.4) is a measurable map from $(\Omega, \mathcal{F})$ to $\mathcal{H}^s$ endowed with the Borel sigma algebra since the partial sums form a Cauchy sequence in $L^2(\Omega; \mathcal{H}^s)$. Thus (4.2.4) endows the space $\mathcal{H}^s(\mathbb{T}^3)$ with a probability measure which is the direct image of p. Let us denote this measure by $\mu_{(u_0, u_1)}$. Then

$$\forall A \subset \mathcal{H}^s, \ \mu_{(u_0, u_1)}(A) = p(\omega \in \Omega : (u_0^\omega, u_1^\omega) \in A).$$

Denote by $\mathcal{M}^s$ the set of measures obtained following this construction:

$$\mathcal{M}^s = \bigcup_{(u_0, u_1) \in \mathcal{H}^s} \{\mu_{(u_0, u_1)}\}.$$

Here are two basic properties of these measures.

Proposition 4.2.2 *For any $s' > s$, if $(u_0, u_1) \notin \mathcal{H}^{s'}$, then*

$$\mu_{(u_0, u_1)}(\mathcal{H}^{s'}) = 0.$$

In other words, the randomisation (4.2.4) does not regularise in the scale of the L^2-based Sobolev spaces (this fact is obvious for the Bernoulli randomisation). Next, if (u_0, u_1) have all their Fourier coefficients different from zero and if $\mathrm{supp}(\theta) = \mathbb{R}$ then $\mathrm{supp}(\mu_{(u_0, u_1)}) = \mathcal{H}^s$. In other words, under these assumptions, for any $(w_0, w_1) \in \mathcal{H}^s$ and any $\epsilon > 0$,

$$\mu_{(u_0, u_1)}(\{(v_0, v_1) \in \mathcal{H}^s : \|(w_0, w_1) - (v_0, v_1)\|_{\mathcal{H}^s} < \epsilon\}) > 0, \qquad (4.2.5)$$

or in yet other words, any set of full $\mu_{(u_0, u_1)}$-measure is dense in $\mathcal{H}^s$.

We have the following global existence and uniqueness result for typical data with respect to an element of $\mathcal{M}^s$.

Theorem 4.2.3 (Existence and Uniqueness) *Let us fix $s \in (0, 1)$ and $\mu \in \mathcal{M}^s$. Then, there exists a full μ measure set $\Sigma \subset \mathcal{H}^s(\mathbb{T}^3)$ such that for every $(v_0, v_1) \in \Sigma$, there exists a unique global solution v of the nonlinear wave equation*

$$(\partial_t^2 - \Delta)v + v^3 = 0, \quad (v(0), \partial_t v(0)) = (v_0, v_1) \qquad (4.2.6)$$

satisfying

$$(v(t), \partial_t v(t)) \in \left(S(t)(v_0, v_1), \partial_t S(t)(v_0, v_1)\right) + C(\mathbb{R}; H^1(\mathbb{T}^3) \times L^2(\mathbb{T}^3)).$$

Furthermore, if we denote by

$$\Phi(t)(v_0, v_1) \equiv (v(t), \partial_t v(t))$$

the flow thus defined, the set Σ is invariant by the map $\Phi(t)$, namely

$$\Phi(t)(\Sigma) = \Sigma, \qquad \forall t \in \mathbb{R}.$$

The next statement gives quantitative bounds on the solutions.

Theorem 4.2.4 (Quantitative Bounds) *Let us fix $s \in (0, 1)$ and $\mu \in \mathcal{M}^s$. Let Σ be the set constructed in Theorem 4.2.3. Then for every $\varepsilon > 0$ there exist $C, \delta > 0$ such that for every $(v_0, v_1) \in \Sigma$, there exists $M > 0$ such that the global solution to (4.2.6) constructed in Theorem 4.2.3 satisfies*

$$v(t) = S(t)\Pi_0^{\perp}(v_0, v_1) + w(t),$$

with

$$\|(w(t), \partial_t w(t))\|_{\mathcal{H}^1(\mathbb{T}^3)} \le C(M + |t|)^{\frac{1-s}{s} + \varepsilon}$$

and

$$\mu((v_0, v_1) : M > \lambda) \le Ce^{-\lambda^{\delta}}.$$

Remark 4.2.5 Recall that Π_0 is the orthogonal projector on the zero Fourier mode and $\Pi_0^{\perp} = \mathrm{Id} - \Pi_0$.

We now further discuss the uniqueness of the obtained solutions. For $s > 1/2$, we have the following statement.

Theorem 4.2.6 (Unique Limit of Smooth Solutions for $s > 1/2$) *Let $s \in (1/2, 1)$. With the notations of the statement of Theorem 4.2.3, let us fix an initial datum $(v_0, v_1) \in \Sigma$ with a corresponding global solution $v(t)$. Let $(v_{0,n}, v_{1,n})_{n=1}^{\infty}$ be a sequence of $\mathcal{H}^1(\mathbb{T}^3)$ such that*

$$\lim_{n \to \infty} \|(v_{0,n} - v_0, v_{1,n} - v_1)\|_{\mathcal{H}^s(\mathbb{T}^3)} = 0.$$

Denote by $v_n(t)$ the solution of the cubic defocusing wave equation with data $(v_{0,n}, v_{1,n})$ defined in Theorem 4.1.2. Then for every $T > 0$,

$$\lim_{n \to \infty} \|(v_n(t) - v(t), \partial_t v_n(t) - \partial_t v(t))\|_{L^{\infty}([0,T]; \mathcal{H}^s(\mathbb{T}^3))} = 0.$$

Thanks to Theorem 4.1.33, we know that for $s \in (0, 1/2)$ the result of Theorem 4.2.6 cannot hold true ! We only have a partial statement.

Theorem 4.2.7 (Unique Limit of Particular Smooth Solutions for $s < 1/2$) *Let $s \in (0, 1/2)$. With the notations of the statement of Theorem 4.2.3, let us fix an initial datum $(v_0, v_1) \in \Sigma$ with a corresponding global solution $v(t)$. Let $(v_{0,n}, v_{1,n})_{n=1}^{\infty}$ be the sequence of $C^{\infty}(\mathbb{T}^3) \times C^{\infty}(\mathbb{T}^3)$ defined as the usual regularisation by convolution, i.e.*

$$v_{0,n} = v_0 \star \rho_n, \qquad v_{1,n} = v_1 \star \rho_n,$$

where $(\rho_n)_{n=1}^{\infty}$ is an approximate identity. Denote by $v_n(t)$ the solution of the cubic defocusing wave equation with data $(v_{0,n}, v_{1,n})$ defined in Theorem 4.1.2. Then for every $T > 0$,

$$\lim_{n \to \infty} \|(v_n(t) - v(t), \partial_t v_n(t) - \partial_t v(t))\|_{L^{\infty}([0,T];\mathcal{H}^s(\mathbb{T}^3))} = 0.$$

Remark 4.2.8 We emphasise that the result of Theorem 4.1.33 applies for the elements of Σ. More precisely, thanks to Theorem 4.1.33, we have that for every $(v_0, v_1) \in \Sigma$ there is a sequence $(v_{0,n}, v_{1,n})_{n=1}^{\infty}$ of elements of $C^{\infty}(\mathbb{T}^3) \times C^{\infty}(\mathbb{T}^3)$ such that

$$\lim_{n \to \infty} \|(v_{0,n} - v_0, v_{1,n} - v_1)\|_{\mathcal{H}^s(\mathbb{T}^3)} = 0$$

but such that if we denote by $v_n(t)$ the solution of the cubic defocusing wave equation with data $(v_{0,n}, v_{1,n})$ defined in Theorem 4.1.2 then for every $T > 0$,

$$\lim_{n \to \infty} \|(v_n(t), \partial_t v_n(t))\|_{L^{\infty}([0,T];\mathcal{H}^s(\mathbb{T}^3))} = \infty.$$

Therefore the choice of the particular regularisation of the initial data in Theorem 4.2.7 is of key importance. It would be interesting to classify the "admissible type of regularisations" allowing to get a statement such as Theorem 4.2.7 .

Remark 4.2.9 We can also see the solutions constructed in Theorem 4.2.3 as the (unique) limit as N tends to infinity of the solutions of the following truncated versions of the cubic defocusing wave equation.

$$(\partial_t^2 - \Delta)S_N u + S_N((S_N u)^3) = 0,$$

where S_N is a Fourier multiplier localising on modes of size $\leq N$. The convergence of a subsequence can be obtained by a compactness argument (cf. [9]). The convergence of the whole sequence however requires strong solutions techniques.

The next question is whether some sort of continuous dependence with respect to the initial data survives in the context of Theorem 4.2.3. In order to state our result concerning the continuous dependence with respect to the initial data, we recall that for any event B (of non vanishing probability) the conditioned probability $p(\cdot|B)$ is the natural probability measure supported by B, defined by

$$p(A|B) = \frac{p(A \cap B)}{p(B)}.$$

We have the following statement.

Theorem 4.2.10 (Conditioned Continuous Dependence) *Let us fix* $s \in (0, 1)$, *let* $A > 0$, *let* $B_A \equiv (V \in \mathcal{H}^s : \|V\|_{\mathcal{H}^s} \leq A)$ *be the closed ball of radius A centered at the origin of* $\mathcal{H}^s$ *and let* $T > 0$. *Let* $\mu \in \mathcal{M}^s$ *and suppose that* θ *(the law of our random variables) is symmetric. Let* $\Phi(t)$ *be the flow of the cubic wave equations defined* μ *almost everywhere in Theorem 4.2.3. Then for* $\varepsilon, \eta > 0$, *we have the bound*

$$\mu \otimes \mu \Big((V, V') \in \mathcal{H}^s \times \mathcal{H}^s : \|\Phi(t)(V) - \Phi(t)(V')\|_{X_T} > \varepsilon \Big|$$

$$\|V - V'\|_{\mathcal{H}^s} < \eta \text{ and } (V, V') \in B_A \times B_A \Big) \leq g(\varepsilon, \eta), \quad (4.2.7)$$

where $X_T \equiv (C([0, T]; \mathcal{H}^s) \cap L^4([0, T] \times \mathbb{T}^3)) \times C([0, T]; H^{s-1})$ *and* $g(\varepsilon, \eta)$ *is such that*

$$\lim_{\eta \to 0} g(\varepsilon, \eta) = 0, \qquad \forall \varepsilon > 0.$$

Moreover, if for $s \in (0, 1/2)$ *we assume in addition that the support of* μ *is the whole* $\mathcal{H}^s$ *(which is true if in the definition of the measure* μ, *we have* $a_i, b_{n,j}, c_{n,j} \neq 0$, $\forall n \in \mathbb{Z}^d$ *and the support of the distribution function of the random variables is* $\mathbb{R}$), *then there exists* $\varepsilon > 0$ *such that for every* $\eta > 0$ *the left hand-side in* (4.2.7) *is positive.*

A probability measure θ on $\mathbb{R}$ is called symmetric if

$$\int_{\mathbb{R}} f(x)d\theta(x) = \int_{\mathbb{R}} f(-x)d\theta(x), \quad \forall f \in L^1(d\theta).$$

A real random variable is called symmetric if its distribution is a symmetric measure on $\mathbb{R}$.

The result of Theorem 4.2.10 is saying that as soon as $\eta \ll \varepsilon$, among the initial data which are η-close to each other, the probability of finding two for which the corresponding solutions to (4.2.1) do not remain ε close to each other, is very small. The last part of the statement is saying that the deterministic version of the uniform continuity property (4.2.7) does not hold and somehow that one cannot get rid of a probabilistic approach in the question concerning the continuous dependence (in $\mathcal{H}^s$, $s < 1/2$) with respect to the data. The ill-posedness result of Theorem 4.1.28 will be of importance in the proof of the last part of Theorem 4.2.10.

4.2.2 Probabilistic Strichartz Estimates

Lemma 4.2.11 *Let* $(l_n(\omega))_{n=1}^\infty$ *be a sequence of real, independent random variables with associated sequence of distributions* $(\theta_n)_{n=1}^\infty$. *Assume that* θ_n *satisfy the property*

$$\exists c > 0 : \forall \gamma \in \mathbb{R}, \forall n \geq 1, \left| \int_{-\infty}^{\infty} e^{\gamma x} d\theta_n(x) \right| \leq e^{c\gamma^2}. \tag{4.2.8}$$

Then there exists $\alpha > 0$ *such that for every* $\lambda > 0$, *every sequence* $(c_n)_{n=1}^\infty \in l^2$ *of real numbers,*

$$p\left(\omega : \left| \sum_{n=1}^{\infty} c_n l_n(\omega) \right| > \lambda \right) \leq 2e^{-\frac{\alpha \lambda^2}{\sum_n c_n^2}}. \tag{4.2.9}$$

As a consequence there exists $C > 0$ *such that for every* $p \geq 2$, *every* $(c_n)_{n=1}^\infty \in l^2$,

$$\left\| \sum_{n=1}^{\infty} c_n l_n(\omega) \right\|_{L^p(\Omega)} \leq C\sqrt{p} \left(\sum_{n=1}^{\infty} c_n^2 \right)^{1/2}. \tag{4.2.10}$$

Remark 4.2.12 The property (4.2.8) is equivalent to assuming that θ_n are of zero mean and assuming that

$$\exists c > 0, C > 0 : \forall \gamma \in \mathbb{R}, \forall n \geq 1, \left| \int_{-\infty}^{\infty} e^{\gamma x} d\theta_n(x) \right| \leq C e^{c\gamma^2}. \tag{4.2.11}$$

Remark 4.2.13 Let us notice that (4.2.8) is readily satisfied if $(l_n(\omega))_{n=1}^\infty$ are standard real Gaussian or standard Bernoulli variables. Indeed in the case of Gaussian

$$\int_{-\infty}^{\infty} e^{\gamma x} d\theta_n(x) = \int_{-\infty}^{\infty} e^{\gamma x} e^{-x^2/2} \frac{dx}{\sqrt{2\pi}} = e^{\gamma^2/2}.$$

In the case of Bernoulli variables one can obtain that (4.2.8) is satisfied by invoking the inequality

$$\frac{e^\gamma + e^{-\gamma}}{2} \leq e^{\gamma^2/2}, \quad \forall \gamma \in \mathbb{R}.$$

More generally, we can observe that (4.2.11) holds if θ_n is compactly supported.

Remark 4.2.14 In the case of Gaussian we can see Lemma 4.2.11 as a very particular case of a L^p smoothing properties of the Hartree-Foch heat flow (see e.g. [44, Section 3] for more details on this issue).

Proof of Lemma 4.2.11 For $t > 0$ to be determined later, using the independence and (4.2.8), we obtain

$$\int_\Omega e^{t \sum_{n \geq 1} c_n l_n(\omega)} dp(\omega) = \prod_{n \geq 1} \int_\Omega e^{t c_n l_n(\omega)} dp(\omega)$$

$$= \prod_{n \geq 1} \int_{-\infty}^{\infty} e^{t c_n x} d\theta_n(x)$$

$$\leq \prod_{n \geq 1} e^{c(t c_n)^2} = e^{(c t^2) \sum_n c_n^2}.$$

Therefore

$$e^{(c t^2) \sum_n c_n^2} \geq e^{t\lambda} \; p\left(\omega : \sum_{n \geq 1} c_n l_n(\omega) > \lambda\right)$$

or equivalently,

$$p\left(\omega : \sum_{n \geq 1} c_n l_n(\omega) > \lambda\right) \leq e^{(c t^2) \sum_n c_n^2} e^{-t\lambda}.$$

We choose t as

$$t \equiv \frac{\lambda}{2c \sum_n c_n^2}.$$

Hence

$$p\left(\omega : \sum_{n \geq 1} c_n l_n(\omega) > \lambda\right) \leq e^{-\frac{\lambda^2}{4c \sum_n c_n^2}}.$$

In the same way (replacing c_n by $-c_n$), we can show that

$$p\left(\omega : \sum_{n \geq 1} c_n l_n(\omega) < -\lambda\right) \leq e^{-\frac{\lambda^2}{4c \sum_n c_n^2}}$$

which completes the proof of (4.2.9). To deduce (4.2.10), we write

$$\left\| \sum_{n=1}^{\infty} c_n l_n(\omega) \right\|_{L^p(\Omega)}^p = p \int_0^{+\infty} p\left(\omega : \left| \sum_{n=1}^{\infty} c_n l_n(\omega) \right| > \lambda\right) \lambda^{p-1} d\lambda$$

$$\leq C p \int_0^{+\infty} \lambda^{p-1} e^{-\frac{c\lambda^2}{\sum_n c_n^2}} d\lambda$$

$$\leq Cp(C\sum_n c_n^2)^{\frac{p}{2}} \int_0^{+\infty} \lambda^{p-1} e^{-\frac{\lambda^2}{2}} d\lambda$$

$$\leq C(Cp\sum_n c_n^2)^{\frac{p}{2}}$$

which completes the proof of Lemma 4.2.11.   □

As a consequence of Lemma 4.2.11, we get the following "probabilistic" Strichartz estimates.

Theorem 4.2.15 *Let us fix* $s \in (0, 1)$ *and let* $\mu \in \mathcal{M}^s$ *be induced via the map* (4.2.4) *from the couple* $(u_0, u_1) \in \mathcal{H}^s$. *Let us also fix* $\sigma \in (0, s]$, $2 \leq p_1 < +\infty$, $2 \leq p_2 \leq +\infty$ *and* $\delta > 1 + \frac{1}{p_1}$. *Then there exists a positive constant* C *such that for every* $p \geq 2$,

$$\left\| \|\langle t \rangle^{-\delta} S(t)(v_0, v_1)\|_{L^{p_1}(\mathbb{R}_t; L^{p_2}(\mathbb{T}^3))} \right\|_{L^p(\mu)} \leq C\sqrt{p} \|(u_0, u_1)\|_{\mathcal{H}^\sigma(\mathbb{T}^3)}. \qquad (4.2.12)$$

As a consequence for every $T > 0$ *and* $p_1 \in [1, \infty)$, $p_2 \in [2, \infty]$,

$$\|S(t)(v_0, v_1)\|_{L^{p_1}([0,T]; L^{p_2}(\mathbb{T}^3))} < \infty, \qquad \mu\text{- almost surely.} \qquad (4.2.13)$$

Moreover, there exist two positive constants C *and* c *such that for every* $\lambda > 0$,

$$\mu\left((v_0, v_1) \in \mathcal{H}^s : \|\langle t \rangle^{-\delta} S(t)(v_0, v_1)\|_{L^{p_1}(\mathbb{R}_t; L^{p_2}(\mathbb{T}^3))} > \lambda\right)$$

$$\leq C \exp\left(-\frac{c\lambda^2}{\|(u_0, u_1)\|^2_{\mathcal{H}^\sigma(\mathbb{T}^3)}}\right). \qquad (4.2.14)$$

Remark 4.2.16 Observe that (4.2.13) applied for $p_2 = \infty$ displays an improvement of 3/2 derivatives with respect to the Sobolev embedding which is stronger than the improvement obtained by the (deterministic) Strichartz estimates (see Remark 4.1.14). The proof of Theorem 4.2.15 exploits the random oscillations of the initial data while the proof of the deterministic Strichartz estimates exploits in a crucial (and subtle) manner the time oscillations of $S(t)$. In the proof of Theorem 4.2.15, we simply neglect these times oscillations.

Remark 4.2.17 In the proof of Theorem 4.2.15, we shall make use of the Sobolev spaces $W^{\sigma,q}(\mathbb{T}^3)$, $\sigma \geq 0$, $q \in (1, \infty)$, defined via the norm

$$\|u\|_{W^{\sigma,q}(\mathbb{T}^3)} = \|(1 - \Delta)^{\sigma/2} u\|_{L^q(\mathbb{T}^3)}.$$

Proof of Theorem 4.2.15 We have that

$$\left\| \|\langle t \rangle^{-\delta} \Pi_0 S(t)(v_0, v_1)\|_{L^{p_1}(\mathbb{R}_t; L^{p_2}(\mathbb{T}^3))} \right\|_{L^p(\mu)}$$

equals

$$\left\| \| \langle t \rangle^{-\delta} (\alpha_0(\omega) a_0 + t \alpha_1(\omega) a_1) \|_{L^{p_1}(\mathbb{R}_t; L^{p_2}(\mathbb{T}^3))} \right\|_{L^p_\omega}. \qquad (4.2.15)$$

A trivial application of Lemma 4.2.11 implies that

$$\| \alpha_j(\omega) \|_{L^p_\omega} \le C \sqrt{p}, \quad j = 0, 1.$$

Therefore, using that $\delta > 1 + 1/p_1$ the expression (4.2.15) can be bounded by

$$(2\pi)^{\frac{3}{p_2}} \left\| \| \langle t \rangle^{-\delta} (\alpha_0(\omega) a_0 + t \alpha_1(\omega) a_1) \|_{L^{p_1}(\mathbb{R}_t)} \right\|_{L^p_\omega} \le C \sqrt{p} (|a_0| + |a_1|).$$

Therefore, it remains to estimate

$$\left\| \| \langle t \rangle^{-\delta} \Pi_0^\perp S(t)(v_0, v_1) \|_{L^{p_1}(\mathbb{R}_t; L^{p_2}(\mathbb{T}^3))} \right\|_{L^p(\mu)}.$$

By a use of the Hölder inequality on $\mathbb{T}^3$, we observe that it suffices to estimate

$$\left\| \| \langle t \rangle^{-\delta} \Pi_0^\perp S(t)(v_0, v_1) \|_{L^{p_1}(\mathbb{R}_t; L^\infty(\mathbb{T}^3))} \right\|_{L^p(\mu)}.$$

Let $q < \infty$ be such that $\sigma > 3/q$. Then by the Sobolev embedding $W^{\sigma,q}(\mathbb{T}^3) \subset C^0(\mathbb{T}^3)$, we have

$$\| \Pi_0^\perp S(t)(v_0, v_1) \|_{L^\infty(\mathbb{T}^3)} \le C \| (1 - \Delta)^{\sigma/2} \Pi_0^\perp S(t)(v_0, v_1) \|_{L^q(\mathbb{T}^3)}.$$

Therefore, we need to estimate

$$\left\| \| \langle t \rangle^{-\delta} (1 - \Delta)^{\sigma/2} \Pi_0^\perp S(t)(v_0, v_1) \|_{L^{p_1}(\mathbb{R}_t; L^q(\mathbb{T}^3))} \right\|_{L^p(\mu)}$$

which equals

$$\left\| \| \langle t \rangle^{-\delta} (1 - \Delta)^{\sigma/2} \Pi_0^\perp S(t)(u_0^\omega, u_1^\omega) \|_{L^{p_1}(\mathbb{R}_t; L^q(\mathbb{T}^3))} \right\|_{L^p_\omega}. \qquad (4.2.16)$$

By using the Hölder inequality in ω, we observe that it suffices to evaluate the last quantity only for $p > \max(p_1, q)$. For such values of p, using the Minkowski inequality, we can estimate (4.2.16) by

$$\left\| \| \langle t \rangle^{-\delta} (1 - \Delta)^{\sigma/2} \Pi_0^\perp S(t)(u_0^\omega, u_1^\omega) \|_{L^p_\omega} \right\|_{L^{p_1}(\mathbb{R}_t; L^q(\mathbb{T}^3))}. \qquad (4.2.17)$$

Now, we can write $(1 - \Delta)^{\sigma/2} \Pi_0^\perp S(t)(u_0^\omega, u_1^\omega)$ as

$$\sum_{n \in \mathbb{Z}_\star^3} \langle n \rangle^\sigma \left(\left(\beta_{n,0}(\omega) b_{n,0} \cos(t|n|) + \beta_{n,1}(\omega) b_{n,1} \frac{\sin(t|n|)}{|n|} \right) \cos(n \cdot x) \right.$$

$$\left. + \left(\gamma_{n,0}(\omega) c_{n,0} \cos(t|n|) + \gamma_{n,1}(\omega) c_{n,1} \frac{\sin(t|n|)}{|n|} \right) \sin(n \cdot x) \right),$$

with

$$\sum_{n \in \mathbb{Z}_\star^3} \langle n \rangle^{2\sigma} \left(|b_{n,0}|^2 + |c_{n,0}|^2 + |n|^{-2}(|b_{n,1}|^2 + |c_{n,1}|^2) \right) \leq C \|(u_0, u_1)\|_{\mathcal{H}^\sigma(\mathbb{T}^3)}^2 .$$

Now using (4.2.10) of Lemma 4.2.11 and the boundedness of sin and cos functions, we obtain that (4.2.17) can be bounded by

$$C \left\| \langle t \rangle^{-\delta} C \sqrt{p} \|(u_0, u_1)\|_{\mathcal{H}^\sigma(\mathbb{T}^3)} \right\|_{L^{p_1}(\mathbb{R}_t; L^q(\mathbb{T}^3))} . \tag{4.2.18}$$

Since $\delta > 1 + 1/p_1$, we can estimate (4.2.18) by

$$C \sqrt{p} \|(u_0, u_1)\|_{\mathcal{H}^\sigma(\mathbb{T}^3)} .$$

This completes the proof of (4.2.12). Let us finally show how (4.2.12) implies (4.2.14). Using the Tchebichev inequality and (4.2.12), we have that

$$\mu \left((v_0, v_1) \in \mathcal{H}^s : \|\langle t \rangle^{-\delta} S(t)(v_0, v_1)\|_{L^{p_1}(\mathbb{R}_t; L^{p_2}(\mathbb{T}^3))} > \lambda \right)$$

is bounded by

$$\lambda^{-p} \left\| \|\langle t \rangle^{-\delta} S(t)(v_0, v_1)\|_{L^{p_1}(\mathbb{R}_t; L^{p_2}(\mathbb{T}^3))} \right\|_{L^p(\mu)}^p \leq \left(C\lambda^{-1} \sqrt{p} \|(u_0, u_1)\|_{\mathcal{H}^\sigma(\mathbb{T}^3)} \right)^p$$

We now choose p as

$$C\lambda^{-1} \sqrt{p} \|(u_0, u_1)\|_{\mathcal{H}^\sigma(\mathbb{T}^3)} = \frac{1}{2} \Leftrightarrow p = \frac{\lambda^2 \|(u_0, u_1)\|_{\mathcal{H}^\sigma(\mathbb{T}^3)}^{-2}}{4C^2},$$

which yields (4.2.14). This completes the proof of Theorem 4.2.15. □

The proof of Theorem 4.2.15 also implies the following statement.

Theorem 4.2.18 *Let us fix $s \in (0, 1)$ and let $\mu \in \mathcal{M}^s$ be induced via the map (4.2.4) from the couple $(u_0, u_1) \in \mathcal{H}^s$. Let us also fix $p \geq 2$, $\sigma \in (0, s]$ and $q < \infty$ such that $\sigma > 3/q$. Then for every $T > 0$,*

$$\|S(t)(v_0, v_1)\|_{L^p([0,T]; W^{\sigma,q}(\mathbb{T}^3))} < \infty, \quad \mu\text{-almost surely.} \tag{4.2.19}$$

4.2.3   Regularisation Effect in the Picard Iteration Expansion

Consider the Cauchy problem

$$(\partial_t^2 - \Delta)u + u^3 = 0, \quad u|_{t=0} = u_0, \ \partial_t u|_{t=0} = u_1, \tag{4.2.20}$$

where (u_0, u_1) is a typical element on the support of $\mu \in \mathcal{M}^s, s \in (0, 1)$. According to the discussion in Sect. 4.1.5 of the previous section, for small times depending on (u_0, u_1), we can hope to represent the solution of (4.2.20) as

$$u = \sum_{j=1}^{\infty} Q_j(u_0, u_1),$$

where Q_j is homogeneous of order j in (u_0, u_1). We have that

$$Q_1(u_0, u_1) = S(t)(u_0, u_1),$$

$$Q_2(u_0, u_1) = 0,$$

$$Q_3(u_0, u_1) = -\int_0^t \frac{\sin((t - \tau)\sqrt{-\Delta})}{\sqrt{-\Delta}} \big(S(\tau)(u_0, u_1)\big)^3 d\tau,$$

etc. We have that μ a.s. $Q_1 \notin H^{\sigma}$ for $\sigma > s$. However, using the probabilistic Strichartz estimates of the previous section, we have that for $T > 0$,

$$\|Q_3(u_0, u_1)\|_{L_T^\infty H^1(\mathbb{T}^3)} \lesssim \|S(t)(u_0, u_1)\|_{L_T^3 L^6(\mathbb{T}^3)}^3 < \infty, \quad \mu\text{-almost surely.}$$

Therefore the second non trivial term in the formal expansion defining the solution is more regular than the initial data ! The strategy will therefore be to write the solution of (4.2.20) as

$$u = Q_1(u_0, u_1) + v,$$

where $v \in H^1$ and solve the equation for v by the methods described in the previous section. In the case of the cubic nonlinearity the deterministic analysis used to solve the equation for v is particularly simple, it is in fact very close to the analysis in the proof of Proposition 4.1.1. For more complicated problems the analysis of the equation for v could involve more advanced deterministic arguments. We refer to [4], where a similar strategy is used in the context of the nonlinear Schrödinger equation and to [16] where it is used in the context of stochastic PDE's.

This argument is not particularly restricted to Q_3. One can imagine situations when for some $m > 3$, Q_m is the first element in the expansion whose regularity fits well in a deterministic analysis. Then we can equally well look for the solutions under the form

$$u = \sum_{j=1}^{m-1} Q_j(u_0, u_1) + v, \tag{4.2.21}$$

and treat v by a deterministic analysis. It is worth noticing that such a situation occurs in the work on parabolic PDE's with a singular random source term [22–24]. In these works in expansions of type (4.2.21) the random initial data (u_0, u_1) should be replaced by the random source term (the white noise). Let us also mention that in the case of parabolic equations the deterministic smoothing comes from elliptic regularity estimates while in the context of the wave equation we basically rely on the smoothing estimate (4.1.5).

4.2.4   The Local Existence Result

Proposition 4.2.19 *Consider the problem*

$$(\partial_t^2 - \Delta)v + (f + v)^3 = 0. \tag{4.2.22}$$

There exists a constant C such that for every time interval $I = [a, b]$ of size 1, every $\Lambda \geq 1$, every $(v_0, v_1, f) \in H^1 \times L^2 \times L^3(I, L^6)$ satisfying

$$\|v_0\|_{H^1} + \|v_1\|_{L^2} + \|f\|_{L^3(I,L^6)}^3 \leq \Lambda$$

there exists a unique solution on the time interval $[a, a + C^{-1}\Lambda^{-2}]$ of (4.2.22) with initial data

$$v(a, x) = v_0(x), \quad \partial_t v(a, x) = v_1(x).$$

Moreover the solution satisfies $\|(v, \partial_t v)\|_{L^\infty([a, a+C^{-1}\Lambda^{-2}], H^1 \times L^2)} \leq C\Lambda$, $(v, \partial_t v)$ is unique in the class $L^\infty([a, a + C^{-1}\Lambda^{-2}], H^1 \times L^2)$ and the dependence in time is continuous.

Proof The proof is very similar to the proof of Proposition 4.1.1. By translation invariance in time, we can suppose that $I = [0, 1]$. We can rewrite the problem as

$$v(t) = S(t)(v_0, v_1) - \int_0^t \frac{\sin((t - \tau)\sqrt{-\Delta})}{\sqrt{-\Delta}}((f(\tau) + v(\tau))^3 d\tau. \tag{4.2.23}$$

Set

$$\Phi_{v_0, v_1, f}(v) \equiv S(t)(v_0, v_1) - \int_0^t \frac{\sin((t - \tau)\sqrt{-\Delta})}{\sqrt{-\Delta}}((f(\tau) + v(\tau))^3 d\tau.$$

Then for $T \in (0, 1]$, using the Sobolev embedding $H^1(\mathbb{T}^3) \subset L^6(\mathbb{T}^3)$, we get

$$\|\Phi_{v_0, v_1, f}(v)\|_{L^\infty([0,T], H^1)}$$

$$\leq C\left(\|v_0\|_{H^1} + \|v_1\|_{L^2} + \int_0^T \|f(\tau)\|_{L^6}^3 d\tau\right) + T \sup_{\tau \in [0,T]} \|v(\tau)\|_{L^6}^3$$

$$\leq C\left(\|v_0\|_{H^1} + \|v_1\|_{L^2} + \|f\|_{L^3(I, L^6)}^3\right) + T \|v\|_{L^\infty([0,T], H^1)}^3.$$

It is now clear that for $T \approx \Lambda^{-2}$ the map $\Phi_{u_0, u_1, f}$ send the ball

$$\{v : \|v\|_{L^\infty([0,T], H^1)} \leq C\Lambda\}$$

into itself. Moreover by a similar argument, we obtain that this map is a contraction on the same ball. Thus we obtain the existence part and the bound on v in H^1. The estimate of $\|\partial_t v\|_{L^2}$ follows by differentiating in t the Duhamel formula (4.2.23). This completes the proof of Proposition 4.2.19. $\square$

4.2.5 Global Existence

In this section, we complete the proof of Theorem 4.2.3. We search v under the form $v(t) = S(t)(v_0, v_1) + w(t)$. Then w solves

$$(\partial_t^2 - \Delta)w + (S(t)(v_0, v_1) + w)^3 = 0, \quad w\mid_{t=0} = 0, \quad \partial_t w \mid_{t=0} = 0. \quad (4.2.24)$$

Thanks to Theorems 4.2.15 and 4.2.18, we have that μ-almost surely,

$$g(t) = \|S(t)(v_0, v_1)\|_{L^6(\mathbb{T}^3)}^3 \in L^1_{loc}(\mathbb{R}_t),$$

$$f(t) = \|S(t)(v_0, v_1)\|_{W^{\sigma, q}(\mathbb{T}^3)} \in L^1_{loc}(\mathbb{R}_t), \quad (4.2.25)$$

$\sigma > 3/q$. The local existence for (4.2.24) follows from Proposition 4.2.19 and the first estimate in (4.2.25). We also deduce from Proposition 4.2.19, that as long as the $H^1 \times L^2$ norm of $(w, \partial_t w)$ remains bounded, the solution w of (4.2.24) exists. Set

$$\mathcal{E}(w(t)) = \frac{1}{2} \int_{\mathbb{T}^3} \left((\partial_t w)^2 + |\nabla_x w|^2 + \frac{1}{2} w^4\right) dx.$$

Using the equation solved by w, we now compute

$$\frac{d}{dt}\mathcal{E}(w(t)) = \int_{\mathbb{T}^3} \left(\partial_t w \partial_t^2 w + \nabla_x \partial_t w \cdot \nabla_x w + \partial_t w \, w^3\right) dx$$

$$= \int_{\mathbb{T}^3} \partial_t w \left(\partial_t^2 w - \Delta w + w^3\right) dx$$

$$= \int_{\mathbb{T}^3} \partial_t w \left(w^3 - (S(t)(v_0, v_1) + w)^3\right) dx.$$

Now, using the Cauchy-Schwarz inequality, the Hölder inequalities and the Sobolev embedding $W^{\sigma,q}(\mathbb{T}^3) \subset C^0(\mathbb{T}^3)$, we can write

$$\frac{d}{dt}\mathcal{E}(w(t)) \leq C\left(\mathcal{E}(w(t))\right)^{1/2} \|w^3 - (S(t)(v_0, v_1) + w)^3\|_{L^2(\mathbb{T}^3)}$$

$$\leq C\left(\mathcal{E}(w(t))\right)^{1/2}$$

$$\times \left(\|S(t)(v_0, v_1)\|^3_{L^6(\mathbb{T}^3)} + \|S(t)(v_0, v_1)\|_{L^\infty(\mathbb{T}^3)} \|w^2\|_{L^2(\mathbb{T}^3)}\right)$$

$$\leq C\left(\mathcal{E}(w(t))\right)^{1/2}$$

$$\times \left(\|S(t)(v_0, v_1)\|^3_{L^6(\mathbb{T}^3)} + \|S(t)(v_0, v_1)\|_{W^{\sigma,q}(\mathbb{T}^3)} \|w^2\|_{L^2(\mathbb{T}^3)}\right)$$

$$\leq C\left(\mathcal{E}(w(t))\right)^{1/2}\left(g(t) + f(t)\left(\mathcal{E}(w(t))\right)^{1/2}\right)$$

and consequently, according to Gronwall inequality and (4.2.25), w exists globally in time.

This completes the proof of the existence and uniqueness part of Theorem 4.2.3. Let us now turn to the construction of an invariant set. Define the sets

$$\Theta \equiv \left\{(v_0, v_1) \in \mathcal{H}^s : \|S(t)(v_0, v_1)\|^3_{L^6(\mathbb{T}^3)} \in L^1_{\text{loc}}(\mathbb{R}_t),\right.$$

$$\left.\|S(t)(v_0, v_1)\|_{W^{\sigma,q}(\mathbb{T}^3)} \in L^1_{\text{loc}}(\mathbb{R}_t)\right\}$$

and $\Sigma \equiv \Theta + \mathcal{H}^1$. Then Σ is of full μ measure for every $\mu \in \mathcal{H}^s$, since so is Θ. We have the following proposition.

Proposition 4.2.20 *Assume that $s > 0$ and let us fix $\mu \in \mathcal{M}^s$. Then, for every $(v_0, v_1) \in \Sigma$, there exists a unique global solution*

$$(v(t), \partial_t v(t)) \in (S(t)(v_0, v_1), \partial_t S(t)(v_0, v_1)) + C(\mathbb{R}; H^1(\mathbb{T}^3) \times L^2(\mathbb{T}^3))$$

of the nonlinear wave equation

$$(\partial_t^2 - \Delta)v + v^3 = 0, \quad (v(0, x), \partial_t v(0, x)) = (v_0(x), v_1(x)). \tag{4.2.26}$$

Moreover for every $t \in \mathbb{R}$, $(v(t), \partial_t v(t)) \in \Sigma$ and thus by the time reversibility Σ is invariant under the flow of (4.2.26).

Proof By assumption, we can write $(v_0, v_1) = (\tilde{v}_0, \tilde{v}_1) + (w_0, w_1)$ with $(\tilde{v}_0, \tilde{v}_1) \in \Theta$ and $(w_0, w_1) \in \mathcal{H}^1$. We search v under the form

$$v(t) = S(t)(\tilde{v}_0, \tilde{v}_1) + w(t).$$

Then w solves

$$(\partial_t^2 - \Delta_{\mathbb{T}^3})w + (S(t)(\tilde{v}_0, \tilde{v}_1) + w)^3 = 0, \quad w\mid_{t=0} = w_0, \quad \partial_t w\mid_{t=0} = w_1.$$

Now, exactly as before, we obtain that

$$\frac{d}{dt}\mathcal{E}(w(t)) \leq C\big(\mathcal{E}(w(t))\big)^{1/2}\big(g(t) + f(t)\big(\mathcal{E}(w(t))\big)^{1/2}\big),$$

where

$$g(t) = \|S(t)(\tilde{v}_0, \tilde{v}_1)\|_{L^6(\mathbb{T}^3)}^3, \quad f(t) = \|S(t)(\tilde{v}_0, \tilde{v}_1)\|_{W^{\sigma,q}(\mathbb{T}^3)}.$$

Therefore thanks to the Gronwall lemma, using that $\mathcal{E}(w(0))$ is well defined, we obtain the global existence for w. Thus the solution of (4.2.26) can be written as

$$v(t) = S(t)(\tilde{v}_0, \tilde{v}_1) + w(t), \quad (w, \partial_t w) \in C(\mathbb{R}; \mathcal{H}^1).$$

Coming back to the definition of Θ, we observe that

$$S(t)(\Theta) = \Theta.$$

Thus $(v(t), \partial_t v(t)) \in \Sigma$.

This completes the proof of Theorem 4.2.3.   □

4.2.6 Unique Limits of Smooth Solutions

In this section, we present the proofs of Theorems 4.2.6 and 4.2.7.

Proof of Theorem 4.2.6 Thanks to Theorem 4.2.3, the Sobolev embeddings and Theorem 4.2.15 we obtain that

$$(v, \partial_t v) \in C(\mathbb{R}; \mathcal{H}^s(\mathbb{T}^3))$$

and

$$v \in L^{p^\star}_{loc}(\mathbb{R}; L^{q^\star}(\mathbb{T}^3)),$$

where $(p^\star, q^\star)$ are as in Corollary 4.1.23 (observe that $q^\star \leq 6$). Once, we have this information the proof of Theorem 4.2.6 follows from Theorem 4.1.22 (here we use the assumption $s > 1/2$) and Corollary 4.1.23. Indeed, let us fix $T > 0$ and let Λ be such that

$$\sup_{0 \leq t \leq T} \|(v(t), \partial_t v(t))\|_{\mathcal{H}^s(\mathbb{T}^3)} < \Lambda - 1.$$

Let $\tau > 0$ be the time of existence associated with Λ in Theorem 4.1.22. We now cover the interval $[0, T]$ with intervals of size τ and using iteratively the continuous dependence statement of Theorem 4.1.22 and the uniqueness statement given by Corollary 4.1.23, we obtain that

$$\lim_{n \to \infty} \|(v_n(t) - v(t), \partial_t v_n(t) - \partial_t v(t))\|_{L^\infty([0,T];\mathcal{H}^s(\mathbb{T}^3))} = 0.$$

This completes the proof of Theorem 4.2.6.   □

We now turn to the proof of Theorem 4.2.7 which is slightly more delicate.

Proof of Theorem 4.2.7 For $(v_0, v_1) \in \Sigma$ we decompose the solution as

$$v(t) = S(t)(v_0, v_1) + w(t), \quad w(0) = 0, \ \partial_t w(0) = 0.$$

Similarly, we decompose the solutions issued from $(v_{0,n}, v_{1,n})$ as

$$v_n(t) = S(t)(v_{0,n}, v_{1,n}) + w_n(t), \quad w_n(0) = 0, \ \partial_t w_n(0) = 0.$$

Using the energy estimates of the previous section, we obtain that

$$\frac{d}{dt}\mathcal{E}(w_n(t)) \leq C\big(\mathcal{E}(w_n(t))\big)^{1/2}\big(g_n(t) + f_n(t)\big(\mathcal{E}(w(t))\big)^{1/2}\big),$$

where

$$g_n(t) = \|S(t)(v_{0,n}, v_{1,n})\|^3_{L^6(\mathbb{T}^3)}, \quad f_n(t) = \|S(t)(v_{0,n}, v_{1,n})\|_{W^{\sigma,q}(\mathbb{T}^3)}.$$

Therefore

$$(\mathcal{E}(w_n(t)))^{1/2} \leq C\Big(\int_0^t g_n(\tau)d\tau\Big)e^{\int_0^t f_n(\tau)d\tau}.$$

Using that

$$S(t)(v_{0,n}, v_{1,n}) = \rho_n \star \big(S(t)(v_0, v_1)\big), \tag{4.2.27}$$

and the fact that $(v_0, v_1) \in \Sigma$, we obtain that

$$\lim_{n \to \infty} \int_0^t g_n(\tau)d\tau = \int_0^t g(\tau)d\tau, \quad \lim_{n \to \infty} \int_0^t f_n(\tau)d\tau = \int_0^t f(\tau)d\tau,$$

where $g(t)$ and $f(t)$ are defined in (4.2.25). Therefore, we obtain that for every $T > 0$ there is $C > 0$ such that for every n,

$$\sup_{0 \le t \le T} \|(w_n(t), \partial_t w_n(t))\|_{\mathcal{H}^1(\mathbb{T}^3)} \le C. \tag{4.2.28}$$

Next, we observe that w and w_n solve the equations

$$(\partial_t^2 - \Delta)w + (S(t)(v_0, v_1) + w)^3 = 0$$

and

$$(\partial_t^2 - \Delta)w_n + (S(t)(v_{0,n}, v_{1,n}) + w_n)^3 = 0.$$

Therefore

$$(\partial_t^2 - \Delta)(w - w_n) = -\big((S(t)(v_0, v_1) + w)^3 - S(t)(v_{0,n}, v_{1,n}) + w_n)^3\big).$$

We multiply the last equation by $\partial_t (w - w_n)$, and by using the Sobolev embedding $H^1(\mathbb{T}^3) \subset L^6(\mathbb{T}^3)$ and the Hölder inequality, we arrive at the bound

$$\frac{d}{dt}\|(w - w_n, \partial_t w - \partial_t w_n)\|_{\mathcal{H}^1(\mathbb{T}^3)}$$

$$\le C\big(\|S(t)(v_0 - v_{0,n}, v_1 - v_{1,n})\|_{L^6(\mathbb{T}^3)} + \|w - w_n\|_{H^1(\mathbb{T}^3)}\big)$$

$$\times \Big(\|S(t)(v_0, v_1)\|^2_{L^6(\mathbb{T}^3)} + \|S(t)(v_{0,n}, v_{1,n})\|^2_{L^6(\mathbb{T}^3)}$$

$$+ \|w\|^2_{H^1(\mathbb{T}^3)} + \|w_n\|^2_{H^1(\mathbb{T}^3)}\Big).$$

Using (4.2.28) and the properties of the solutions obtained in Theorem 4.2.3, we obtain

$$\frac{d}{dt}\|(w - w_n, \partial_t w - \partial_t w_n)\|_{\mathcal{H}^1(\mathbb{T}^3)}$$

$$\le C\big(\|S(t)(v_0 - v_{0,n}, v_1 - v_{1,n})\|_{L^6(\mathbb{T}^3)} + \|w - w_n\|_{H^1(\mathbb{T}^3)}\big)$$

$$\times \Big(\|S(t)(v_0, v_1)\|^2_{L^6(\mathbb{T}^3)} + \|S(t)(v_{0,n}, v_{1,n})\|^2_{L^6(\mathbb{T}^3)} + C\Big).$$

The last inequality implies the following bound for $t \in [0, T]$,

$$
\|(w(t) - w_n(t), \partial_t w(t) - \partial_t w_n(t))\|_{\mathcal{H}^1}
$$

$$
\leq C \int_0^t \|S(\tau)(v_0 - v_{0,n}, v_1 - v_{1,n})\|_{L^6}
$$

$$
\times \left(\|S(\tau)(v_0, v_1)\|_{L^6}^2 + \|S(\tau)(v_{0,n}, v_{1,n})\|_{L^6}^2 + C \right) d\tau
$$

$$
\times \exp \left(\int_0^t (\|S(\tau)(v_0, v_1)\|_{L^6}^2 + \|S(\tau)(v_{0,n}, v_{1,n})\|_{L^6}^2 + C) d\tau \right). \quad (4.2.29)
$$

More precisely, we used that if $x(t) \geq 0$ satisfies the differential inequality

$$
\dot{x}(t) \leq C z(t)(y(t) + x(t)), \quad x(0) = 0,
$$

for some $z(t) \geq 0$ and $y(t) \geq 0$ then

$$
x(t) \leq C \int_0^t y(\tau) z(\tau) d\tau \exp \left(\int_0^t z(\tau) d\tau \right).
$$

Coming back to (4.2.29) and using the Hölder inequality, we get for $t \in [0, T]$,

$$
\|(w(t) - w_n(t), \partial_t w(t) - \partial_t w_n(t))\|_{\mathcal{H}^1}
$$

$$
\leq C \|S(t)(v_0 - v_{0,n}, v_1 - v_{1,n})\|_{L_T^2 L^6}
$$

$$
\times \left(\|S(t)(v_0, v_1)\|_{L_T^4 L^6}^2 + \|S(t)(v_{0,n}, v_{1,n})\|_{L_T^4 L^6}^2 + C \right)
$$

$$
\times \exp \left(\int_0^t (\|S(\tau)(v_0, v_1)\|_{L^6}^2 + \|S(\tau)(v_{0,n}, v_{1,n})\|_{L^6}^2 + C) d\tau \right). \quad (4.2.30)
$$

Recalling (4.2.27), we obtain that for $1 < p < \infty$,

$$
\lim_{n \to \infty} \int_0^T \|S(\tau)(v_0 - v_{0,n}, v_1 - v_{1,n})\|_{L^6(\mathbb{T}^3)}^p d\tau = 0.
$$

Therefore (4.2.30) implies that

$$
\lim_{n \to \infty} \|(w(t) - w_n(t), \partial_t w(t) - \partial_t w_n(t))\|_{L^\infty([0,T];\mathcal{H}^1(\mathbb{T}^3))} = 0.
$$

Recall that

$$
v(t) = S(t)(v_0, v_1) + w(t), \quad v_n(t) = S(t)(v_{0,n}, v_{1,n}) + w_n(t).
$$

Using once again (4.2.27) and

$$\partial_t S(t)(v_{0,n}, v_{1,n}) = \rho_n \star \left(\partial_t S(t)(v_0, v_1)\right)$$

we get

$$\lim_{n\to\infty} \|(S(t)(v_0, v_1) - S(t)(v_{0,n}, v_{1,n}),$$

$$\partial_t S(t)(v_0, v_1) - \partial_t S(t)(v_{0,n}, v_{1,n}))\|_{L^\infty([0,T];\mathcal{H}^s(\mathbb{T}^3))} = 0$$

and consequently

$$\lim_{n\to\infty} \|(v(t) - v_n(t), \partial_t v(t) - \partial_t v_n(t))\|_{L^\infty([0,T];\mathcal{H}^s(\mathbb{T}^3))} = 0.$$

This completes the proof of Theorem 4.2.7.   □

Remark 4.2.21 In the proof of Theorem 4.2.7, we essentially used that the regularisation by convolution works equally well in H^s and L^p ($p < \infty$) and that it commutes with the Fourier multipliers such as the free evolution $S(t)$. Any other regularisation respecting these two properties would produce smooth solutions converging to the singular dynamics constructed in Theorem 4.2.3.

4.2.7 Conditioned Large Deviation Bounds

In this section, we prove conditioned large deviation bounds which are the main tool in the proof of Theorem 4.2.10.

Proposition 4.2.22 *Let $\mu \in \mathcal{M}^s$, $s \in (0, 1)$ and suppose that the real random variable with distribution θ, involved in the definition of μ is symmetric. Then for $\delta > 1 + \frac{1}{p_1}$, $2 \le p_1 < \infty$ and $2 \le p_2 \le \infty$ there exist positive constants c, C such that for every positive $\varepsilon, \lambda, \Lambda$ and A,*

$$\mu \otimes \mu\Big(((v_0, v_1), (v_0', v_1')) \in \mathcal{H}^s \times \mathcal{H}^s :$$

$$\|\langle t\rangle^{-\delta} S(t)(v_0 - v_0', v_1 - v_1')\|_{L^{p_1}(\mathbb{R}_t; L^{p_2}(\mathbb{T}^3))} > \lambda$$

$$or \ \|\langle t\rangle^{-\delta} S(t)(v_0 + v_0', v_1 + v_1')\|_{L^{p_1}(\mathbb{R}_t; L^{p_2}(\mathbb{T}^3))}$$

$$> \Lambda\Big|\|(v_0 - v_0', v_1 - v_1')\|_{\mathcal{H}^s(\mathbb{T}^3)} \le \varepsilon$$

$$and \ \|(v_0 + v_0', v_1 + v_1')\|_{\mathcal{H}^s(\mathbb{T}^3)} \le A\Big) \le C\Big(e^{-c\frac{\lambda^2}{\varepsilon^2}} + e^{-c\frac{\Lambda^2}{A^2}}\Big).$$

$$(4.2.31)$$

We shall make use of the following elementary lemmas.

Lemma 4.2.23 *For $j = 1, 2$, let E_j be two Banach spaces endowed with measures μ_j. Let $f : E_1 \times E_2 \to \mathbb{C}$ and $g_1, g_2 : E_2 \to \mathbb{C}$ be three measurable functions. Then*

$$\mu_1 \otimes \mu_2 \Big((x_1, x_2) \in E_1 \times E_2 : |f(x_1, x_2)| > \lambda \Big| |g_1(x_2)| \le \varepsilon, \ |g_2(x_2)| \le A \Big)$$

$$\le \sup_{x_2 \in E_2, |g_1(x_2)| \le \varepsilon, |g_2(x_2)| \le A} \mu_1 (x_1 \in E_1 : |f(x_1, x_2)| > \lambda),$$

where by sup *we mean the essential supremum.*

Lemma 4.2.24 *Let g_1 and g_2 be two independent identically distributed real random variables with symmetric distribution. Then $g_1 \pm g_2$ have symmetric distributions. Moreover if h is a Bernoulli random variable independent of g_1 then hg_1 has the same distribution as g_1.*

Proof of Proposition 4.2.22 Define

$$\mathcal{E} \equiv \mathbb{R} \times \mathbb{R}^{\mathbb{Z}_*^3} \times \mathbb{R}^{\mathbb{Z}_*^3},$$

equipped with the natural Banach space structure coming from the l^∞ norm. We endow $\mathcal{E}$ with a probability measure μ_0 defined via the map

$$\omega \mapsto \Big(k_0(\omega), \big(l_n(\omega) \big)_{n \in \mathbb{Z}_*^3}, \big(h_n(\omega) \big)_{n \in \mathbb{Z}_*^3} \Big),$$

where (k_0, l_n, h_n) is a system of independent Bernoulli variables.
  For $h = \big(x, (y_n)_{n \in \mathbb{Z}_*^3}, (z_n)_{n \in \mathbb{Z}_*^3} \big) \in \mathcal{E}$ and

$$u(x) = a + \sum_{n \in \mathbb{Z}_*^3} \Big(b_n \cos(n \cdot x) + c_n \sin(n \cdot x) \Big),$$

we define the operation $\odot$ by

$$h \odot u \equiv ax + \sum_{n \in \mathbb{Z}_*^3} \Big(b_n y_n \cos(n \cdot x) + c_n z_n \sin(n \cdot x) \Big).$$

Let us first evaluate the quantity

$$\mu \otimes \mu \Big(((v_0, v_1), (v_0', v_1')) \in \mathcal{H}^s \times \mathcal{H}^s :$$

$$\| \langle t \rangle^{-\delta} S(t)(v_0 - v_0', v_1 - v_1') \|_{L^{p_1}(\mathbb{R}_t; L^{p_2}(\mathbb{T}^3))} > \lambda \Big|$$

$$\| (v_0 - v_0', v_1 - v_1') \|_{\mathcal{H}^s(\mathbb{T}^3)} \le \varepsilon$$

$$\text{and } \| (v_0 + v_0', v_1 + v_1') \|_{\mathcal{H}^s(\mathbb{T}^3)} \le A \Big). \qquad (4.2.32)$$

Observe that, thanks to Lemma 4.2.24, (4.2.32) equals

$$\mu \otimes \mu \otimes \mu_0 \otimes \mu_0 \Big(((v_0, v_1), (v_0', v_1'), (h_0, h_1)) \in \mathcal{H}^s \times \mathcal{H}^s \times \mathcal{E} \times \mathcal{E} :$$

$$\|\langle t \rangle^{-\delta} S(t)(h_0 \odot (v_0 - v_0'), h_1 \odot (v_1 - v_1'))\|_{L^{p_1}(\mathbb{R}_t; L^{p_2}(\mathbb{T}^3))} > \lambda \Big|$$

$$\|(h_0 \odot (v_0 - v_0'), h_1 \odot (v_1 - v_1'))\|_{\mathcal{H}^s(\mathbb{T}^3)} \leq \varepsilon$$

$$\text{and } \|(h_0 \odot (v_0 + v_0'), h_1 \odot (v_1 + v_1'))\|_{\mathcal{H}^s(\mathbb{T}^3)} \leq A \Big). \tag{4.2.33}$$

Since the $H^s(\mathbb{T}^3)$ norm of a function f depends only on the absolute value of its Fourier coefficients, we deduce that (4.2.33) equals

$$\mu \otimes \mu \otimes \mu_0 \otimes \mu_0 \Big(((v_0, v_1), (v_0', v_1'), (h_0, h_1)) \in \mathcal{H}^s \times \mathcal{H}^s \times \mathcal{E} \times \mathcal{E} :$$

$$\|\langle t \rangle^{-\delta} S(t)(h_0 \odot (v_0 - v_0'), h_1 \odot (v_1 - v_1'))\|_{L^{p_1}(\mathbb{R}_t; L^{p_2}(\mathbb{T}^3))} > \lambda \Big|$$

$$\|(v_0 - v_0', v_1 - v_1')\|_{\mathcal{H}^s(\mathbb{T}^3)} \leq \varepsilon \text{ and } \|(v_0 + v_0', v_1 + v_1')\|_{\mathcal{H}^s(\mathbb{T}^3)} \leq A \Big). \tag{4.2.34}$$

We now apply Lemma 4.2.23 with $\mu_1 = \mu_0 \otimes \mu_0$ and $\mu_2 = \mu \otimes \mu$ to get that (4.2.34) is bounded by

$$\sup_{\|(v_0 - v_0', v_1 - v_1')\|_{\mathcal{H}^s(\mathbb{T}^3)} \leq \varepsilon} \mu_0 \otimes \mu_0 \Big((h_0, h_1) \in \mathcal{E} \times \mathcal{E} :$$

$$\|\langle t \rangle^{-\delta} S(t)(h_0 \odot (v_0 - v_0'), h_1 \odot (v_1 - v_1'))\|_{L^{p_1}(\mathbb{R}_t; L^{p_2}(\mathbb{T}^3))} > \lambda \Big). \tag{4.2.35}$$

We now apply Theorem 4.2.15 (with Bernoulli variables) to obtain that (4.2.32) is bounded by $C \exp(-c\frac{\lambda^2}{\varepsilon^2})$. A very similar argument gives that

$$\mu \otimes \mu \Big(((v_0, v_1), (v_0', v_1')) \in \mathcal{H}^s \times \mathcal{H}^s :$$

$$\|\langle t \rangle^{-\delta} S(t)(v_0 + v_0', v_1 + v_1')\|_{L^{p_1}(\mathbb{R}_t; L^{p_2}(\mathbb{T}^3))} > \Lambda \Big|$$

$$\|(v_0 - v_0', v_1 - v_1')\|_{\mathcal{H}^s(\mathbb{T}^3)} \leq \varepsilon$$

$$\text{and } \|(v_0 + v_0', v_1 + v_1')\|_{\mathcal{H}^s(\mathbb{T}^3)} \leq A \Big)$$

is bounded by $C \exp(-c\frac{\Lambda^2}{A^2})$. This completes the proof of Proposition 4.2.22. $\qquad \square$

4.2.8   End of the Proof of the Conditioned Continuous Dependence

In this section, we complete the proof of Theorem 4.2.10. According to (a variant of) Proposition 4.2.22, we have that for any

$$2 \le p_1 < +\infty, \ 2 \le p_2 \le +\infty, \ \delta > 1 + \frac{1}{p_1}, \ \eta \in (0, 1),$$

one has

$$\mu \otimes \mu\Big((V_0, V_1) \in \mathcal{H}^s \times \mathcal{H}^s : \ \|\langle t \rangle^{-\delta} S(t)(V_0 - V_1)\|_{L^{p_1}(\mathbb{R}_t; L^{p_2}(\mathbb{T}^3))} > \eta^{\frac{1}{2}}$$

$$\text{or } \|\langle t \rangle^{-\delta} S(t)(V_0)\|_{L^{p_1}(\mathbb{R}_t; L^{p_2}(\mathbb{T}^3))} > \log\log\log(\eta^{-1})$$

$$\text{or } \|\langle t \rangle^{-\delta} S(t)(V_1)\|_{L^{p_1}(\mathbb{R}_t; L^{p_2}(\mathbb{T}^3))} > \log\log\log(\eta^{-1})\Big|$$

$$\|V_0 - V_1\|_{\mathcal{H}^s(\mathbb{T}^3)} < \eta \ \text{ and } \ \|V_j\|_{\mathcal{H}^s(\mathbb{T}^3)} \le A, \ j = 0, 1\Big) \longrightarrow 0,$$

as $\eta \to 0$. Therefore, we can also suppose that

$$\|\langle t \rangle^{-\delta} S(t)(V_0 - V_1)\|_{L^{p_1}(\mathbb{R}_t; L^{p_2}(\mathbb{T}^3))} \le \eta^{\frac{1}{2}} \qquad (4.2.36)$$

and

$$\|\langle t \rangle^{-\delta} S(t)(V_j)\|_{L^{p_1}(\mathbb{R}_t; L^{p_2}(\mathbb{T}^3))} \le \log\log\log(\eta^{-1}), \ j = 0, 1, \qquad (4.2.37)$$

when we estimate the needed conditional probability.

We therefore need to estimate the difference of two solutions under the assumptions (4.2.36) and (4.2.37), in the regime $\eta \ll 1$. Let

$$v_j(t) = S(t)(V_j) + w_j(t), \quad j = 0, 1$$

be two solutions of the cubic wave equation with data V_j. We thus have

$$(w_j(0), \partial_t w_j(0)) = (0, 0).$$

Applying the energy estimate, performed several times in this section, for $j = 0, 1$, we get the bound

$$\frac{d}{dt} \mathcal{E}^{1/2}(w_j(t)) \le C\Big(\|S(t)(V_j)\|^3_{L^6(\mathbb{T}^3)} + \|S(t)(V_j)\|_{L^\infty(\mathbb{T}^3)} \mathcal{E}^{1/2}(w_j(t))\Big),$$

and therefore, under the assumptions (4.2.36) and (4.2.37), for $t \in [0, T]$ one has

$$\mathcal{E}^{1/2}(w_j(t)) \leq C_T \, e^{C_T \log\log\log(\eta^{-1})} (\log\log\log(\eta^{-1}))^3 \qquad (4.2.38)$$

$$\leq C_T [\log(\eta^{-1})]^{\frac{1}{20}},$$

where here and in the sequel we denote by C_T different constants depending only on T (but independent of η).

We next estimate the difference $w_0 - w_1$. Using the equations solved by w_0, w_1, we infer that

$$\frac{d}{dt} \| w_0(t, \cdot) - w_1(t, \cdot) \|_{\mathcal{H}^1(\mathbb{T}^3)}^2$$

$$\leq 2 \left| \int_{\mathbb{T}^3} \partial_t(w_0(t, x) - w_1(t, x))(\partial_t^2 - \Delta)(w_0(t, x) - w_1(t, x)) dx \right|$$

$$\leq C \| w_0(t, \cdot) - w_1(t, \cdot) \|_{\mathcal{H}^1(\mathbb{T}^3)}$$

$$\| (w_0 + S(t)(V_0))^3 - (w_1 + S(t)(V_1))^3 \|_{L^2(\mathbb{T}^3)}, \qquad (4.2.39)$$

where for shortness we denote $\|(u, \partial_t u)\|_{\mathcal{H}^1}$ simply by $\|u\|_{\mathcal{H}^1}$.

Thanks to (4.2.39) and the Sobolev embedding $H^1(\mathbb{T}^3) \subset L^6(\mathbb{T}^3)$, we get that

$$\frac{d}{dt} \| w_0(t, \cdot) - w_1(t, \cdot) \|_{\mathcal{H}^1(\mathbb{T}^3)}$$

is bounded by

$$C \Big(\| w_0(t, \cdot) - w_1(t, \cdot) \|_{\mathcal{H}^1(\mathbb{T}^3)} + \| S(t)(V_0 - V_1) \|_{L^6(\mathbb{T}^3)} \Big)$$

$$\Big(\| w_0(t, \cdot) \|_{H^1(\mathbb{T}^3)}^2 + \| w_1(t, \cdot) \|_{H^1(\mathbb{T}^3)}^2$$

$$+ \| S(t)(V_0) \|_{L^6(\mathbb{T}^3)}^2 + \| S(t)(V_1) \|_{L^6(\mathbb{T}^3)}^2 \Big).$$

Therefore, using (4.2.38) and the Gronwall lemma, under the assumptions (4.2.36) and (4.2.37), for $t \in [0, T]$,

$$\| w_0(t, \cdot) - w_1(t, \cdot) \|_{\mathcal{H}^1(\mathbb{T}^3)} \leq C_T \eta^{\frac{1}{2}} [\log(\eta^{-1})]^{\frac{1}{10}} e^{C_T [\log(\eta^{-1})]^{\frac{1}{10}}}$$

$$\leq C_T \eta^{\frac{1}{4}}.$$

In particular by the Sobolev embedding

$$\| w_0 - w_1 \|_{L^4([0, T] \times \mathbb{T}^3)} \leq C_T \eta^{\frac{1}{4}},$$

and therefore under the assumption (4.2.36),

$$\|v_0 - v_1\|_{L^4([0,T] \times \mathbb{T}^3)} \leq C_T \eta^{\frac{1}{4}}.$$

In summary, we obtained that for a fixed $\varepsilon > 0$, the $\mu \otimes \mu$ measure of V_0, V_1 such that

$$\|\Phi(t)(V_0) - \Phi(t)(V_1)\|_{X_T} > \varepsilon$$

under the conditions (4.2.36), (4.2.37) and $\|V_0 - V_1\|_{\mathcal{H}^s} < \eta$ is zero, as far as $\eta > 0$ is sufficiently small. Therefore, we obtain that the left hand side of (4.2.7) tends to zero as $\eta \to 0$. This ends the proof of the first part of Theorem 4.2.10.

For the second part of the proof of Theorem 4.2.10, we argue by contradiction. Suppose thus that for every $\varepsilon > 0$ there exist $\eta > 0$ and Σ of full $\mu \otimes \mu$ measure such that

$$\forall (V, V') \in \Sigma \cap (B_A \times B_A), \ \|V - V'\|_{\mathcal{H}^s} < \eta$$

$$\implies \|\Phi(t)(V) - \Phi(t)(V')\|_{X_T} < \varepsilon.$$

Let us apply the previous affirmation with $\varepsilon = 1/n, n = 1, 2, 3 \ldots$ which produces full measure sets $\Sigma(n)$. Set

$$\Sigma_1 \equiv \bigcap_{n=1}^{\infty} \Sigma(n).$$

Then Σ_1 is of full $\mu \otimes \mu$ measure and we have that

$$\forall \varepsilon > 0, \ \exists \eta > 0, \ \forall (V, V') \in \Sigma_1 \cap (B_A \times B_A),$$

$$\|V - V'\|_{\mathcal{H}^s} < \eta \implies \|\Phi(t)(V) - \Phi(t)(V')\|_{X_T} < \varepsilon. \quad (4.2.40)$$

Next for $V \in \mathcal{H}^s$ we define $\mathcal{A}(V) \subset \mathcal{H}^s$ by

$$\mathcal{A}(V) \equiv \{V' \in \mathcal{H}^s : (V, V') \in \Sigma_1\}.$$

According to Fubini Theorem, there exists $\mathcal{E} \subset \mathcal{H}^s$ a set of full μ measure such that for every $V \in \mathcal{E}$ the set $\mathcal{A}(V)$ is a full μ measure.

We are going to extend $\Phi(t)$ to a uniformly continuous map on B_A. For that purpose, we first extend $\Phi(t)$ to a uniformly continuous map on dense set of B_A. Let $\{(V_j)_{j \in \mathbb{N}}\}$ be a dense set of B_A for the $\mathcal{H}^s$ topology. For $j \in \mathbb{N}$, we can construct by induction a sequence $(V_{j,n})$ such that

$$V_{j,n} \in B_A \cap \mathcal{E} \cap \bigcap_{m < n} \mathcal{A}(V_{j,m}) \cap \bigcap_{l < j, q \in \mathbb{N}} \mathcal{A}(V_{l,q}), \quad \|V_{j,n} - V_j\|_{\mathcal{H}^s} < 1/n.$$

Indeed, the induction assumption guarantees that the set

$$\mathcal{E} \cap \bigcap_{m<n} \mathcal{A}(V_{j,m}) \bigcap_{l<j,q\in\mathbb{N}} \mathcal{A}(V_{l,q})$$

has measure 1 (as an intersection of sets of measure 1) and consequently is dense. Notice that by construction, we have

$$(V_{k,n}, V_{l,m}) \in \Sigma_1, \forall k < l, \forall n, m \in \mathbb{N}, \text{ and } \forall k = l, n < m. \tag{4.2.41}$$

Using (4.2.41) for $k = l$, we obtain according to (4.2.40) that for any fixed k, the sequence $\Phi(t)(V_{k,n})_{n\in\mathbb{N}}$ is a Cauchy sequence in X_T and we can define $\overline{\Phi(t)}(V_j)$ as its limit. Using again (4.2.41), for $k \neq l$, we see according to (4.2.40) that the map $\overline{\Phi(t)}$ is uniformly continuous on the set $\{(V_j)_{j\in\mathbb{N}}\}$. Therefore $\Phi(t)$ can be extended by density to a uniformly continuous map, on the whole B_A. Let us denote by $\overline{\Phi(t)}$ the extension of $\Phi(t)$ to B_A. We therefore have

$$\forall \varepsilon > 0, \ \exists \eta > 0, \ \forall V, V' \in B_A,$$

$$\|V - V'\|_{\mathcal{H}^s} < \eta \implies \|\overline{\Phi(t)}(V) - \overline{\Phi(t)}(V')\|_{X_T} < \varepsilon. \tag{4.2.42}$$

We have the following lemma.

Lemma 4.2.25 *For $V \in (C^\infty(\mathbb{T}^3) \times C^\infty(\mathbb{T}^3)) \cap B_A$, we have that $\overline{\Phi(t)}(V) = (u, u_t)$, where u is the unique classical solution on $[0, T]$ of*

$$(\partial_t^2 - \Delta)u + u^3 = 0, \quad (u(0), \partial_t u(0)) = V.$$

Proof Let us first show that first component of

$$\overline{\Phi(t)}(V) \equiv (\overline{\Phi_1(t)}(V), \overline{\Phi_2(t)}(V))$$

is a solution of the cubic wave equation. Observe that by construction, necessarily $\overline{\Phi_2(t)}(V) = \partial_t \overline{\Phi_1(t)}(V)$ in the distributional sense (in $\mathcal{D}'((0, T) \times \mathbb{T}^3)$).

Again by construction, we have that

$$V = \lim_{n\to\infty} V_n,$$

in $\mathcal{H}^s$ where V_n are such that

$$(\partial_t^2 - \Delta)(\Phi_1(t)(V_n)) + (\Phi_1(t)(V_n))^3 = 0, \tag{4.2.43}$$

with the notation $\Phi(t) = (\Phi_1(t), \Phi_2(t))$. In addition,

$$\overline{\Phi(t)}(V) = \lim_{n\to\infty} \Phi(t)(V_n),$$

in X_T. We therefore have that

$$(\partial_t^2 - \Delta)(\overline{\Phi_1(t)}(V)) = \lim_{n \to \infty} (\partial_t^2 - \Delta)(\Phi_1(t)(V_n)),$$

in the distributional sense. Moreover, coming back to the definition of X_T, we also obtain that

$$(\overline{\Phi_1(t)}(V))^3 = \lim_{n \to \infty} (\Phi_1(t)(V_n))^3,$$

in $L^{4/3}([0, T] \times \mathbb{T}^3)$. Therefore, passing into the limit $n \to \infty$ in ((4.2.43)), we obtain that $\overline{\Phi_1(t)}(V)$ solves the cubic wave equation (with data V). Moreover, since $(\overline{\Phi_1(t)}(V))^3 \in L^{4/3}([0, T] \times \mathbb{T}^3)$, it also satisfies the Duhamel formulation of the equation.

Let us denote by $u(t)$, $t \in [0, T]$ the classical solution of

$$(\partial_t^2 - \Delta)u + u^3 = 0, \quad (u(0), \partial_t u(0)) = V,$$

defined by Theorem 4.1.2. Set $v \equiv \overline{\Phi_1(t)}(V)$. Since our previous analysis has shown that v is a solution of the cubic wave equation, we have that

$$(\partial_t^2 - \Delta)(u - v) + u^3 - v^3 = 0, \quad (u(0), \partial_t u(0)) = (0, 0). \tag{4.2.44}$$

We now invoke the $L^4 - L^{4/3}$ non homogenous estimates for the three dimensional wave equation. Namely, thanks to Theorem 4.1.21, we have that there exists a constant (depending on T) such that for every interval $I \subset [0, T]$, the solution of the wave equation

$$(\partial_t^2 - \Delta)w = F, \quad (u(0), \partial_t u(0)) = (0, 0)$$

satisfies

$$\|w\|_{L^4(I \times \mathbb{T}^3)} \le C \|F\|_{L^{4/3}(I \times \mathbb{T}^3)}. \tag{4.2.45}$$

Applying (4.2.45) in the context of (4.2.44) together with the Hölder inequality yields the bound

$$\|u - v\|_{L^4(I \times \mathbb{T}^3)} \le C \big(\|u\|_{L^4(I \times \mathbb{T}^3)}^2 + \|v\|_{L^4(I \times \mathbb{T}^3)}^2\big) \|u - v\|_{L^4(I \times \mathbb{T}^3)}. \tag{4.2.46}$$

Since $u, v \in L^4(I \times \mathbb{T}^3)$, we can find a partition of intervals $I_1, \ldots, I_l$ of $[0, T]$ such that

$$C \big(\|u\|_{L^4(I_j \times \mathbb{T}^3)}^2 + \|v\|_{L^4(I_j \times \mathbb{T}^3)}^2\big) < \frac{1}{2}, \quad j = 1, \ldots, l.$$

We now apply (4.2.46) with $I = I_j$, $j = 1, \ldots, l$ to conclude that $u = v$ on I_1, then on I_2 and so on up to I_l which gives that $u = v$ on $[0, T]$. Thus $u = \overline{\Phi_1(t)}(V)$ and therefore also $\partial_t u = \overline{\Phi_2(t)}(V)$. This completes the proof of Lemma 4.2.25. □

It remains now to apply Lemma 4.2.25 to the sequence of smooth data in the statement of Theorem 4.1.28 to get a contradiction with (4.2.42). More precisely, if (U_n) is the sequence involved in the statement of Theorem 4.1.28, the result of Theorem 4.1.28 affirms that $\overline{\Phi(t)}(U_n)$ tends to infinity in $L^\infty([0, T]; \mathcal{H}^s)$ while (4.2.42) affirms that the same sequence tends to zero in the same space $L^\infty([0, T]; \mathcal{H}^s)$. This completes the proof of Theorem 4.2.10.

4.2.9 Extensions to More General Nonlinearities

In the remarkable work by Oh-Pocovnicu [35, 36] (based on the previous contributions [2, 38]) it is shown that the result of Theorem 4.2.3 can be extended to the energy critical equation

$$(\partial_t^2 - \Delta)v + v^5 = 0.$$

This equation is H^1 critical and the data is a typical element with respect to $\mu \in \mathcal{M}^s$, $s > 1/2$. We refer also to [31, 43] for extensions of Theorem 4.2.3 to nonlinearities between cubic and quintic.

4.2.10 Notes

For the case $s = 0$ and the proof of the quantitative bounds displayed by Theorem 4.2.4, we refer to [8]. For the proof of Proposition 4.2.2, we refer to [6, Appendix B] and [8, Appenidix B2]. The probabilistic part of our analysis only relies on linear bounds such as Lemma 4.2.11. In other situations multi-linear versions of these bounds are of importance (see [4, 13, 33]). The above mentioned work by Oh-Pocovnicu relies on a much more complicated deterministic analysis (such as the concentration compactness) and also on a significant extension of the probabilistic energy bound used in the proof of Theorem 4.2.3.

Our starting point and main motivation toward the probabilistic well-posedness results presented in this section was the ill-posedness result of Theorem 4.1.28 of the previous section. As already mentioned the method of proof has some similarities with the earlier work [16] or with the even earlier work of Bourgain [4] on the invariance of the Gibbs measure associated with the nonlinear Schrödinger equation

$$(i\partial_t + \Delta)u = |u|^2 u, \tag{4.2.47}$$

posed on the two dimensional torus. The main purpose of [4] is to show the invariance of the Gibbs measure and as a byproduct one gets the global existence and uniqueness of solutions of (4.2.47) with a suitable random data belonging a.s. to $H^{-\varepsilon}(\mathbb{T}^2)$ for every $\varepsilon > 0$ but missing a.s. $L^2(\mathbb{T}^2)$. In the time of writing of [4] statement such as Theorem 4.1.28 or Theorem 4.1.33 were not known in the context of (4.2.47). In the recent work [34], the analogue of Theorems 4.1.28 and 4.1.33 in the context of (4.2.47) is obtained. Most likely, the analysis of [4] can be adapted in order to get the analogue of Theorem 4.2.7 in the context of (4.2.47). As a consequence, it looks that we can see from the same view point (4.2.47) with data on the support of the Gibbs measure and the cubic defocusing wave equation with random data of super-critical regularity presented in these lectures. We plan to address this issue in a future work.

4.3 Random Data Global Well-Posedness with Data of Supercritical Regularity via Invariant Measures

In the previous section, we presented a method to construct global in time solutions for the cubic defocusing wave equation posed on the three dimensional torus with random data of supercritical regularity $(H^s(\mathbb{T}^3) \times H^{s-1}(\mathbb{T}^3), s \in (0, 1/2))$. These solutions are unique in a suitable sense and depend continuously (in a conditional sense) on the initial data. The method we used is based on a local in time result showing that even if the data is of supercritical regularity, we can find a local solution written as "free evolution" (keeping the Sobolev regularity of the initial data) plus "a remainder of higher regularity". The term of higher regularity is then regular enough to allow us to deal it with the deterministic methods to treat the equation. The globalisation was then done by establishing an energy bound for the remainder in a probabilistic manner, here of course the energy conservation law is the key structure allowing to perform the analysis. Moreover, we have shown that the problem is ill-posed with data of this supercritical regularity and this in turn implied the impossibility to see the constructed flow as the unique extension of the regular solutions flow.

   In this section, we will show another method for global in time solutions for a defocusing wave equation with data of supercritical regularity. The construction of local solutions will be based on the same principle as in the previous section, i.e. we shall again see the solution as a "free evolution" plus "a remainder of higher regularity". However the globalisation will be done by a different argument (due to Bourgain [3, 4]) based on exploiting the invariance of the Gibbs measure associated with the equation. The Gibbs measure is constructed starting from the energy conservation law and therefore this energy conservation law is again the key structure allowing to perform the global in time analysis. This method of globalisation by invariant measures is working only for very particular choice of the initial data and in this sense it is much less general than the method presented in the previous section. On the other hand the method based on exploiting invariant measures gives a strong macroscopic information about the constructed flow, namely one has a precise information on the measure evolution along the time. The method presented in the previous section gives essentially no information about the evolution in time under the constructed flow of the measures in $\mathcal{M}^s$. We shall come back to this issue in the next section.

Our model to present the method of globalisation via invariant measures will be the radial nonlinear wave equation posed on the unit ball of $\mathbb{R}^3$, with Dirichlet boundary conditions. Let Θ be the unit ball of $\mathbb{R}^3$. Consider the nonlinear wave equation with Dirichlet boundary condition posed on Θ,

$$(\partial_t^2 - \Delta)w + |w|^\alpha w = 0, \quad (w, \partial_t w)|_{t=0} = (f_1, f_2), \quad \alpha > 0, \qquad (4.3.1)$$

subject to Dirichlet boundary conditions

$$u\,|_{\mathbb{R}_t \times \partial\Theta} = 0,$$

with radial real valued initial data (f_1, f_2).

We now make some algebraic manipulations on (4.3.1) allowing to write it as a first order equation in t. Set $u \equiv w + i\sqrt{-\Delta}^{-1}\partial_t w$. Observe that Δ^{-1} is well-defined because 0 is not in the spectrum of the Dirichlet Laplacian. Then we have that u solves the equation

$$(i\partial_t - \sqrt{-\Delta})u - \sqrt{-\Delta}^{-1}\left(|\text{Re}(u)|^\alpha \text{Re}(u)\right) = 0, \quad u|_{t=0} = u_0, \qquad (4.3.2)$$

with $u|_{\mathbb{R}\times\partial\Theta} = 0$, where $u_0 = f_1 + i\sqrt{-\Delta}^{-1} f_2$. We consider (4.3.2) for data in the (complex) Sobolev spaces $H_{rad}^s(\Theta)$ of radial functions.

Equation (4.3.2) is (formally) an Hamiltonian equation on $L^2(\Theta)$ with Hamiltonian,

$$\frac{1}{2}\|\sqrt{-\Delta}(u)\|_{L^2(\Theta)}^2 + \frac{1}{\alpha+2}\|\text{Re}(u)\|_{L^{\alpha+2}(\Theta)}^{\alpha+2} \qquad (4.3.3)$$

which is (formally) conserved by the flow of (4.3.2).

Let us next discuss the measure describing the initial data set. For $s < 1/2$, we define the measure μ on $H_{rad}^s(\Theta)$ as the image measure under the map from a probability space $(\Omega, \mathcal{A}, p)$ to $H_{rad}^s(\Theta)$ equipped with the Borel sigma algebra, defined by

$$\omega \longmapsto \sum_{n=1}^\infty \frac{h_n(\omega) + il_n(\omega)}{n\pi} e_n, \qquad (4.3.4)$$

where $((h_n, l_n))_{n=1}^\infty$ is a sequence of independent standard real Gaussian random variables. In (4.3.4), the functions $(e_n)_{n=1}^\infty$ are the radial eigenfunctions of the

Dirichlet Laplacian on Θ, associated with eigenvalues $(\pi n)^2$. The eigenfunctions e_n have the following explicit form

$$e_n(r) = \frac{\sin(n\pi r)}{r}, \quad 0 \le r \le 1.$$

They are the analogues of $\cos(n \cdot x)$ and $\sin(n \cdot x)$, $n \in \mathbb{Z}^3$ used in the analysis on $\mathbb{T}^3$ in the previous section. One has that $\mu(H_{rad}^{1/2}(\Theta)) = 0$. By the method described in the previous section one may show that (4.3.2) is ill-posed in $H_{rad}^s(\Theta)$ for $s < \frac{3}{2} - \frac{2}{\alpha}$. Therefore for $\alpha > 2$ the map (4.3.4) describes functions of supercritical Sobolev regularity (i.e. $H_{rad}^s(\Theta)$ with s smaller than $\frac{3}{2} - \frac{2}{\alpha}$). The situation is therefore similar to the analysis of the cubic defocusing wave equation on $\mathbb{T}^3$ with data in $H^s \times H^{s-1}$, $s < 1/2$ considered in the previous section. As in the previous section, we can still get global existence and uniqueness for (4.3.2), almost surely with respect to μ.

Theorem 4.3.1 *Let $s < 1/2$. Suppose that $\alpha < 3$. Let us fix a real number p such that $\max(4, 2\alpha) < p < 6$. Then there exists a full μ measure set $\Sigma \subset H_{rad}^s(\Theta)$ such that for every $u_0 \in \Sigma$ there exists a unique global solution of (4.3.2)*

$$u \in C(\mathbb{R}, H_{rad}^s(\Theta)) \cap L_{loc}^p(\mathbb{R}_t; L^p(\Theta)).$$

The solution can be written as

$$u(t) = S(t)(u_0) + v(t),$$

where $S(t) = e^{-it\sqrt{-\Delta}}$ is the free evolution and $v(t) \in H_{rad}^\sigma(\Theta)$ for some $\sigma > 1/2$. Moreover

$$\|u(t)\|_{H^s(\Theta)} \le C(s)\left(\log(2 + |t|)\right)^{\frac{1}{2}}.$$

The proof of Theorem 4.3.1 is based on the following local existence result.

Proposition 4.3.2 *For a given positive number $\alpha < 3$ we choose a real number p such that $\max(4, 2\alpha) < p < 6$. Then we fix a real number σ by $\sigma = \frac{3}{2} - \frac{4}{p}$. There exist $C > 0$, $c \in (0, 1]$, $\gamma > 0$ such that for every $R \ge 1$ if we set $T = cR^{-\gamma}$ then for every radially symmetric u_0 satisfying*

$$\|S(t)u_0\|_{L^p((0,2)\times\Theta)} \le R$$

there exists a unique solution u of (4.3.2) such that

$$u(t) = S(t)u_0 + v(t)$$

with $v \in X_T^\sigma$ (the Strichartz spaces defined in the previous section). Moreover

$$\|v\|_{X_T^\sigma} \le CR.$$

In particular, since $S(t)$ is 2 periodic and thanks to the Strichartz estimates,

$$\sup_{t \in [-T,T]} \|S(\tau)u(t)\|_{L^p(\tau \in (0,2); L^p(\Theta))} \le CR.$$

In addition, if $u_0 \in H^s(\Theta)$ (and thus $s < \sigma$) then

$$\|u(t)\|_{H^s(\Theta)} \le \|S(t)u_0\|_{H^s(\Theta)} + \|v(t)\|_{H^s(\Theta)} \le \|u_0\|_{H^s(\Theta)} + CR.$$

Using probabilistic Strichartz estimates for $S(t)$ as we did in the previous section, we can deduce the following corollary of Proposition 4.3.2.

Proposition 4.3.3 *Under the assumptions of Proposition 4.3.2 there is a set Σ of full μ measure such that for every $u_0 \in \Sigma$ there is $T > 0$ and a unique solution of (4.3.2) on $[0, T]$ in*

$$C([0, T], H^s_{rad}(\Theta)) \cap L^p_{loc}(\mathbb{R}_t; L^p(\Theta)).$$

Moreover for every $T \le 1$ there is a set $\Sigma_T \subset \Sigma$ such that

$$\mu(\Sigma_T) \ge 1 - Ce^{-c/T^\delta}, \quad c > 0, \delta > 0$$

and such that for every $u_0 \in \Sigma_T$ the time of existence is at least T.

Let us next define the Gibbs measures associated with (4.3.2). Using [1, Theorem 4], we have that for $\alpha < 4$ the quantity

$$\|\sum_{n=1}^{\infty} \frac{h_n(\omega) + il_n(\omega)}{n\pi} e_n\|_{L^{\alpha+2}(\Theta)} \tag{4.3.5}$$

is finite almost surely. Moreover the restriction $\alpha < 4$ is optimal because for $\alpha = 4$ the quantity (4.3.5) is infinite almost surely. Therefore, for $\alpha < 4$, we can define a nontrivial measure ρ as the image measure on $H^s_{rad}(\Theta)$ by the map (4.3.4) of the measure

$$\exp\left(-\frac{1}{\alpha+2}\|\sum_{n=1}^{\infty} \frac{h_n(\omega)}{n\pi} e_n\|_{L^{\alpha+2}(\Theta)}^{\alpha+2}\right) dp(\omega). \tag{4.3.6}$$

The measure ρ is the Gibbs measures associated with (4.3.2) and it can be formally seen as

$$\exp\left(-\frac{1}{2}\|\sqrt{-\Delta}(u)\|_{L^2(\Theta)}^2 - \frac{1}{\alpha+2}\|\text{Re}(u)\|_{L^{\alpha+2}(\Theta)}^{\alpha+2}\right) du,$$

where a renormalisation of

$$\exp\left(-\frac{1}{2}\|\sqrt{-\Delta}(u)\|_{L^2(\Theta)}^2\right)du,$$

corresponds to the measure μ and

$$\exp\left(-\frac{1}{\alpha+2}\|\mathrm{Re}(u)\|_{L^{\alpha+2}(\Theta)}^{\alpha+2}\right),$$

corresponds to the density in (4.3.6). Thanks to the conservation of the Hamiltonian (4.3.3), the measure ρ is expected to be invariant under the flow of (4.3.2). This expectation is also supported by the fact that the vector field defining (4.3.2) is (formally) divergence free. This fact follows again from the Hamiltonian structure of (4.3.2).

Observe that if a Borel set $A \subset H^s(\Theta)$ is of full ρ measure then A is also of full μ measure. Therefore, it suffices to solve (4.3.2) globally in time for u_0 in a set of full ρ measure.

We now explain how the local existence result of Proposition 4.3.2 can be combined with invariant measure considerations in order to get global existence of the solution. The details can be found in [7]. Consider a truncated version of (4.3.2)

$$(i\partial_t - \sqrt{-\Delta})u - S_N\left(\sqrt{-\Delta}^{-1}\left(|S_N\mathrm{Re}(u)|^\alpha S_N\mathrm{Re}(u)\right)\right) = 0, \qquad (4.3.7)$$

where S_N is a suitable "projector" tending to the identity as N goes to infinity. Let us denote by $\Phi_N(t)$ the flow of (4.3.7). This flow is well-defined for a fixed N because for frequencies $\gg N$ it is simply the linear flow and for the remaining frequencies one can use that (4.3.7) has the preserved energy

$$\frac{1}{2}\|\sqrt{-\Delta}(u)\|_{L^2(\Theta)}^2 + \frac{1}{\alpha+2}\|S_N\mathrm{Re}(u)\|_{L^{\alpha+2}(\Theta)}^{\alpha+2}. \qquad (4.3.8)$$

The energy (4.3.8) allows us to define an approximated Gibbs measure ρ_N. One has that ρ_N is invariant under $\Phi_N(t)$ by the Liouville theorem and the invariance of complex Gaussians under rotations (for the frequencies $\gg N$). In addition, ρ_N converges in a strong sense to ρ as $N \to \infty$.

Let us fix $T \gg 1$ and a small $\epsilon > 0$. Our goal is to find a set of ρ residual measure $< \epsilon$ such that for initial data in this set we can solve (4.3.2) up to time T.

The local existence theory implies that as far as

$$\|S(t)u\|_{L^p(\tau\in(0,2);L^p(\Theta))} \le R, \quad R \ge 1 \qquad (4.3.9)$$

we can define the solution of the true equation with datum u for times of order $R^{-\gamma}$, $\gamma > 0$, the bound (4.3.9) is propagated and moreover *on the interval of existence this solution is the limit as $N \to \infty$ of the solutions of the truncated equation* (4.3.7) *with the same datum.*

Our goal is to show that with a suitably chosen $R = R(T, \varepsilon)$ we can propagate the bound (4.3.9) for the solutions of the approximated equation (4.3.7) (for $N \gg 1$) up to time T for initial data in a set of residual ρ measure $\lesssim \varepsilon$.

For $R > 1$, we define the set B_R as

$$B_R = \{u : \|S(t)u\|_{L^p(\tau \in (0,2); L^p(\Theta))} \leq R\}.$$

As mentioned the (large) number R will be determined depending on T and ε. Thanks to the probabilistic Strichartz estimates for $S(t)$, we have the bound

$$\rho(B_R^c) < e^{-\kappa R^2} \tag{4.3.10}$$

for some $\kappa > 0$. Let $\tau \approx R^{-\gamma}$ be the local existence time associated to R given by Proposition 4.3.2. Define the set B by

$$B = \bigcap_{k=0}^{[T/\tau]} \Phi_N(-k\tau)(B_R). \tag{4.3.11}$$

Thanks to the local theory, we can propagate (4.3.9) for data in B up to time T. On the other hand, using the invariance of ρ_N under $\Phi_N(t)$ and (4.3.10), we obtain that

$$\rho_N(B^c) \lesssim TR^\gamma e^{-\kappa R^2}.$$

We now choose R so that

$$TR^\gamma e^{-\kappa R^2} \sim \varepsilon.$$

In other words

$$R \sim \left(\log\left(\frac{T}{\varepsilon}\right)\right)^{\frac{1}{2}}.$$

This fixes the value of R. With this choice of R, $\rho(B^c) < \varepsilon$, provided $N \gg 1$. With this value of R the set B defined by (4.3.11) is such that for data in B we have the bound (4.3.9) up to time T on a set of residual ρ measure $< \varepsilon$. Now, we can pass to the limit $N \to \infty$ thanks to the above mentioned consequence of the local theory and hence defining the solution of the true equation (4.3.2) up to time T for data in a set of ρ residual measure $< \varepsilon$.

We now apply the last conclusion with $T = 2^j$ and $\varepsilon/2^j$. This produces a set $\Sigma_{j,\varepsilon}$ such that $\rho((\Sigma_{j,\varepsilon})^c) < \varepsilon/2^j$ an for $u_0 \in \Sigma_{j,\varepsilon}$ we can solve (4.3.2) up to time 2^j. We next set

$$\Sigma_\varepsilon = \bigcap_{j=1}^{\infty} \Sigma_{j,\varepsilon}.$$

Clearly, we have $\rho((\Sigma_\varepsilon)^c) < \varepsilon$ and for $u_0 \in \Sigma_\varepsilon$, we can define a global solution of (4.3.2). Finally

$$\Sigma = \bigcup_{j=1}^{\infty} \Sigma_{2-j}$$

is a set of full ρ measure on which we can define globally the solutions of (4.3.2). The previous construction also keeps enough information allowing to get the claimed uniqueness property.

Remark 4.3.4 In [5], a part of the result of Theorem 4.3.1 was extended to $\alpha < 4$ which is the full range of the definition of the measure ρ.

Remark 4.3.5 The previous discussion has shown that we have two methods to globalise the solutions in the context of random data well-posedness for the non-linear wave equation. The one of the previous section is based on energy estimates while the method of this section is based on invariant measures considerations. It is worth mentioning that these two methods are also employed in the context of singular stochastic PDE's. More precisely in [32] the globalisation is done via the (more flexible) method of energy estimates while in [25] one globalises by exploiting invariant measures considerations.

4.4   Quasi-Invariant Measures

4.4.1   Introduction

4.4.1.1   Motivation

In Sect. 4.2, for each $s \in (0, 1)$ we introduced a family of measures $\mathcal{M}^s$ on the Sobolev space $\mathcal{H}^s(\mathbb{T}^3) = H^s(\mathbb{T}^3) \times H^{s-1}(\mathbb{T}^3)$. Then for each $\mu \in \mathcal{M}^s$, we succeeded to define a unique global flow $\Phi(t)$ of the cubic defocusing wave equation a.s. with respect to μ. This result is of interest for the solvability of the Cauchy problem associated with the cubic defocusing wave equation for data in $\mathcal{H}^s(\mathbb{T}^3)$, especially for $s < 1/2$ because for these regularities this Cauchy problem is ill-posed in the Hadamard sense in $\mathcal{H}^s(\mathbb{T}^3)$. On the other hand the methods of Sect. 4.2 give no information about the transport by $\Phi(t)$ of the measures in $\mathcal{M}^s$, even for large s. Of course, $\mathcal{M}^s$ can be defined for any $s \in \mathbb{R}$ and for $s \geq 1$ the global existence a.s. with respect to an element of $\mathcal{M}^s$ follows from Theorem 4.1.2. The question of the transport of the measures of $\mathcal{M}^s$ under $\Phi(t)$ is of interest in the context of the macroscopic description of the flow of the cubic defocusing wave equation. It is no longer only a low regularity issue and the answer of this question is a priori not clear at all for regular solutions either.

On the other hand, in Sect. 4.3, we constructed a very particular (Gaussian) measure μ on the Sobolev spaces of radial functions on the unit disc of $\mathbb{R}^3$ such that a.s. with respect to this measure the nonlinear defocusing wave equation with nonlinear term $|u|^\alpha u$, $\alpha \in (2, 3)$ has a well defined dynamics. The typical Sobolev regularity on the support of this measure is supercritical and thus again this result is of interest concerning the individual behaviour of the trajectories. This result is also of interest concerning the macroscopic description of the flow because, we can also prove by the methods of Sect. 4.3 that the transported measure by the flow is absolutely continuous with respect to μ. Unfortunately, the method of Sect. 4.3 is only restricted to a very particular initial distribution with data of low regularity.

Motivated by the previous discussion, a natural question to ask is what can be said for the transport of the measures of $\mathcal{M}^s$ under the flow of the cubic defocusing wave equation. In this section we discuss some recent progress on this question.

4.4.1.2 Statement of the Result

Consider the cubic defocusing wave equation

$$(\partial_t^2 - \Delta)u + u^3 = 0, \tag{4.4.1}$$

where $u : \mathbb{R} \times \mathbb{T}^d \to \mathbb{R}$. We rewrite (4.4.1) as the first order system

$$\partial_t u = v, \quad \partial_t v = \Delta u - u^3. \tag{4.4.2}$$

As we already know, if (u, v) is a smooth solution of (4.4.2) then

$$\frac{d}{dt} H(u(t), v(t)) = 0,$$

where

$$H(u, v) = \frac{1}{2} \int_{\mathbb{T}^d} \left(v^2 + |\nabla u|^2 \right) + \frac{1}{4} \int_{\mathbb{T}^d} u^4. \tag{4.4.3}$$

Thanks to Theorem 4.1.2, for $d \leq 3$ the Cauchy problem associated with (4.4.2) is globally well-posed in $\mathcal{H}^s(\mathbb{T}^d) = H^s(\mathbb{T}^d) \times H^{s-1}(\mathbb{T}^d)$, $s \geq 1$. Denote by $\Phi(t)$: $\mathcal{H}^s(\mathbb{T}^d) \to \mathcal{H}^s(\mathbb{T}^d)$ the resulting flow. As we already mentioned, we are interested in the statistical description of $\Phi(t)$. Let $\mu_{s,d}$ be the measure formally defined by

$$d\mu_{s,d} = Z_{s,d}^{-1} e^{-\frac{1}{2} \|(u,v)\|_{\mathcal{H}^{s+1}}^2} du\, dv$$

or

$$d\mu_{s,d} = Z_{s,d}^{-1} \prod_{n \in \mathbb{Z}^2} e^{-\frac{1}{2} \langle n \rangle^{2(s+1)} |\widehat{u}_n|^2} e^{-\frac{1}{2} \langle n \rangle^{2s} |\widehat{v}_n|^2} \, d\widehat{u}_n d\widehat{v}_n ,$$

where $\widehat{u}_n$ and $\widehat{v}_n$ denote the Fourier transforms of u and v respectively. Recall that $\langle n \rangle = (1 + |n|^2)^{\frac{1}{2}}$.

Rigorously one can define the Gaussian measure $\mu_{s,d}$ as the induced probability measure under the map

$$\omega \longmapsto (u^\omega(x), v^\omega(x))$$

with

$$u^\omega(x) = \sum_{n \in \mathbb{Z}^d} \frac{g_n(\omega)}{\langle n \rangle^{s+1}} e^{in \cdot x}, \quad v^\omega(x) = \sum_{n \in \mathbb{Z}^d} \frac{h_n(\omega)}{\langle n \rangle^s} e^{in \cdot x} . \qquad (4.4.4)$$

In (4.4.4), $(g_n)_{n \in \mathbb{Z}^d}$, $(h_n)_{n \in \mathbb{Z}^d}$ are two sequences of "standard" complex Gaussian random variables, such that $g_n = \overline{g_{-n}}$, $h_n = \overline{h_{-n}}$ and such that $\{g_n, h_n\}$ are independent, modulo the central symmetry. The measures $\mu_{s,d}$ can be seen as special cases of the measures in $\mathcal{M}^s$ considered in Sect. 4.2. The partial sums of the series in (4.4.4) are a Cauchy sequence in $L^2(\Omega; \mathcal{H}^\sigma(\mathbb{T}^d))$ for every $\sigma < s + 1 - \frac{d}{2}$ and therefore one can see $\mu_{s,d}$ as a probability measure on $\mathcal{H}^\sigma$ for a fixed $\sigma < s + 1 - \frac{d}{2}$. Therefore, thanks to the results of Sect. 4.2, for $d \leq 3$, the flow $\Phi(t)$ can be extended $\mu_{s,d}$ almost surely, provided $s > \frac{d}{2} - 1$. We have the following result.

Theorem 4.4.1 *Let $s \geq 0$ be an integer. Then the measure $\mu_{s,1}$ is quasi-invariant under the flow of (4.4.2).*

We recall that given a measure space (X, μ), we say that μ is *quasi-invariant* under a transformation $T : X \to X$ if the transported measure $T_* \mu = \mu \circ T^{-1}$ and μ are equivalent, i.e. mutually absolutely continuous with respect to each other. The proof of Theorem 4.4.1 is essentially contained in the analysis of [45].

For $d = 2$ the situation is much more complicated. Recently in [37], we were able to prove the following statement.

Theorem 4.4.2 *Let $s \geq 2$ be an even integer. Then the measure $\mu_{s,2}$ is quasi-invariant under the flow of (4.4.2).*

We expect that by using the methods of Sect. 4.2, one can extend the result of Theorem 4.4.2 to all $s > 0$, not necessarily an integer.

It would be interesting to decide whether one can extend the result of Theorem 4.4.2 to the three dimensional case. It could be that the type of renormalisations employed in the context of singular stochastic PDE's or the QFT become useful in this context.

From now on we consider $d = 2$ and we denote $\mu_{s,2}$ simply by μ_s.

4.4.1.3 Relation to Cameron-Martin Type Results

In probability theory, there is an extensive literature on the transport property of
Gaussian measures under linear and nonlinear transformations. The statements of
Theorems 4.4.1 and 4.4.2 can be seen as such kind of results for the nonlinear
transformation defined by the flow map of the cubic defocusing wave equation. The
most classical result concerning the transport property of Gaussian measures is the
result of Cameron-Martin [10] giving an optimal answer concerning the shifts. The
Cameron-Martin theorem in the context of the measures μ_s is saying that for a fixed
$(h_1, h_2) \in \mathcal{H}^\sigma, \sigma < s$, the transport of μ_s under the shift

$$(u, v) \longmapsto (u, v) + (h_1, h_2),$$

is absolutely continuous with respect to μ_s if and only if $(h_1, h_2) \in \mathcal{H}^{s+1}$.

If we denote by $S(t)$ the free evolution associated with (4.4.2) then for $(u, v) \in$
$\mathcal{H}^\sigma$, we classically have that the flow of the nonlinear wave equation can be
decomposed as

$$\Phi(t)(u, v) = S(t)\big((u, v) + (h_1, h_2)\big), \tag{4.4.5}$$

where $(h_1, h_2) = (h_1(u, v), h_2(u, v)) \in \mathcal{H}^{\sigma+1}$. In other word there is one derivative
smoothing and no more. Of course, if $\sigma < s$ then $\sigma + 1 < s + 1$ and therefore
the result of Theorem 4.4.2 represents a statement displaying fine properties of the
vector field generating $\Phi(t)$. More precisely if in (4.4.5) $(h_1, h_2) \in \mathcal{H}^{\sigma+1}$ were
fixed (independent of (u, v)) then the transported measures would *not* be absolutely
continuous with respect to μ_s !

Let us next compare the result of Theorem 4.4.2 with a result of Ramer [39]. For
$\sigma < s$, let us consider an invertible map Ψ on $\mathcal{H}^\sigma(\mathbb{T}^2)$ of the form

$$\Psi(u, v) = (u, v) + F(u, v),$$

where $F : \mathcal{H}^\sigma(\mathbb{T}^2) \to \mathcal{H}^{s+1}(\mathbb{T}^2)$. Under some more assumptions, the most
important being that

$$DF(u, v) : \mathcal{H}^{s+1}(\mathbb{T}^2) \to \mathcal{H}^{s+1}(\mathbb{T}^2)$$

is a Hilbert-Schmidt map, the analysis of [39] implies that μ_s is quasi-invariant
under Ψ. A typical example for the F is

$$F(u, v) = \varepsilon(1 - \Delta)^{-1-\delta}(u^2, v^2), \quad \delta > 0, \ |\varepsilon| \ll 1,$$

i.e. 2-smoothing is needed in order to ensure the Hilbert-Schmidt assumption.
Therefore the approach of Ramer is far from being applicable in the context of the
flow map of the nonlinear wave equation because for the nonlinear wave equation
there is only 1-smoothing.

Let us finally discuss the Cruzeiro generalisation of the Cameron-Martin theorem. In [15], Ana Bela Cruzeiro considered a general equation of the form

$$\partial_t u = X(u), \tag{4.4.6}$$

where X is an infinite dimensional vector field. She proved that μ_s would be quasi-invariant under the flow of (4.4.6) if we suppose a number of assumptions, the most important being of type:

$$\int_{H^\sigma(\mathbb{T}^2)} e^{\operatorname{div}(X(u))} d\mu_s(u) < \infty. \tag{4.4.7}$$

The problem is how to check the abstract assumption (4.4.7) for concrete examples. Very roughly speaking the result of Theorem 4.4.2 aims to verify assumptions of type (4.4.7) "in practice".

4.4.2 Elements of the Proof

In this section, we present some of the key steps in the proof of Theorem 4.4.2.

4.4.2.1 An Equivalent Gaussian Measure

Since the quadratic part of (4.4.3) does not control the L^2 norm of u, we will prove the quasi-invariance for the equivalent measure $\widetilde{\mu}_s$ defined as the induced probability measure under the map

$$\omega \in \Omega \longmapsto (u^\omega(x), v^\omega(x))$$

with

$$u^\omega(x) = \sum_{n \in \mathbb{Z}^2} \frac{g_n(\omega)}{(1 + |n|^2 + |n|^{2s+2})^{\frac{1}{2}}} e^{in \cdot x}, \quad v^\omega(x) = \sum_{n \in \mathbb{Z}^2} \frac{h_n(\omega)}{(1 + |n|^{2s})^{\frac{1}{2}}} e^{in \cdot x}.$$

Formally $\widetilde{\mu}_s$ can be seen as

$$Z^{-1} e^{-\frac{1}{2} \int v^2 - \frac{1}{2} \int (D^s v)^2 - \frac{1}{2} \int u^2 - \frac{1}{2} \int |\nabla u|^2 - \frac{1}{2} \int (D^{s+1} u)^2} \, du \, dv,$$

where

$$D \equiv \sqrt{-\Delta}.$$

As we shall see below, the expression

$$\frac{1}{2}\int_{\mathbb{T}^2} v^2 + \frac{1}{2}\int_{\mathbb{T}^2} (D^s v)^2 + \frac{1}{2}\int_{\mathbb{T}^2} u^2 + \frac{1}{2}\int_{\mathbb{T}^2} |\nabla u|^2 + \frac{1}{2}\int_{\mathbb{T}^2} (D^{s+1} u)^2 \qquad (4.4.8)$$

is the main part of the quadratic part of the renormalised energy in the context of the nonlinear wave equation (4.4.2). Using the result of Kakutani [26], we can show that for $s > 1/2$ the Gaussian measures μ_s and $\widetilde{\mu}_s$ are equivalent.

4.4.2.2 The Renormalised Energies

Consider the truncated wave equation

$$\partial_t u = v, \qquad \partial_t v = \Delta u - \pi_N((\pi_N u)^3), \qquad (4.4.9)$$

where π_N is the Dirichlet projector on frequencies $n \in \mathbb{Z}^2$ such that $|n| \le N$. If (u, v) is a solution of (4.4.9) then

$$\partial_t \left[\frac{1}{2} \int_{\mathbb{T}^2} (D^s v_N)^2 + \frac{1}{2} \int_{\mathbb{T}^2} (D^{s+1} u_N)^2 \right] = \int_{\mathbb{T}^2} (D^{2s} v_N)(-u_N^3),$$

where $(u_N, v_N) = (\pi_N u, \pi_N v)$. Clearly $\partial_t u_N = v_N$. Observe that for $s = 0$, we recover the conservation of the truncated energy $H_N(u, v)$, defined by

$$H_N(u, v) \equiv H(\pi_N u, \pi_N v).$$

For $s \ge 2$, an even integer, using the Leibniz rule, we get

$$\int_{\mathbb{T}^2} (D^{2s} v_N)(-u_N^3) = -3 \int_{\mathbb{T}^2} D^s v_N \, D^s u_N \, u_N^2$$

$$+ \sum_{\substack{|\alpha|+|\beta|+|\gamma|=s \\ |\alpha|,|\beta|,|\gamma|<s}} c_{\alpha,\beta,\gamma} \int_{\mathbb{T}^2} D^s v_N \, \partial^\alpha u_N \partial^\beta u_N \partial^\gamma u_N,$$

for some unessential constants $c_{\alpha,\beta,\gamma}$.

It will be convenient in the sequel to suppose that the integration on $\mathbb{T}^2$ is done with respect to a probability measure. Therefore the integrations will be done with respect to the Lebesgue measure multiplied by $(2\pi)^{-2}$. We can write

$$-3\int_{\mathbb{T}^2} D^s v_N \, D^s u_N \, u_N^2 = -\frac{3}{2}\partial_t\left[\int_{\mathbb{T}^2}(D^s u_N)^2 u_N^2\right] + 3\int_{\mathbb{T}^2}(D^s u_N)^2 \, v_N \, u_N$$

$$= -\frac{3}{2}\partial_t\left[\int_{\mathbb{T}^2}\Pi_0^\perp[(D^s u_N)^2]\,\Pi_0^\perp[u_N^2]\right] + 3\int_{\mathbb{T}^2}\Pi_0^\perp[(D^s u_N)^2]\,\Pi_0^\perp[v_N \, u_N]$$

$$-\frac{3}{2}\partial_t\left[\int_{\mathbb{T}^2}(D^s u_N)^2\int_{\mathbb{T}^2}u_N^2\right] + 3\int_{\mathbb{T}^2}(D^s u_N)^2\int_{\mathbb{T}^2}v_N \, u_N, \qquad (4.4.10)$$

where $\Pi_0^\perp$ is again the projector on the nonzero frequencies, i.e.

$$(\Pi_0^\perp(f))(x) = f(x) - \int_{\mathbb{T}^2} f(y)dy.$$

The last two terms on the right-hand side of (4.4.10) are problematic because

$$\lim_{N\to\infty}\mathbb{E}_{\tilde{\mu}_s}\left[\int_{\mathbb{T}^2}(D^s \pi_N u)^2\right] = +\infty.$$

Therefore, we need to use a renormalisation in the definitions of the energies. Define σ_N by

$$\sigma_N = \mathbb{E}_{\tilde{\mu}_s}\left[\int_{\mathbb{T}^2}(D^s \pi_N u)^2\right] = \sum_{\substack{n\in\mathbb{Z}^2 \\ |n|\leq N}}\frac{|n|^{2s}}{1 + |n|^2 + |n|^{2s+2}} \sim \log N.$$

Then, we have

$$-\frac{3}{2}\partial_t\left[\int_{\mathbb{T}^2}(D^s u_N)^2\int_{\mathbb{T}^2}u_N^2\right] + 3\int_{\mathbb{T}^2}(D^s u_N)^2\int v_N \, u_N$$

$$= -\frac{3}{2}\partial_t\left[\left(\int_{\mathbb{T}^2}(D^s u_N)^2 - \sigma_N\right)\int_{\mathbb{T}^2}u_N^2\right] + 3\left(\int_{\mathbb{T}^2}(D^s u_N)^2 - \sigma_N\right)\int_{\mathbb{T}^2}v_N \, u_N.$$

Now, the term

$$\int_{\mathbb{T}^2}(D^s u_N)^2 - \sigma_N$$

is a good term because thanks to Wiener chaos estimates, we have the bound

$$\left\|\int_{\mathbb{T}^2}(D^s \pi_N u)^2 - \sigma_N\right\|_{L^p(d\tilde{\mu}_s(u,v))} \leq Cp,$$

where the constant C is independent of p and N. We define $\tilde{H}_{s,N}(u, v)$ by

$$\tilde{H}_{s,N}(u, v) = \frac{1}{2} \int_{\mathbb{T}^2} (D^s v)^2 + \frac{1}{2} \int_{\mathbb{T}^2} (D^{s+1} u)^2 + \frac{3}{2} \int_{\mathbb{T}^2} (D^s u)^2 u^2 - \frac{3}{2} \sigma_N \int_{\mathbb{T}^2} u^2 .$$

We can summarise the previous analysis as follows: if (u, v) is a solution of (4.4.9) then

$$\partial_t \tilde{H}_{s,N}(u_N, v_N) = 3 \int_{\mathbb{T}^2} \Pi_0^\perp [(D^s u_N)^2] \, \Pi_0^\perp [v_N u_N]$$

$$+ \sum_{\substack{|\alpha|+|\beta|+|\gamma|=s \\ |\alpha|,|\beta|,|\gamma|<s}} c_{\alpha,\beta,\gamma} \int_{\mathbb{T}^2} D^s v_N \, \partial^\alpha u_N \partial^\beta u_N \partial^\gamma u_N$$

$$+ 3 \left(\int_{\mathbb{T}^2} (D^s u_N)^2 - \sigma_N \right) \int_{\mathbb{T}^2} v_N u_N . \qquad (4.4.11)$$

All terms in the right hand-side of (4.4.11) are suitable for a perturbative analysis. We finally define the full modified energy $H_{s,N}(u, v)$ as

$$H_{s,N}(u, v) = \tilde{H}_{s,N}(u, v) + H(u, v) + \frac{1}{2} \int_{\mathbb{T}^2} u^2 ,$$

where H is defined by (4.4.3). The quadratic part of $H_{s,N}$ (except the renormalisation term which is morally quartic) is now given by (4.4.8). Therefore in order to prove the quasi-invariance it will be of crucial importance to study the variation in time of $H_{s,N}$. Here is the main quantitative bound used in the proof of Theorem 4.4.2.

Theorem 4.4.3 *Let $s \geq 2$ be an even integer and let us denote by $\Phi_N(t)$ the flow of*

$$\partial_t u = v, \qquad \partial_t v = \Delta u - \pi_N((\pi_N u)^3) .$$

Then for every $r > 0$ there is a constant C such that for every $p \geq 2$ and every $N \geq 1$,

$$\left(\int_{H_N(u,v) \leq r} \left| \partial_t H_{s,N}(\pi_N \Phi_N(t)(u, v))|_{t=0} \right|^p d\tilde{\mu}_s(u, v) \right)^{\frac{1}{p}} \leq Cp .$$

4.4.2.3 On the Proof of Theorem 4.4.3

Using Eq. (4.4.9), we have that

$$\partial_t H_{s,N}(u_N, v_N) = \partial_t \tilde{H}_{s,N}(u_N, v_N) + \int_{\mathbb{T}^2} u_N v_N .$$

Therefore, coming back to (4.4.11), we obtain

$$\partial_t \tilde{H}_{s,N}(\pi_N \Phi_N(t)(u, v))|_{t=0} = \int_{\mathbb{T}^2} \pi_N u \pi_N v + Q_1(u, v) + Q_2(u, v) + Q_3(u, v),$$

where

$$Q_1(u, v) = 3 \int_{\mathbb{T}^2} \Pi_0^{\perp}[(D^s \pi_N u)^2] \Pi_0^{\perp}[\pi_N v \, \pi_N u],$$

$$Q_2(u, v) = \sum_{\substack{|\alpha|+|\beta|+|\gamma|=s \\ |\alpha|,|\beta|,|\gamma|<s}} c_{\alpha,\beta,\gamma} \int_{\mathbb{T}^2} D^s \pi_N v \, \partial^\alpha \pi_N u \partial^\beta \pi_N u \partial^\gamma \pi_N u,$$

$$Q_3(u, v) = 3 \left(\int_{\mathbb{T}^2} (D^s u_N)^2 - \sigma_N \right) \int_{\mathbb{T}^2} \pi_N v \, \pi_N u.$$

Let us first consider

$$\int_{\mathbb{T}^2} \pi_N u \pi_N v . \tag{4.4.12}$$

We need to estimate (4.4.12) under the restriction

$$\int_{\mathbb{T}^2} (|\nabla \pi_N u|^2 + (\pi_N v)^2 + \frac{1}{2}(\pi_N u)^4) \leq 2r. \tag{4.4.13}$$

Using the compactness of $\mathbb{T}^2$, one can see that under the restriction (4.4.13),

$$\left| \int_{\mathbb{T}^2} \pi_N u \pi_N v \right| \leq \|\pi_N u\|_{L^2(\mathbb{T}^2)} \|\pi_N v\|_{L^2(\mathbb{T}^2)} \leq C \|\pi_N u\|_{L^4(\mathbb{T}^2)} \|\pi_N v\|_{L^2(\mathbb{T}^2)} \leq C r^{\frac{3}{4}}.$$

Let us next consider $Q_3(u, v)$. For $r > 0$, we define $\mu_{s,r,N}$ as

$$d\mu_{s,r,N}(u, v) = \chi_{H_N(u,v) \leq r} \, d\tilde{\mu}_s(u, v),$$

where $\chi_{H_N(u,v) \leq r}$ stays for the characteristic function of the set

$$\{(u, v) : H_N(u, v) \leq r\}.$$

The goal is to show that

$$\|Q_3(u, v)\|_{L^p(d\mu_{s,r,N}(u,v))} \leq Cp,$$

with a constant C independent of N and p. Since we already checked that under (4.4.13),

$$\left| \int_{\mathbb{T}^2} \pi_N v \, \pi_N u \right| \le C_r,$$

we obtain that

$$\|Q_3(u,v)\|_{L^p(d\mu_{s,r,N}(u,v))} \le C_r \left\| \int_{\mathbb{T}^2} (D^s \pi_N u)^2 - \sigma_N \right\|_{L^p(d\mu_{s,r,N}(u,v))}$$

$$\le C_r \left\| \int_{\mathbb{T}^2} (D^s \pi_N u)^2 - \sigma_N \right\|_{L^p(d\widetilde{\mu}_s(u,v))}.$$

On the other hand

$$\left\| \int_{\mathbb{T}^2} (D^s \pi_N u)^2 - \sigma_N \right\|_{L^p(d\widetilde{\mu}_s(u,v))} = \left\| \sum_{\substack{n \in \mathbb{Z}^2 \\ |n| \le N}} \frac{(|g_n(\omega)|^2 - 1)|n|^{2s}}{1 + |n|^2 + |n|^{2s+2}} \right\|_{L^p(\Omega)}$$

and by using Wiener chaos estimates, we have

$$\left\| \sum_{\substack{n \in \mathbb{Z}^2 \\ |n| \le N}} \frac{(|g_n(\omega)|^2 - 1)|n|^{2s}}{1 + |n|^2 + |n|^{2s+2}} \right\|_{L^p(\Omega)} \le Cp \left\| \sum_{\substack{n \in \mathbb{Z}^2 \\ |n| \le N}} \frac{(|g_n(\omega)|^2 - 1)|n|^{2s}}{1 + |n|^2 + |n|^{2s+2}} \right\|_{L^2(\Omega)} \le Cp$$

which provides the needed bound for $Q_3(u,v)$.

The analysis of

$$Q_1(u,v) = 3 \int_{\mathbb{T}^2} \Pi_0^\perp [(D^s \pi_N u)^2] \, \Pi_0^\perp [\pi_N v \, \pi_N u]$$

is the most delicate part of the analysis and relies on subtle multi-linear arguments. The analysis of $Q_2(u,v)$ follows similar lines.

Basically, we are allowed to have outputs as

$$\|D^\sigma u\|_{L^\infty(\mathbb{T}^2)}, \quad \sigma < s$$

with a loss $\sqrt{p}$ and $H_N(u,v)$ with no loss in p. The outputs $H_N(u,v)$ follow from deterministic analysis and thus have no loss in p but they are regularity consuming.

We observe that a naive Hölder inequality approach clearly fails. A purely probabilistic argument based on Wiener chaos estimates fails because the output power of p is too large. The basic strategy is therefore to perform a multi-scale analysis redistributing properly the derivative losses by never having more then quadratic weight of the contribution of the Wiener chaos estimate.

When analysing the 4-linear expression defining $Q_1(u, v)$, we suppose that

$$D^s \pi_N u, \quad D^s \pi_N u, \quad \pi_N v, \quad \pi_N u$$

are localised at dyadic frequencies N_1, N_2, N_3, N_4 respectively.

We first consider the case when $N_4 \gtrsim (\max(N_1, N_2))^{\frac{1}{100}}$. In this case we exchange some regularity of $D^s \pi_N u$ with this of $\pi_N u$ and we perform the naive linear analysis.

Therefore, in the analysis of $Q_1(u, v)$ we can suppose that

$$N_4 \ll (\max(N_1, N_2))^{\frac{1}{100}}.$$

In this case, we have that

$$\max(N_1, N_2) \sim \max(N_j, \ j = 1, 2, 3, 4).$$

By symmetry, we can suppose that $N_1 = \max(N_1, N_2)$. We next consider the case

$$N_3 \ll N_1^{1-a}, \quad a = a(s) \ll 1.$$

In this case, we perform a bi-linear Wiener chaos estimate and we have some gain of regularity in the localisation of $\Pi_0^\perp [(D^s \pi_N u)^2]$. Finally, we consider the case

$$N_1 \sim \max(N_j, \ j = 1, 2, 3, 4), \quad N_4 \ll (\max(N_1, N_2))^{\frac{1}{100}}, \quad N_3 \gtrsim N_1^{1-a}$$

In this case, we perform a tri-linear Wiener chaos estimate and we have enough gain of regularity in the localisation of

$$\Pi_0^\perp [(J^s \pi_N u)^2] \pi_N v.$$

This essentially explains the argument leading to the key estimate of Theorem 4.4.3. We refer to [37] for the details.

4.4.2.4   On the Soft Analysis

We can observe that

$$H_{s,N}(u, v) = (4.4.8) + \frac{3}{2} \int_{\mathbb{T}^2} (D^s u)^2 u^2 - \frac{3}{2} \sigma_N \int_{\mathbb{T}^2} u^2 + \int_{\mathbb{T}^2} u^4.$$

By classical arguments from QFT, we can define

$$\lim_{N \to \infty} \left(\frac{3}{2} \int (D^s \pi_N u)^2 (\pi_N u)^2 - \frac{3}{2} \sigma_N \int (\pi_N u)^2 \right)$$

in $L^p(d\widetilde{\mu}_s(u, v))$, $p < \infty$. Denote this limit by $R(u)$. Essentially speaking, once we have the key estimate, we study the quasi-invariance of

$$\chi_{H(u,v)\le r}\, e^{-R(u)-\int u^4}\, d\widetilde{\mu}_s(u, v) \qquad (4.4.14)$$

by soft analysis techniques.

Let us finally explain the importance of the loss p in the key estimate of Theorem 4.4.3. Denote by $x(t)$ the measure evolution of a set having zero measure with respect to (4.4.14). Essentially speaking, using the key estimate and the arguments introduced in [46, 47], we obtain that $x(t)$ satisfy the estimate

$$\dot{x}(t) \le Cp(x(t))^{1-\frac{1}{p}}, \quad x(0) = 0. \qquad (4.4.15)$$

Integrating the last estimate leads to $x(t) \le (Ct)^p$. Taking the limit $p \to \infty$, we infer that $x(t) = 0$ for $0 \le t < 1/C$. Since C is an absolute constant, we can iterate the argument and show that $x(t)$ is vanishing. Observe that this argument would not work if in (4.4.15), we have p^α, $\alpha > 1$ instead of p. In order to make the previous reasoning rigorous, we need to use some more or less standard approximation arguments. We refer to [45] and [37] for the details of such type of reasoning.

Acknowledgements I am grateful to Leonardo Tolomeo, Tadahiro Oh and Yuzhao Wang for their remarks on the manuscript. I am very grateful to Chenmin Sun for pointing our an error in a previous version of Lemma 4.1.32. I am particularly indebted to Nicolas Burq and Tadahiro Oh since this text benefitted from the discussions we had on the topics discussed in the lectures. I am grateful to Franco Flandoli and Massimiliano Gubinelli for inviting me to give these lectures.

References

1. A. Ayache, N. Tzvetkov, L^p properties of Gaussian random series. Trans. Am. Math. Soc. **360**, 4425–4439 (2008)
2. A. Benyi, T. Oh, O. Pocovnicu, On the probabilistic Cauchy theory of the cubic nonlinear Schrödinger equation on $\mathbb{R}^d$, $d \ge 3$. Trans. Am. Math. Soc. Ser. B **2**, 1–50 (2015)
3. J. Bourgain, Periodic nonlinear Schrödinger equation and invariant measures. Commun. Math. Phys. **166**, 1–26 (1994)
4. J. Bourgain, Invariant measures for the 2d-defocusing nonlinear Schrödinger equation. Commun. Math. Phys. **176**, 421–445 (1996)
5. J. Bourgain, A. Bulut, Invariant Gibbs measure evolution for the radial nonlinear wave equation on the 3D ball. J. Funct. Anal. **266**, 2319–2340 (2014)
6. N. Burq, N. Tzvetkov, Random data Cauchy theory for supercritical wave equations I. Local theory. Invent. Math. **173**, 449–475 (2008)
7. N. Burq, N. Tzvetkov, Random data Cauchy theory for supercritical wave equations II. A global existence result. Invent. Math. **173**, 477–496 (2008)
8. N. Burq, N. Tzvetkov, Probabilistic well-posedness for the cubic wave equation. J. Eur. Math. Soc. **16**, 1–30 (2014)

9. N. Burq, L. Thomann, N. Tzvetkov, Global infinite energy solutions for the cubic wave equation. Bull. Soc. Math. Fr. **143**, 301–313 (2015)
10. R.H. Cameron, W.T. Martin, Transformation of Wiener integrals under translations. Ann. Math. **45**, 386–396 (1944)
11. M. Christ, A. Kiselev, Maximal functions associated to filtrations. J. Funct. Anal. **179**, 409–425 (2001)
12. M. Christ, J. Colliander, T. Tao, Ill-posedness for nonlinear Schrödinger and wave equations, Preprint, November 2003
13. J. Colliander, T. Oh, Almost sure local well-posedness of the cubic NLS below L^2. Duke Math. J. **161**, 367–414 (2012)
14. J. Colliander, M. Keel, G. Staffilani, H. Takaoka, T. Tao, Almost conservation laws and global rough solutions to a nonlinear Schrödinger equation. Math. Res. Lett. **9**, 659–682 (2002)
15. A.B. Cruzeiro, Equations différentielles sur l'espace de Wiener et formules de Cameron-Martin non linéaires. J. Funct. Anal. **54**, 206–227 (1983)
16. G. Da Prato, A. Debussche, Strong solutions to the stochastic quantization equations. Ann. Probab. **31**, 1900–1916 (2003)
17. L. Farah, F. Rousset, N. Tzvetkov, Oscillatory integrals and global well-posedness for the 2D Boussinesq equation. Bull. Braz. Math. Soc. **43**, 655–679 (2012)
18. I. Gallagher, F. Planchon, On global solutions to a defocusing semi-linear wave equation. Rev. Mat. Iberoamericana **19**, 161–177 (2003)
19. J. Ginibre, G. Velo, Generalized Strichartz inequalities for the wave equation. J. Funct. Anal. **133**, 50–68 (1995)
20. M. Grillakis, Regularity and asymptotic behaviour of the wave equation with a critical non linearity. Ann. Math. **132**, 485–509 (1990)
21. M. Grillakis, Regularity for the wave equation with a critical non linearity. Commun. Pures Appl. Math. **45**, 749–774 (1992)
22. M. Gubinelli, P. Imkeller, P. Perkowski, Paracontrolled distributions and singular PDEs. Forum Math. varPi **3**, e6, 75 pp. (2015)
23. M. Hairer, Solving the KPZ equation. Ann. Math. **178**, 559–664 (2013)
24. M. Hairer, A theory of regularity structures. Invent. Math. **198**, 269–504 (2014)
25. M. Hairer, K. Matetski, Discretisations of rough stochastic PDEs. Ann. Probab. **46**, 1651–1709 (2018)
26. S. Kakutani, On equivalence of infinite product measures. Ann. Math. **49**, 214–224 (1948)
27. C. Kenig, G. Ponce, L. Vega, Global well-posedness for semi-linear wave equations. Commun. Partial Differ. Equ. **25**, 1741–1752 (2000)
28. G. Lebeau, Perte de régularité pour les équation d'ondes sur-critiques. Bull. Soc. Math. Fr. **133**, 145–157 (2005)
29. E. Lieb, M. Loss, *Analysis*. Graduate Studies in Mathematics, vol. 14 (American Mathematical Society, Providence, 2001)
30. H. Lindblad, C. Sogge, On existence and scattering with minimal regularity for semilinear wave equations. J. Funct. Anal. **130**, 357–426 (1995)
31. J. Lührmann, D. Mendelson, Random data Cauchy theory for nonlinear wave equations of power-type on $\mathbb{R}^3$. Commun. Partial Differ. Equ. **39**, 2262–2283 (2014)
32. J.C. Mourrat, H. Weber, The dynamic Φ_3^4 model comes down from infinity. Commun. Math. Phys. **356**, 673–753 (2017)
33. A. Nahmod, G. Staffilani, Almost sure well-posedness for the periodic 3D quintic nonlinear Schrödinger equation below the energy space. J. Eur. Math. Soc. **17**, 1687–1759 (2015)
34. T. Oh, A remark on norm inflation with general initial data for the cubic nonlinear Schrödinger equations in negative Sobolev spaces. Funkcial. Ekvac. **60**, 259–277 (2017)
35. T. Oh, O. Pocovnicu, Probabilistic global well-posedness of the energy-critical defocusing quintic nonlinear wave equation on $\mathbb{R}^3$. J. Math. Pures Appl. **105**, 342–366 (2016)
36. T. Oh, O. Pocovnicu, A remark on almost sure global well-posedness of the energy-critical defocusing nonlinear wave equations in the periodic setting. Tohoku Math. J. **69**, 455–481 (2017)

37. T. Oh, N. Tzvetkov, Quasi-invariant Gaussian measures for the two-dimensional defocusing cubic nonlinear wave equation. J. Eur. Math. Soc. (to appear)
38. O. Pocovnicu, Almost sure global well-posedness for the energy-critical defocusing nonlinear wave equation on $\mathbb{R}^d$, $d = 4$ and 5. J. Eur. Math. Soc. **19**, 2521–2575 (2017)
39. R. Ramer, On nonlinear transformations of Gaussian measures. J. Funct. Anal. **15**, 166–187 (1974)
40. T. Roy, Adapted linear-nonlinear decomposition and global well-posedness for solutions to the defocusing cubic wave equation on $\mathbb{R}^3$. Discrete Contin. Dynam. Syst. A **24**, 1307–1323 (2009)
41. J. Shatah, M. Struwe, Regularity results for nonlinear wave equations. Ann. Math. **138**, 503–518 (1993)
42. J. Shatah, M. Struwe, Well-posedness in the energy space for semi-linear wave equations with critical growth. Int. Math. Res. Not. **1994**, 303–309 (1994)
43. C. Sun, B. Xia, Probabilistic well-posedness for supercritical wave equation on $\mathbb{T}^3$. Ill. J. Math. **60**, 481–503 (2016)
44. N. Tzvetkov, Construction of a Gibbs measure associated to the periodic Benjamin-Ono equation. Probab. Theory Relat. Fields **146**, 481–514 (2010)
45. N. Tzvetkov, Quasi-invariant Gaussian measures for one dimensional Hamiltonian PDE's. Forum Math. Sigma **3**, e28, 35 pp. (2015)
46. N. Tzvetkov, N. Visciglia, Invariant measures and long-time behavior for the Benjamin-Ono equation. Int. Math. Res. Not. **2014**, 4679–4714 (2014)
47. N. Tzvetkov, N. Visciglia, Invariant measures and long time behaviour for the Benjamin-Ono equation II. J. Math. Pures Appl. **103**, 102–141 (2015)
48. B. Xia, Equations aux dérivées partielles et aléa, PhD thesis, University of Paris Sud, July 2016

LECTURE NOTES IN MATHEMATICS

 Springer

Editors in Chief: J.-M. Morel, B. Teissier;

Editorial Policy

1. Lecture Notes aim to report new developments in all areas of mathematics and their applications – quickly, informally and at a high level. Mathematical texts analysing new developments in modelling and numerical simulation are welcome.

   Manuscripts should be reasonably self-contained and rounded off. Thus they may, and often will, present not only results of the author but also related work by other people. They may be based on specialised lecture courses. Furthermore, the manuscripts should provide sufficient motivation, examples and applications. This clearly distinguishes Lecture Notes from journal articles or technical reports which normally are very concise. Articles intended for a journal but too long to be accepted by most journals, usually do not have this "lecture notes" character. For similar reasons it is unusual for doctoral theses to be accepted for the Lecture Notes series, though habilitation theses may be appropriate.

2. Besides monographs, multi-author manuscripts resulting from SUMMER SCHOOLS or similar INTENSIVE COURSES are welcome, provided their objective was held to present an active mathematical topic to an audience at the beginning or intermediate graduate level (a list of participants should be provided).

   The resulting manuscript should not be just a collection of course notes, but should require advance planning and coordination among the main lecturers. The subject matter should dictate the structure of the book. This structure should be motivated and explained in a scientific introduction, and the notation, references, index and formulation of results should be, if possible, unified by the editors. Each contribution should have an abstract and an introduction referring to the other contributions. In other words, more preparatory work must go into a multi-authored volume than simply assembling a disparate collection of papers, communicated at the event.

3. Manuscripts should be submitted either online at www.editorialmanager.com/lnm to Springer's mathematics editorial in Heidelberg, or electronically to one of the series editors. Authors should be aware that incomplete or insufficiently close-to-final manuscripts almost always result in longer refereeing times and nevertheless unclear referees' recommendations, making further refereeing of a final draft necessary. The strict minimum amount of material that will be considered should include a detailed outline describing the planned contents of each chapter, a bibliography and several sample chapters. Parallel submission of a manuscript to another publisher while under consideration for LNM is not acceptable and can lead to rejection.

4. In general, **monographs** will be sent out to at least 2 external referees for evaluation.

   A final decision to publish can be made only on the basis of the complete manuscript, however a refereeing process leading to a preliminary decision can be based on a pre-final or incomplete manuscript.

   Volume Editors of **multi-author works** are expected to arrange for the refereeing, to the usual scientific standards, of the individual contributions. If the resulting reports can be forwarded to the LNM Editorial Board, this is very helpful. If no reports are forwarded or if other questions remain unclear in respect of homogeneity etc, the series editors may wish to consult external referees for an overall evaluation of the volume.

5. Manuscripts should in general be submitted in English. Final manuscripts should contain at least 100 pages of mathematical text and should always include

   – a table of contents;
   – an informative introduction, with adequate motivation and perhaps some historical remarks: it should be accessible to a reader not intimately familiar with the topic treated;
   – a subject index: as a rule this is genuinely helpful for the reader.
   – For evaluation purposes, manuscripts should be submitted as pdf files.

6. Careful preparation of the manuscripts will help keep production time short besides ensuring satisfactory appearance of the finished book in print and online. After acceptance of the manuscript authors will be asked to prepare the final LaTeX source files (see LaTeX templates online: https://www.springer.com/gb/authors-editors/book-authors-editors/manuscriptpreparation/5636) plus the corresponding pdf- or zipped ps-file. The LaTeX source files are essential for producing the full-text online version of the book, see http://link.springer.com/bookseries/304 for the existing online volumes of LNM). The technical production of a Lecture Notes volume takes approximately 12 weeks. Additional instructions, if necessary, are available on request from lnm@springer.com.

7. Authors receive a total of 30 free copies of their volume and free access to their book on SpringerLink, but no royalties. They are entitled to a discount of 33.3 % on the price of Springer books purchased for their personal use, if ordering directly from Springer.

8. Commitment to publish is made by a *Publishing Agreement*; contributing authors of multiauthor books are requested to sign a *Consent to Publish form*. Springer-Verlag registers the copyright for each volume. Authors are free to reuse material contained in their LNM volumes in later publications: a brief written (or e-mail) request for formal permission is sufficient.

Addresses:
Professor Jean-Michel Morel, CMLA, École Normale Supérieure de Cachan, France
E-mail: moreljeanmichel@gmail.com

Professor Bernard Teissier, Equipe Géométrie et Dynamique,
Institut de Mathématiques de Jussieu – Paris Rive Gauche, Paris, France
E-mail: bernard.teissier@imj-prg.fr

Springer: Ute McCrory, Mathematics, Heidelberg, Germany,
E-mail: lnm@springer.com

Printed in the United States
By Bookmasters